Genomic Characterization of Antimicrobial Resistance and Evolution Mechanism of Bacteria

Genomic Characterization of Antimicrobial Resistance and Evolution Mechanism of Bacteria

Guest Editors

Daniel Gyamfi Amoako

Linda Bester

MDPI

Basel • Beijing • Wuhan • Barcelona • Belgrade • Novi Sad • Cluj • Manchester

Guest Editors
Daniel Gyamfi Amoako
Department of Pathobiology
University of Guelph
Guelph
Canada

Linda Bester
College of Health Sciences
University of KwaZulu-Natal
Durban
South Africa

Editorial Office
MDPI AG
Grosspeteranlage 5
4052 Basel, Switzerland

This is a reprint of the Special Issue, published open access by the journal *Antibiotics* (ISSN 2079-6382), freely accessible at: https://www.mdpi.com/journal/antibiotics/special_issues/99G9X8VQ4O.

For citation purposes, cite each article independently as indicated on the article page online and as indicated below:

Lastname, A.A.; Lastname, B.B. Article Title. *Journal Name* **Year**, *Volume Number*, Page Range.

ISBN 978-3-7258-5307-6 (Hbk)
ISBN 978-3-7258-5308-3 (PDF)
https://doi.org/10.3390/books978-3-7258-5308-3

Contents

About the Editors . . . . **vii**

Preface . . . . **ix**

Daniel Gyamfi Amoako and Linda Antionette Bester
Genomic Characterization of Antimicrobial Resistance and Evolution Mechanism of Bacteria
Reprinted from: *Antibiotics* **2025**, *14*, 945, https://doi.org/10.3390/antibiotics14090945 . . . . . **1**

Shuan Er, Yichen Ding, Linda Wei Lin Tan, Yik Ying Teo, Niranjan Nagarajan and Henning Seedorf
Comparative Genomics of DH5α-Inhibiting *Escherichia coli* Isolates from Feces of Healthy Individuals Reveals Common Co-Occurrence of Bacteriocin Genes with Virulence Factors and Antibiotic Resistance Genes
Reprinted from: *Antibiotics* **2025**, *14*, 860, https://doi.org/10.3390/antibiotics14090860 . . . . . **4**

Mako Watanabe, Ryuichi Nakano, Keizo Yamamoto, Akiyo Nakano, Yuki Suzuki, Kai Saito, et al.
Unique Regulation of Sed-1 β-Lactamase in *Citrobacter sedlakii*: Insights on Resistance to Third-Generation Cephalosporin
Reprinted from: *Antibiotics* **2025**, *14*, 823, https://doi.org/10.3390/antibiotics14080823 . . . . . **11**

Sasini Jayaweera, Pondpan Suwanthada, David Atomanyi Barnes, Charlotte Poussier, Tomoyasu Nishimura, Naoki Hasegawa, et al.
Investigation of WQ-3810, a Fluoroquinolone with a High Potential Against Fluoroquinolone-Resistant *Mycobacterium avium*
Reprinted from: *Antibiotics* **2025**, *14*, 704, https://doi.org/10.3390/antibiotics14070704 . . . . . **28**

Nada A. Fahmy, Sumin Karna, Angel Bhusal, Ajran Kabir, Erdal Erol and Yosra A. Helmy
Multidrug Resistance and Virulence Traits of *Salmonella enterica* Isolated from Cattle: Genotypic and Phenotypic Insights
Reprinted from: *Antibiotics* **2025**, *14*, 689, https://doi.org/10.3390/antibiotics14070689 . . . . . **41**

Shima E. Abdalla, Linda A. Bester, Akebe L. K. Abia, Mushal Allam, Arshad Ismail, Sabiha Y. Essack and Daniel G. Amoako
Genomic Insights of Antibiotic-Resistant *Escherichia coli* Isolated from Intensive Pig Farming in South Africa Using 'Farm-to-Fork' Approach
Reprinted from: *Antibiotics* **2025**, *14*, 446, https://doi.org/10.3390/antibiotics14050446 . . . . . **64**

Natalia Belkova, Uliana Nemchenko, Elizaveta Klimenko, Nadezhda Smurova, Raisa Zugeeva, Marina Sukhoreva, et al.
Resistance of *Pseudomonas aeruginosa* to Antibiotics During Long-Term Persistence in Patients with Cystic Fibrosis
Reprinted from: *Antibiotics* **2025**, *14*, 302, https://doi.org/10.3390/antibiotics14030302 . . . . . **88**

Kanako Ishihara, Suzuka Someno, Kaoru Matsui, Chisato Nakazawa, Takahiro Abe, Hayato Harima, et al.
Determination of Antimicrobial Resistance Megaplasmid-Like pESI Structures Contributing to the Spread of *Salmonella* Schwarzengrund in Japan
Reprinted from: *Antibiotics* **2025**, *14*, 288, https://doi.org/10.3390/antibiotics14030288 . . . . . **101**

Deneke Wolde, Tadesse Eguale, Girmay Medhin, Aklilu Feleke Haile, Haile Alemayehu, Adane Mihret, et al.
Genomic Characterization of Extended-Spectrum β-Lactamase-Producing and Third-Generation Cephalosporin-Resistant *Escherichia coli* Isolated from Stools of Primary Healthcare Patients in Ethiopia
Reprinted from: *Antibiotics* **2024**, *13*, 851, https://doi.org/10.3390/antibiotics13090851 . . . . . **118**

Bakoena A. Hetsa, Jonathan Asante, Joshua Mbanga, Arshad Ismail, Akebe L. K. Abia, Daniel G. Amoako and Sabiha Y. Essack
Genomic Characterization of Methicillin-Resistant and Methicillin-Susceptible *Staphylococcus aureus* Implicated in Bloodstream Infections, KwaZulu-Natal, South Africa: A Pilot Study
Reprinted from: *Antibiotics* **2024**, *13*, 796, https://doi.org/10.3390/antibiotics13090796 . . . . . **137**

Thunchanok Yaikhan, Sirikan Suwannasin, Kamonnut Singkhamanan, Sarunyou Chusri, Rattanaruji Pomwised, Monwadee Wonglapsuwan and Komwit Surachat
Genomic Characterization of Multidrug-Resistant Enterobacteriaceae Clinical Isolates from Southern Thailand Hospitals: Unraveling Antimicrobial Resistance and Virulence Mechanisms
Reprinted from: *Antibiotics* **2024**, *13*, 531, https://doi.org/10.3390/antibiotics13060531 . . . . . **157**

Kinga Tóth, Ivelina Damjanova, Levente Laczkó, Lilla Buzgó, Virág Lesinszki, Erika Ungvári, et al.
Genomic Epidemiology of C2/H30Rx and C1-M27 Subclades of *Escherichia coli* ST131 Isolates from Clinical Blood Samples in Hungary
Reprinted from: *Antibiotics* **2024**, *13*, 363, https://doi.org/10.3390/antibiotics13040363 . . . . . **171**

About the Editors

Daniel Gyamfi Amoako

Daniel Gyamfi Amoako is a researcher at the Department of Pathobiology at the University of Guelph, Canada, and the College of Health Sciences at the University of KwaZulu-Natal, South Africa. He has a longstanding interest in microbial bioinformatics; his research applies genome sequencing and data analytics to better understand emerging pathogens and their mechanisms of resistance and virulence. He has been involved in many large-scale bacterial whole-genome sequencing projects and actively translates microbial genomic data into genomic epidemiology for the benefit of the public, private, food, animal, and environmental health sectors (One-Health Context) globally. He has been working with several organizations such as the Africa Pathogen Genomics Initiative (Africa PGI)/Africa CDC, Centre for Respiratory Diseases and Meningitis at the National Institute for Communicable Diseases, SEQAFRICA-Fleming Fund, Network for Genomic Surveillance in South Africa (NGS-SA), COG-Train/COVID-19 Genomics UK Consortium, Mobilome Ontology (MOBIO) working group/Canada, RSV genotyping consortium-ECDC-WHO and more, spearheading training and implementation of bioinformatics worldwide. To date, Amoako has authored or co-authored over 120 publications in leading peer-reviewed journals, with over 9000 citations and an H-index of 40. Additionally, he has successfully supervised several postgraduate students, contributing to the advancement of scientific knowledge in his field.

Linda Bester

Linda Bester serves as the Deputy Chair of the Animal Research Ethics Committee at the University of KwaZulu-Natal. She is employed full-time at the Biomedical Resource Unit, which is a laboratory animal facility located on the Westville campus. In 2023, she received an NRF rating of C2 and collaborated as a member of the Africa CDC Pathogen Genomic Initiative (PGI) Foodborne Disease Resistance Genomic Surveillance Focus Group, endorsed by the African Union, tackling pressing diarrheal health issues across the continent. Her research interests include foodborne bacteria, zoonosis and the dissemination of bacteria within specific ecological systems. Linda has supervised and co-supervised several PhD and MSc students. Her research foundation is rooted in antibiotic resistance surveillance in bacteria from food animals, and it has expanded to focus on the spread of bacteria within certain ecological environments. Additionally, her professional background has cultivated a strong interest in laboratory diagnostic tools and protocols.

Preface

"There is no better time than the present to act with wisdom." This adage reminds us that decisive action, though often delayed, remains urgent when the stakes are high. In recent years, the World Health Organization and its Tripartite partners have repeatedly emphasized that antimicrobial resistance (AMR) is one of the greatest threats to global health, food security and sustainable development. The release of the 2024 Global Antimicrobial Resistance and Use Surveillance System (GLASS) report further reinforced the urgent need for robust surveillance and coordinated responses. By calling for stronger commitments across human, animal and environmental health, these global initiatives echo the very principles that guided this collection.

The Special Issue titled "Genomic Characterization of Antimicrobial Resistance and Evolution Mechanism of Bacteria" brings together original research articles and reviews that deepen our understanding of the genetic determinants driving bacterial AMR. At a time when resistance to frontline antibiotics threatens both human and animal health, genomics offers an indispensable framework for decoding resistance mechanisms, evolutionary pathways, and the mobilization strategies of resistance genes. The scope of this volume spans diverse bacterial taxa and ecological settings, from clinical isolates and livestock production systems to environmental reservoirs, reflecting the One Health dimensions of AMR. The contributing studies explore molecular epidemiology, plasmid and mobile genetic element dynamics, resistance–virulence linkages and the evolutionary plasticity of bacterial genomes. Collectively, these works illustrate the power of next-generation sequencing and bioinformatics in mapping resistance landscapes and informing strategies for risk assessment, surveillance, and mitigation.

The aim of this compilation is not only to provide a consolidated view of current genomic insights but also to stimulate new collaborations across disciplines. By presenting perspectives from Africa, Asia, Europe and the Americas, it underscores the global scale of the AMR challenge while offering context-specific findings that advance both science and policy. Equally important, this collection reflects the dedication of its authors researchers from diverse disciplines, institutions and regions whose collective expertise enriches the volume. Their contributions demonstrate the value of international collaboration in advancing genomic research and highlight the shared responsibility of the global scientific community in tackling AMR. This work is addressed to microbiologists, molecular epidemiologists, genomic scientists and healthcare professionals, as well as students and early-career researchers seeking to understand the evolutionary basis of resistance. We hope that the diversity of studies presented here will serve both as a resource for ongoing research and as an inspiration for innovative approaches to combating AMR.

Daniel Gyamfi Amoako and Linda Bester
Guest Editors

antibiotics

MDPI

Editorial

Genomic Characterization of Antimicrobial Resistance and Evolution Mechanism of Bacteria

Daniel Gyamfi Amoako [1,2,*] and Linda Antionette Bester [1,3,*]

1 College of Health Sciences, University of KwaZulu-Natal, Durban 4000, South Africa
2 Department of Pathobiology, University of Guelph, Guelph, ON N1G 2W1, Canada
3 Department of Biochemistry and Microbiology, Faculty of Science, Engineering and Agriculture, University of Venda, Thohoyandou 0950, South Africa
* Correspondence: amoakodg@gmail.com (D.G.A.); besterl@ukzn.ac.za (L.A.B.)

Antimicrobial resistance (AMR) continues to rank among the most pressing global health threats, frequently referred to as a *"silent pandemic"* that undermines decades of progress in infectious disease control while jeopardizing both human and animal health [1]. The relentless emergence of resistant pathogens is fueled by the overuse and misuse of antimicrobials in clinical, veterinary, and agricultural settings, coupled with the remarkable genetic adaptability of bacteria via mechanisms like horizontal gene transfer. In this context, the genomic era has opened unprecedented avenues for discovery. Whole-genome sequencing and metagenomics now enable researchers to move beyond phenotypic observations to dissect the molecular blueprints of resistance determinants. These technologies illuminate how resistance determinants, virulence factors, and other mobile genetic elements (MGEs) interact to shape pathogen fitness and epidemiology across human, animal, and environmental reservoirs [2]. Furthermore, the integration of real-time genomic surveillance combining diagnostic microbiology, pathogen sequence data, and One Health collaboration offers timely insights for clinical practice and public health policies aimed at mitigating AMR spread [3].

This Special Issue assembles eleven diverse contributions from 97 authors, collectively enriching our comprehension of the molecular, evolutionary, and ecological dimensions of AMR. The studies included span various bacterial taxa, clinical and agricultural contexts, and geographic regions, highlighting both universal trends and local particularities in resistance evolution.

Several studies emphasize the critical role of One Health strategies in monitoring resistance across the farm-to-fork continuum and primary healthcare settings. Abdalla et al. employed genomic surveillance in South African intensive pig farming systems, revealing a broad spectrum of plasmid-borne resistance and virulence genes disseminated throughout the production chain from farms to transport and abattoirs, underscoring potential public health risks. Complementarily, Wolde et al. characterized extended-spectrum β-lactamase (ESBL)-producing *Escherichia coli* isolates from Ethiopian primary healthcare patients, uncovering notable genetic diversity, high-risk sequence types, and transferable plasmids, highlighting the mobility of resistomes within community health settings.

Comprehensive genomic and phenotypic analyses reveal the coexistence of AMR genes and virulence factors across multiple bacterial pathogens. Fahmy et al. investigated *Salmonella enterica* isolates from cattle, identifying multidrug resistance, robust biofilm formation, and diverse virulence determinants that enhance persistence and zoonotic potential. Hetsa et al. studied bloodstream infections caused by *Staphylococcus aureus* in South Africa, encompassing methicillin-resistant and susceptible strains, and reported complex resistance profiles coupled with variable virulence gene content, raising significant treatment challenges. Likewise, Yaikhan et al. examined clinical *Enterobacteriaceae* isolates from Thailand, unveiling multidrug resistance alongside novel bacterial species and diverse plasmid-mediated resistance mechanisms. In parallel, Er et al. analyzed bacteriocin-producing *E. coli* from healthy individuals, discovering strains co-harboring virulence and resistance elements, cautioning that such gut colonizers could serve as cryptic reservoirs of AMR.

The evolutionary plasticity of bacterial resistance is vividly illustrated in studies dissecting MGEs and genomic adaptations. Ishihara et al. traced the dissemination of *pESI-like* megaplasmids in *Salmonella Schwarzengrund*, demonstrating how structural variations in these large plasmids drive the spread of multidrug resistance among foodborne pathogens. In Hungary, Tóth et al. examined *E. coli* ST131 bloodstream isolates, revealing dynamic chromosomal integration of β-lactamase genes and plasmid rearrangements underpinning the global success of this high-risk clone. Watanabe et al. provided mechanistic insights into the regulation of *Sed-1* β-lactamase in *Citrobacter sedlakii*, showing that single-point mutations and regulatory disruptions can transform otherwise susceptible strains into extended-spectrum β-lactamase producers.

The clinical ramifications of persistent AMR and therapeutic challenges are reflected in research addressing chronic infections and novel antimicrobials. Belkova et al. investigated *Pseudomonas aeruginosa* isolates persisting in cystic fibrosis patients under long-term therapy, revealing complex resistance trajectories, including instances of partial re-sensitization, highlighting the dynamic nature of resistance adaptation in chronic infections. Jayaweera et al. evaluated the novel fluoroquinolone WQ-3810 against resistant *Mycobacterium avium* strains, demonstrating significantly enhanced efficacy compared to existing agents and promising synergy in combination therapy.

The contributions in this Special Issue reveal AMR as both a scientific challenge and a global systems crisis. Evidence from agricultural, clinical, and community settings shows how the mobility of resistance determinants, often linked to virulence, reshapes bacterial populations and undermines health. Genomics is indispensable for uncovering hidden pathways of dissemination and persistence. Beyond documenting resistance, these studies underscore two priorities: advancing novel therapeutics and embedding genomic surveillance within One Health frameworks. Achieving impact requires global commitment to data sharing, harmonized surveillance, and policies that translate genomic insights into coordinated interventions aimed at curbing the trajectory of the AMR crisis.

Conflicts of Interest: The authors declare no conflict of interest.

List of Contributions:

1. Er, S.; Ding, Y.; Tan, L.W.L.; Teo, Y.Y.; Nagarajan, N.; Seedorf, H. Comparative Genomics of DH5α-Inhibiting *Escherichia coli* Isolates from Feces of Healthy Individuals Reveals Common Co-Occurrence of Bacteriocin Genes with Virulence Factors and Antibiotic Resistance Genes. *Antibiotics* **2025**, *14*, 860. https://doi.org/10.3390/antibiotics14090860.
2. Watanabe, M.; Nakano, R.; Yamamoto, K.; Nakano, A.; Suzuki, Y.; Saito, K.; Nakashima, S.; Endo, K.; Narita, K.; Yano, H. Unique Regulation of Sed-1 β-Lactamase in *Citrobacter sedlakii*: Insights on Resistance to Third-Generation Cephalosporin. *Antibiotics* **2025**, *14*, 823. https://doi.org/10.3390/antibiotics14080823.
3. Jayaweera, S.; Suwanthada, P.; Barnes, D.A.; Poussier, C.; Nishimura, T.; Hasegawa, N.; Nishiuchi, Y.; Thapa, J.; Gordon, S.V.; Kim, H.; et al. Investigation of WQ-3810, a Fluoroquinolone

with a High Potential against Fluoroquinolone-Resistant *Mycobacterium avium*. *Antibiotics* **2025**, *14*, 704. https://doi.org/10.3390/antibiotics14070704.

4. Fahmy, N.A.; Karna, S.; Bhusal, A.; Kabir, A.; Erol, E.; Helmy, Y.A. Multidrug Resistance and Virulence Traits of *Salmonella enterica* Isolated from Cattle: Genotypic and Phenotypic Insights. *Antibiotics* **2025**, *14*, 689. https://doi.org/10.3390/antibiotics14070689.
5. Abdalla, S.E.; Bester, L.A.; Abia, A.L.K.; Allam, M.; Ismail, A.; Essack, S.Y.; Amoako, D.G. Genomic Insights of Antibiotic-Resistant *Escherichia coli* Isolated from Intensive Pig Farming in South Africa Using 'Farm-to-Fork' Approach. *Antibiotics* **2025**, *14*, 446. https://doi.org/10.3390/antibiotics14050446.
6. Belkova, N.; Nemchenko, U.; Klimenko, E.; Smurova, N.; Zugeeva, R.; Sukhoreva, M.; Sinkov, V.; Savilov, E. Resistance of *Pseudomonas aeruginosa* to Antibiotics during Long-Term Persistence in Patients with Cystic Fibrosis. *Antibiotics* **2025**, *14*, 302. https://doi.org/10.3390/antibiotics14030302.
7. Ishihara, K.; Someno, S.; Matsui, K.; Nakazawa, C.; Abe, T.; Harima, H.; Omatsu, T.; Ozawa, M.; Iwabuchi, E.; Asai, T. Determination of Antimicrobial Resistance Megaplasmid-Like pESI Structures Contributing to the Spread of *Salmonella Schwarzengrund* in Japan. *Antibiotics* **2025**, *14*, 288. https://doi.org/10.3390/antibiotics14030288.
8. Wolde, D.; Eguale, T.; Medhin, G.; Haile, A.F.; Alemayehu, H.; Mihret, A.; Pirs, M.; Strašek Smrdel, K.; Avberšek, J.; Kušar, D.; et al. Genomic Characterization of Extended-Spectrum β-Lactamase-Producing and Third-Generation Cephalosporin-Resistant *Escherichia coli* Isolated from Stools of Primary Healthcare Patients in Ethiopia. *Antibiotics* **2024**, *13*, 851. https://doi.org/10.3390/antibiotics13090851.
9. Hetsa, B.A.; Asante, J.; Mbanga, J.; Ismail, A.; Abia, A.L.K.; Amoako, D.G.; Essack, S.Y. Genomic Characterization of Methicillin-Resistant and Methicillin-Susceptible *Staphylococcus aureus* Implicated in Bloodstream Infections, KwaZulu-Natal, South Africa: A Pilot Study. *Antibiotics* **2024**, *13*, 796. https://doi.org/10.3390/antibiotics13090796.
10. Yaikhan, T.; Suwannasin, S.; Singkhamanan, K.; Chusri, S.; Pomwised, R.; Wonglapsuwan, M.; Surachat, K. Genomic Characterization of Multidrug-Resistant *Enterobacteriaceae* Clinical Isolates from Southern Thailand Hospitals: Unraveling Antimicrobial Resistance and Virulence Mechanisms. *Antibiotics* **2024**, *13*, 531. https://doi.org/10.3390/antibiotics13060531.
11. Tóth, K.; Damjanova, I.; Laczkó, L.; Buzgó, L.; Lesinszki, V.; Ungvári, E.; Jánvári, L.; Hanczvikkel, A.; Tóth, Á.; Szabó, D. Genomic Epidemiology of C2/H30Rx and C1-M27 Subclades of *Escherichia coli* ST131 Isolates from Clinical Blood Samples in Hungary. *Antibiotics* **2024**, *13*, 363. https://doi.org/10.3390/antibiotics13040363.

References

1. Antimicrobial resistance: A silent pandemic. *Nat. Commun.* **2024**, *15*, 6198. [CrossRef] [PubMed]
2. Waddington, C.; Carey, M.E.; Boinett, C.J.; Higginson, E.; Veeraraghavan, B.; Baker, S. Exploiting genomics to mitigate the public health impact of antimicrobial resistance. *Genome Med.* **2022**, *14*, 15. [CrossRef] [PubMed]
3. Sherry, N.L.; Lee, J.Y.H.; Giulieri, S.G.; Connor, C.H.; Horan, K.; Lacey, J.A.; Lane, C.R.; Carter, G.P.; Seemann, T.; Egli, A.; et al. Genomics for antimicrobial resistance—Progress and future directions. *Antimicrob. Agents Chemother.* **2025**, *69*, e0108224. [CrossRef] [PubMed]

antibiotics

MDPI

Brief Report

Comparative Genomics of DH5α-Inhibiting *Escherichia coli* Isolates from Feces of Healthy Individuals Reveals Common Co-Occurrence of Bacteriocin Genes with Virulence Factors and Antibiotic Resistance Genes

Shuan Er [1,†], Yichen Ding [1,†], Linda Wei Lin Tan [2], Yik Ying Teo [2,3], Niranjan Nagarajan [4,5] and Henning Seedorf [1,6,*]

1 Temasek Life Sciences Laboratory, 1 Research Link, Singapore 117604, Singapore; shuan_er@yahoo.com.sg (S.E.); yichending@hotmail.com (Y.D.)
2 Saw Swee Hock School of Public Health, National University of Singapore, 12 Science Drive 2, Singapore 117549, Singapore; linda_tan@nus.edu.sg (L.W.L.T.); ephtyy@nus.edu.sg (Y.Y.T.)
3 Department of Statistics and Data Science, National University of Singapore, Singapore 117546, Singapore
4 Genome Institute of Singapore, A*STAR, Singapore 138672, Singapore; nagarajann@gis.a-star.edu.sg
5 Yong Loo Lin School of Medicine, National University of Singapore, Singapore 117597, Singapore
6 Department of Biological Sciences, National University of Singapore, Singapore 117558, Singapore
* Correspondence: henning@tll.org.sg
† These authors contributed equally to this work.

Abstract: Background/Objectives: The presence of multi-drug-resistant (MDR) bacteria in healthy individuals poses a significant public health concern, as these strains may contribute to or even facilitate the dissemination of antibiotic resistance genes (ARGs) and virulence factors. In this study, we investigated the genomic features of antimicrobial-producing *Escherichia coli* strains from the gut microbiota of healthy individuals in Singapore. **Methods**: Using a large-scale screening approach, we analyzed 3107 *E. coli* isolates from 109 fecal samples for inhibitory activity against *E. coli* DH5α and performed whole-genome sequencing on 37 representative isolates. **Results**: Our findings reveal genetically diverse strains, with isolates belonging to five phylogroups (A, B1, B2, D, and F) and 23 unique sequence types (STs). Bacteriocin gene clusters were widespread (92% of isolates carried one or more bacteriocin gene clusters), with colicins and microcins dominating the profiles. Notably, we identified an *hcp-et3-4* gene cluster encoding an effector linked to a Type VI secretion system. Approximately 40% of the sequenced isolates were MDR, with resistance for up to eight antibiotic classes in one strain (strain D96). Plasmids were the primary vehicles for ARG dissemination, but chromosomal resistance determinants were also detected. Additionally, over 55% of isolates were classified as potential extraintestinal pathogenic *E. coli* (ExPEC), raising concerns about their potential pathogenicity outside the intestinal tract. **Conclusions**: Our study highlights the co-occurrence of bacteriocin genes, ARGs, and virulence genes in gut-residing *E. coli*, underscoring their potential role in shaping microbial dynamics and antibiotic resistance. While bacteriocin-producing strains show potential as probiotic alternatives, careful assessment of their safety and genetic stability is necessary for therapeutic applications.

Keywords: bacteriocin; antimicrobial resistance; multi-drug resistance; extraintestinal pathogenic *E. coli* (ExPEC); gut microbiome

1. Introduction

The emergence of multi-drug-resistant (MDR) microorganisms is posing serious threats to global health care and raising calls for novel approaches to prevent the spread of infectious diseases. The rising prevalence of MDR strains in healthy subjects is, in this regard, particularly worrisome, as such individuals could contribute to the dissemination of resistance genes into populations and/or into pathogenic bacteria. In previous studies, we reported a high abundance of a multi-drug-resistant *E. coli* strain, 94EC, in the feces of a healthy subject that had not been treated with antibiotics prior to sampling [1,2]. This strain harbored resistance genes for seven different classes of antibiotics, including last-line antibiotics, such as tigecycline and colistin [3]. It was noted that 94EC colonies displayed antimicrobial activity against other Enterobacteriaceae strains on agar plates (previously unpublished). The underlying causes for the antimicrobial activity and high abundance in the host remained unclear, and whole-genome sequencing of the strain was therefore performed in order to analyze the genome for potential contributing features.

Genome analysis revealed the presence of several bacteriocin gene clusters within the strain (previously unpublished). Bacteriocins are known to have important roles in ecology and microbial population dynamics and have also been of interest for applications in biotechnological processes or as potential alternatives for antibiotic therapeutics [4–6]. However, their co-occurrence with antibiotic resistance genes and their role in the spread of antibiotic resistance or in stabilizing the resistome are less well understood. Such combinations may influence microbial fitness, strain persistence, and the spread of resistance, but, as of yet, their ecological and clinical implications are unclear.

To address this knowledge gap, we performed a broader screen for bacteriocinogenic *E. coli* strains in fecal samples from healthy individuals, identifying isolates that exhibited antimicrobial activity against *E. coli* DH5α. The main objective of this study was to investigate the genomic features of bacteriocin-producing *E. coli* from the gut microbiota of non-clinical subjects, with a particular focus on the co-occurrence of bacteriocin genes, ARGs, and virulence factors. In doing so, we sought to assess their potential implications for gut microbial ecology and the dissemination of antibiotic resistance.

2. Results

We screened 3107 colonies from 109 fecal samples of healthy individuals in the Singapore Integrative Omics Study for antimicrobial activity against *E. coli* DH5α using a large-scale spot-on-lawn assay (Figure 1). Colonies were initially selected on MacConkey agar to enrich for gram-negative bacteria. One representative inhibitory isolate was chosen per positive sample. The previously described MDR strain 94EC was included due to its clinical relevance. In total, 38 *E. coli* isolates were selected for whole-genome comparative analysis. The presence of bacteriocin genes, antibiotic resistance genes, and virulence factors was predicted as described in the Section 4.

The *E. coli* isolates (Figure 2) were predicted to belong to five distinct phylogroups. The detected phylogroups were A (15/38, 39.5%) and D (8/38, 21.1%), followed by B1 (6/38, 15.8%), B2 (5/38, 13.2%), and then F (4/38, 10.5%). All 38 isolates were assigned to 23 unique sequence types (STs) and one unassigned (D96). Of note, ST10 was the most prevalent (8 isolates), followed by ST69 (4 isolates); among the remaining STs, five were doubletons and 17 were singletons (including one unassigned). The core genome phylogenetic tree illustrates the distribution of phylogroups and STs. Specifically, phylogroups B2 and F fall into a distinct group (Group I) comprising nine isolates belonging to seven unique STs while A, B1 and D make up another group (Group II) consisting of 29 isolates belonging to 17 unique STs.

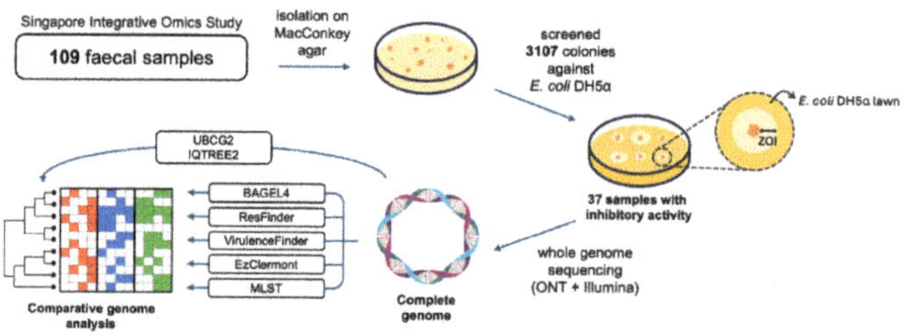

Figure 1. Schematic overview of workflow: fecal samples were screened on MacConkey agar and then against *E. coli* DH5α for inhibitory activity. Whole-genome sequencing and comparative genomic analysis were carried out for positive isolates.

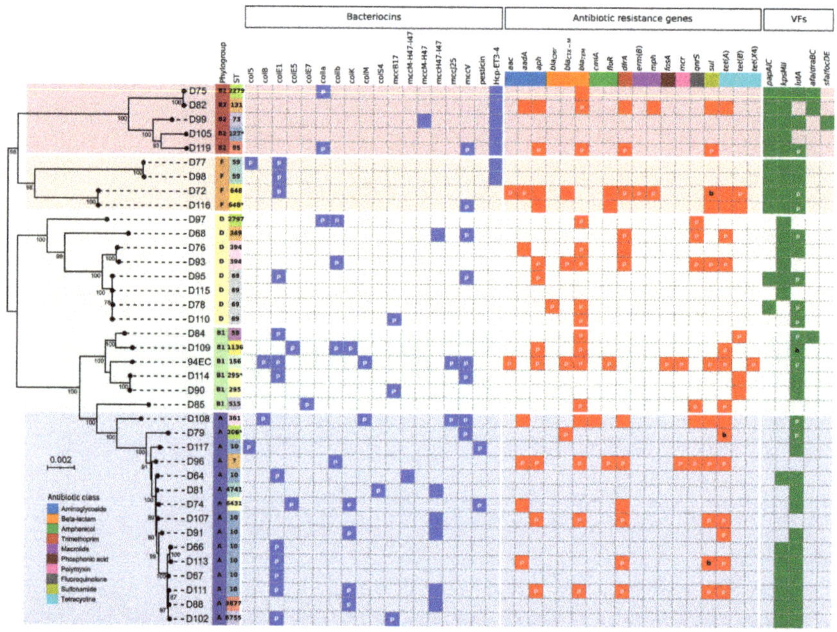

Figure 2. Heatmap illustrating distribution of genomic factors among *E. coli* isolates. Presence of bacteriocin, antibiotic resistance, and virulence factor gene clusters are marked by colored tiles of blue, red, and green, respectively. Non-solid tiles indicate absence of genes. Genomic location of genes is represented by letters: none—chromosome, p—plasmid, and b—both chromosome and plasmid. Maximum likelihood phylogenetic tree was constructed using concatenated alignment of 81 core genes supported by 1000 ultrafast bootstrap replicates. Only bootstrap values ≥ 70% are shown. Scale bar: 0.002 nucleotide substitutions per site.

Bacteriocin gene cluster prediction revealed a vast spread of bacteriocin types among 35 out of the 38 isolates (92%). A total of 17 bacteriocin variants were identified, of which tenwere colicins, six are microcins and one pesticin. Of the 38 isolates, seven harbored 3–5 bacteriocin variants, 28 harbored 1–2 variants, and three (found exclusively in phylogroup D) did not possess any known bacteriocins despite producing inhibitory activity against *E. coli* DH5α growth. ColE1 was the most frequently occurring bacteriocin, identified in 13 out of the 38 isolates (34.2%). The majority of ColE1 carriers belonged to phylogroups A, B1, and F and were absent from phylogroup B2, and only one out of nine phylogroup D isolates possessed this bacteriocin.

Interestingly, we identified a chromosomal *hcp-et3-4* gene cluster in seven out of the 38 isolates. Although the *hcp-et3-4* gene product is an effector of the Type VI secretion system (T6SS), it contains the Pyocin S (ET3) and the Colicin-DNase (ET4) bacteriocin domains [7]; hence, we classified this gene cluster as a bacteriocin variant for the scope of this study. We found that all isolates harboring *hcp-et3-4* were restricted to Group I—phylogroup B2 had five isolates and phylogroup F had two isolates. Among the other chromosomally encoded gene clusters, we also identified the siderophore-microcins MccM-H47 and MccM-H47–I47. Of note, the MccM-H47–I47 cluster was the most prevalent, and it was exclusively carried by the Group II isolates—phylogroup A had six isolates and phylogroup D had one isolate. The other two clusters were each present in a single isolate.

The ARG profiles of the isolates can provide valuable insights into the gut antibiotic resistance landscape of healthy individuals in Singapore. The prediction of ARG clusters identified resistance to 12 unique classes of antibiotics—the most prevalent being tetracycline and beta-lactam (each 18/38, 47.4%), followed by aminoglycoside (15/38, 39.5%), trimethoprim (12/38, 31.6%), sulfonamide (11/38, 28.9%), and fluoroquinolone (6/38, 15.8%). About one-third of all isolates were predicted to not carry any known ARGs.

Strains are considered multi-drug resistant (MDR) if they are resistant to at least three distinct classes of antibiotics [8]. Our analysis revealed that nearly 40% of isolates (15/38) were MDR. Most prominently, D96 was predicted to be resistant to eight classes; D72, D108, and 94EC were resistant to seven classes; and D82 and D93 were resistant to six classes. A majority of ARGs were located on plasmids (68.7%); 28.4% were located on the chromosome only, and 2.9% were found on both. Interestingly, eight isolates were predicted to have most, if not all, ARGs located on the chromosome, out of which half of them were MDR. This presents an opportunity for investigating the evolution of the intrinsic (chromosomal) resistome, which is necessary for predicting the likelihood of emergence of antibiotic resistance in bacterial populations [9].

In addition to bacteriocin gene and ARG profiling, the diversity of virulence factors (VFs) among these isolates was also elucidated. On average, the Group I isolates carried substantially more VFs than the Group II isolates (Group I: 33.2 ± 6.9 VFs; Group II: 22.9 ± 5.7 VFs). The isolates of phylogroups B2 and F (Group I) have frequently been implicated as extraintestinal pathogenic *E. coli* (ExPEC) in humans, capable of causing infections outside of the gastrointestinal tract. In our dataset, we found that 55.3% (21/38) of the isolates were considered ExPEC ($\geq$2 ExPEC VF markers [10]). A further breakdown revealed that two isolates (D75 and D99) carried four ExPEC markers; seven isolates carried three ExPEC markers; and 12 isolates harbored two ExPEC markers. Indeed, it was observed that all Group I isolates were ExPEC strains and possessed $\geq$3 ExPEC markers (except D105), whereas slightly less than half of the Group II isolates were ExPEC, and all of these Group II isolates carried $\leq$2 markers, except D95 (phylogroup D)—which carried 3 markers. ExPEC strains are of clinical significance, as they can cause a myriad of non-intestinal infections in the human body and have the ability to transmit resistance genes to other pathogenic bacteria. Our study showed a relatively high proportion of ExPEC strains present in the feces of healthy human subjects. Furthermore, about one-third of these were MDR, thereby raising concerns of potential problematic infections by these opportunistic pathogens.

3. Discussion

Here, we show evidence for a diverse profile of DH5α-inhibiting *E. coli* in a healthy Singaporean cohort. These strains were made up of at least 23 unique STs belonging to five phylogroups, with phylogroup A being the most prevalent. Our findings contrast with a previous study conducted on clinical samples that reported a higher prevalence in phylogroup B2 *E. coli* [8]. A possible reason for this observation could be the difference in sample sources—isolates in this study were derived from fecal samples of healthy subjects, whereas Šmajs et al. analyzed isolates from urinary tract infections (UTIs). Phylogroup B2 *E. coli* are often associated with UTIs, and Mcc-H47 has been suggested to be an important determinant for facilitating the colonization and subsequent emergence of phylogroup B2 strains from the intestinal reservoir [9,10]. Hence, this may explain the higher incidence of Mcc-H47 observed in clinical UTI samples compared with our dataset.

Most of the obtained isolates (92%) carried bacteriocins. Our findings are concordant with the fact that colicins are typically plasmid encoded, while microcins can be either plasmid or chromosomally encoded [7]. Furthermore, a large proportion of them were MDR and ExPEC strains based on bioinformatic prediction. Approximately 70% of the detected ARGs were located on plasmids that could then act as molecular vehicles to spread resistance genes to highly virulent pathogens, thus posing a threat to treatment using conventional antibiotics. Further experimental research will be necessary to investigate the in vivo impact of the co-occurrence of bacteriocin genes, ARGs, and virulence factors in these strains. In parallel, it will also be important to assess whether these isolates have the potential to suppress multi-drug-resistant bacteria under in vivo conditions. Such findings would support recent suggestions that bacteriocins could serve as viable antimicrobial agents [4]. A key advantage lies in the specificity of the bacteriocins that the isolates produce that can selectively target pathogenic bacteria (typically close relatives of the bacteriocin producer); thus, the narrow antimicrobial spectrum minimizes disruption to the surrounding microbiota. For instance, Mortzfeld et al. showed that Mcc-I47 exhibits specific inhibitory activity against Enterobacteriaceae strains, and its potency is comparable to commonly prescribed antimicrobials [11]. The targeted antimicrobial activity of bacteriocins, combined with their origin as commensal gut microbes, makes the isolates potentially suitable as probiotic candidates. However, safety and efficacy must be ensured by selecting isolates devoid of ARGs and can also be further optimized by genetic engineering to remove undesirable genomic elements. Lastly, since our analysis was conducted on isolates from healthy individuals, it will be essential to compare these findings with *E. coli* strains isolated from clinical settings to better understand their potential implications for human health.

4. Materials and Methods

4.1. Isolation and Screening

Fecal samples from 109 individuals from the 2018 Singapore Integrative Omics Study were collected and processed as described previously [1–3]. Fecal matter was inoculated in Luria–Bertani (LB) broth for 3 h of shaking at 200 rpm at 37 °C and then plated onto MacConkey agar and incubated overnight at 37 °C. Colonies ($n = 3107$) were picked and spotted onto a lawn of *Escherichia coli* DH5α, which was chosen due to its non-pathogenic and non-bacteriocin-producing qualities. The strain is highly susceptible to bacteriocins produced by other strains. The plate was incubated at 37 °C for 18 h before checking for the zone of inhibition of *Escherichia coli* DH5α. Colonies of *Escherichia coli* DH5α served as negative control, with 94EC serving as positive control. Colonies that had a visible zone of inhibition larger than 0.5–1 mm were selected for further processing (see supporting Figure S1 for an example). DNA extraction and whole-genome sequencing were performed for 38 isolates that exhibited inhibitory activity.

4.2. DNA Extraction and Sequencing

Genomic DNA was extracted using Qiagen Genomic-tip 20/G according to the manufacturer's protocol. DNA libraries were prepared using an ONT rapid barcoding kit (SQK-RBK004) and sequenced on MinION R9.4.1 flow cells. Sequenced reads were then basecalled using guppy on high accuracy mode. Genomic DNA was also sequenced on NovaSeq 6000 PE150 by a third-party vendor (NovogeneAIT) to generate short reads. Hybrid genome assembly was performed using Unicycler with default parameters [12]. Complete plasmid sequences were obtained for the majority of isolates with circularized assemblies; where this was not possible, contigs carrying plasmid replicon sequences were used to infer plasmid origin [13].

4.3. Bioinformatics

The presence of bacteriocin gene clusters was predicted with a BAGEL4 webserver [14]. Antibiotic resistance and virulence factor gene clusters were predicted using ResFinder v4.5.0 and VirulenceFinder v2.0.5 at 90% identity and coverage with other default parameters [15–17]. Isolates were considered ExPEC when $\geq$2 ExPEC VF markers were detected [10]. Phylogroups and sequence types were determined with EzClermont and MLST v2.0.9 using default parameters [18,19]. The phylogenetic tree was inferred by maximum likelihood using IQ-TREE v2.3.6 supported with 1000 replications approximated by an ultrafast bootstrap on a concatenated multiple sequence alignment of 81 single-copy core genes generated from a UBCG2 pipeline using default parameters [20–22]. The heatmap was generated in Python 3 using the pandas, seaborn, and matplotlib modules.

Supplementary Materials: The following supporting information can be downloaded at: https://www.mdpi.com/article/10.3390/antibiotics14090860/s1, Figure S1: Representative image of spot-on-lawn assay. Colonies of inhibitory strains were picked and spotted onto a lawn of *Escherichia coli* DH5α, which was chosen as it is non-pathogenic and also non-bacteriocin producing.

Author Contributions: Conceptualization, S.E., Y.D. and H.S.; methodology, S.E. and Y.D.; software, S.E. and Y.D.; validation, S.E. and Y.D.; formal analysis S.E. and Y.D.; investigation, S.E., Y.D. and H.S.; resources, L.W.L.T., Y.Y.T., N.N. and H.S.; data curation, S.E. and Y.D.; writing—original draft preparation, S.E.; writing—review and editing, S.E., Y.D., Y.Y.T., N.N. and H.S.; visualization, S.E.; supervision, H.S.; project administration, Y.Y.T., N.N. and H.S. All authors have read and agreed to the published version of the manuscript.

Funding: This research received no external funding.

Institutional Review Board Statement: This study was conducted in compliance with the Declaration of Helsinki and national and institutional standards. The collection of samples for this study was approved under the National University of Singapore Institutional Review Board code H-17–026.

Informed Consent Statement: Informed consent was obtained from all subjects involved in the study.

Data Availability Statement: Data of sequenced genomes are available under NCBI BioProject ID PRJNA1258756.

Acknowledgments: The experimental work for this study was supported by Temasek Life Sciences Laboratory.

Conflicts of Interest: The authors declare no conflict of interest.

References

1. Ding, Y.; Er, S.; Tan, A.; Gounot, J.-S.; Saw, W.-Y.; Tan, L.W.L.; Teo, Y.Y.; Nagarajan, N.; Seedorf, H. Comparison of *tet*(X4)-containing contigs assembled from metagenomic sequencing data with plasmid sequences of isolates from a cohort of healthy subjects. *Microbiol. Spectr.* **2024**, *12*, e0396923. [CrossRef]

2. Ding, Y.; Saw, W.-Y.; Tan, L.W.L.; Moong, D.K.N.; Nagarajan, N.; Teo, Y.Y.; Seedorf, H. Emergence of tigecycline- and eravacycline-resistant Tet(X4)-producing Enterobacteriaceae in the gut microbiota of healthy Singaporeans. *J. Antimicrob. Chemother.* **2020**, *75*, 3480–3484. [CrossRef]
3. Ding, Y.; Saw, W.-Y.; Tan, L.W.L.; Moong, D.K.N.; Nagarajan, N.; Teo, Y.Y.; Seedorf, H. Extended-Spectrum β-Lactamase-Producing and *mcr-1*-Positive *Escherichia coli* from the Gut Microbiota of Healthy Singaporeans. *Appl. Environ. Microbiol.* **2021**, *87*, e0048821. [CrossRef]
4. Cotter, P.D.; Ross, R.P.; Hill, C. Bacteriocins—A viable alternative to antibiotics? *Nat. Rev. Microbiol.* **2013**, *11*, 95–105. [CrossRef]
5. Heilbronner, S.; Krismer, B.; Brötz-Oesterhelt, H.; Peschel, A. The microbiome-shaping roles of bacteriocins. *Nat. Rev. Microbiol.* **2021**, *19*, 726–739. [CrossRef] [PubMed]
6. Riley, M.A.; Wertz, J.E. Bacteriocins: Evolution, ecology, and application. *Annu. Rev. Microbiol.* **2002**, *56*, 117–137. [CrossRef] [PubMed]
7. Ma, J.; Pan, Z.; Huang, J.; Sun, M.; Lu, C.; Yao, H. The Hcp proteins fused with diverse extended-toxin domains represent a novel pattern of antibacterial effectors in type VI secretion systems. *Virulence* **2017**, *8*, 1189–1202. [CrossRef]
8. Magiorakos, A.P.; Srinivasan, A.; Carey, R.B.; Carmeli, Y.; Falagas, M.E.; Giske, C.G.; Harbarth, S.; Hindler, J.F.; Kahlmeter, G.; Olsson-Liljequist, B.; et al. Multidrug-resistant, extensively drug-resistant and pandrug-resistant bacteria: An international expert proposal for interim standard definitions for acquired resistance. *Clin. Microbiol. Infect.* **2012**, *18*, 268–281. [CrossRef]
9. Martínez, J.L.; Baquero, F.; Andersson, D.I. Predicting antibiotic resistance. *Nat. Rev. Microbiol.* **2007**, *5*, 958–965. [CrossRef] [PubMed]
10. Johnson, J.R.; Murray, A.C.; Gajewski, A.; Sullivan, M.; Snippes, P.; Kuskowski, M.A.; Smith, K.E. Isolation and molecular characterization of nalidixic acid-resistant extraintestinal pathogenic *Escherichia coli* from retail chicken products. *Antimicrob. Agents Chemother.* **2003**, *47*, 2161–2168. [CrossRef]
11. Chakraborty, A.; Saralaya, V.; Adhikari, P.; Shenoy, S.; Baliga, S.; Hegde, A. Characterization of *Escherichia coli* Phylogenetic Groups Associated with Extraintestinal Infections in South Indian Population. *Ann. Med. Health Sci. Res.* **2015**, *5*, 241–246. [CrossRef]
12. Wick, R.R.; Judd, L.M.; Gorrie, C.L.; Holt, K.E. Unicycler: Resolving bacterial genome assemblies from short and long sequencing reads. *PLoS Comput. Biol.* **2017**, *13*, e1005595. [CrossRef]
13. Carattoli, A.; Zankari, E.; García-Fernández, A.; Voldby Larsen, M.; Lund, O.; Villa, L.; Møller Aarestrup, F.; Hasman, H. In silico detection and typing of plasmids using PlasmidFinder and plasmid multilocus sequence typing. *Antimicrob. Agents Chemother.* **2014**, *58*, 3895–3903. [CrossRef]
14. van Heel, A.J.; de Jong, A.; Song, C.; Viel, J.H.; Kok, J.; Kuipers, O.P. BAGEL4: A user-friendly web server to thoroughly mine RiPPs and bacteriocins. *Nucleic Acids Res.* **2018**, *46*, W278–W281. [CrossRef]
15. Bortolaia, V.; Kaas, R.S.; Ruppe, E.; Roberts, M.C.; Schwarz, S.; Cattoir, V.; Philippon, A.; Allesoe, R.L.; Rebelo, A.R.; Florensa, A.F.; et al. ResFinder 4.0 for predictions of phenotypes from genotypes. *J. Antimicrob. Chemother.* **2020**, *75*, 3491–3500. [CrossRef]
16. Joensen, K.G.; Scheutz, F.; Lund, O.; Hasman, H.; Kaas, R.S.; Nielsen, E.M.; Aarestrup, F.M. Real-Time Whole-Genome Sequencing for Routine Typing, Surveillance, and Outbreak Detection of Verotoxigenic *Escherichia coli*. *J. Clin. Microbiol.* **2014**, *52*, 1501–1510. [CrossRef] [PubMed]
17. Tetzschner, A.M.M.; Johnson, J.R.; Johnston, B.D.; Lund, O.; Scheutz, F. In Silico Genotyping of Escherichia coli Isolates for Extraintestinal Virulence Genes by Use of Whole-Genome Sequencing Data. *J. Clin. Microbiol.* **2020**, *58*, e01269-20. [CrossRef]
18. Larsen, M.V.; Cosentino, S.; Rasmussen, S.; Friis, C.; Hasman, H.; Marvig, R.L.; Jelsbak, L.; Sicheritz-Pontén, T.; Ussery, D.W.; Aarestrup, F.M.; et al. Multilocus Sequence Typing of Total-Genome-Sequenced Bacteria. *J. Clin. Microbiol.* **2012**, *50*, 1355–1361. [CrossRef]
19. Waters, N.R.; Abram, F.; Brennan, F.; Holmes, A.; Pritchard, L. Easy phylotyping of *Escherichia coli* via the EzClermont web app and command-line tool. *Access Microbiol.* **2020**, *2*, e000143. [CrossRef] [PubMed]
20. Hoang, D.T.; Chernomor, O.; von Haeseler, A.; Minh, B.Q.; Vinh, L.S. UFBoot2: Improving the Ultrafast Bootstrap Approximation. *Mol. Biol. Evol.* **2018**, *35*, 518–522. [CrossRef] [PubMed]
21. Kim, J.; Na, S.-I.; Kim, D.; Chun, J. UBCG2: Up-to-date bacterial core genes and pipeline for phylogenomic analysis. *J. Microbiol.* **2021**, *59*, 609–615. [CrossRef] [PubMed]
22. Minh, B.Q.; Schmidt, H.A.; Chernomor, O.; Schrempf, D.; Woodhams, M.D.; von Haeseler, A.; Lanfear, R. IQ-TREE 2: New Models and Efficient Methods for Phylogenetic Inference in the Genomic Era. *Mol. Biol. Evol.* **2020**, *37*, 1530–1534. [CrossRef] [PubMed]

antibiotics

MDPI

Article

Unique Regulation of Sed-1 β-Lactamase in *Citrobacter sedlakii*: Insights on Resistance to Third-Generation Cephalosporin

Mako Watanabe [1,†], Ryuichi Nakano [1,*,†], Keizo Yamamoto [2], Akiyo Nakano [1], Yuki Suzuki [1], Kai Saito [1], Satoko Nakashima [3], Kentaro Endo [3], Kazuya Narita [3] and Hisakazu Yano [1]

1 Department of Microbiology and Infectious Diseases, Nara Medical University, 840 Shijo-cho, Kashihara 6348521, Nara, Japan
2 Department of Chemistry, Nara Medical University, 88 Shijo-cho, Kashihara 6340813, Nara, Japan
3 Division of Central Clinical Laboratory, Iwate Medical University Hospital, 1-1-1 Idaidori, Yahaba-cho, Shiwa-gun 0283694, Iwate, Japan
* Correspondence: rnakano@naramed-u.ac.jp
† These authors contributed equally to this work.

Abstract: Background: The *Citrobacter* genus harbors class C (AmpC) and class A β-lactamases. *Citrobacter freundii* produces an inducible AmpC β-lactamase controlled by the LysR-type transcriptional regulator AmpR and cytosolic amidase AmpD. *Citrobacter sedlakii* produces the class A β-lactamase Sed-1, whose expression is believed to be regulated by the transcriptional regulator SedR and AmpD. **Objectives:** *C. sedlakii* NR2807, isolated in Japan, is resistant to third-generation cephalosporins and displays extended-spectrum β-lactamase characteristics. Here, we sought to understand the mechanism for successful resistance to third-generation cephalosporins by investigating the regulators controlling Sed-1 production. **Methods:** Plasmids containing $bla_{\text{Sed-1}}$ and *sedR* (pCR2807) or truncated *sedR* (pCR2807ΔSedR) were constructed and introduced into *Escherichia coli*. Antibiotic-resistant mutants of NR2807 were obtained, and enzyme kinetics were assessed. **Results:** The AmpD mutant (pCR2807/ML4953) showed an 8-fold increase in cefotaxime MIC and an 8.46-fold increase in Sed-1 activity compared to the wild-type (pCR2807/ML4947). However, induction of pCR2807/ML4947 also led to a 1.32-fold higher Sed-1 activity, indicating semi-inducibility. Deletion of *sedR* (pCR2807ΔSedR/ML4947) led to a 4-fold decrease in cefotaxime MIC and 1.93-fold lower Sed-1 activity, confirming SedR as an activator. While wild-type *C. sedlakii* ATCC51115 is susceptible to third-generation cephalosporins, the AmpD mutation in NR2807 led to Sed-1 overproduction and resistance to this class of antibiotics. Finally, mutagenesis revealed that amino acid substitution in Sed-1 conferred resistance to ceftazidime and extended-spectrum β-lactamase characteristics. **Conclusions:** Sed-1 producers, though usually susceptible to third-generation cephalosporins, may develop extended-spectrum β-lactamase traits due to AmpD or Sed-1 mutations, thereby requiring careful monitoring.

Keywords: *Citrobacter*; Sed-1 β-lactamase; regulation; expression; ESBL; mutant

1. Introduction

Antimicrobial resistance represents a significant global public health concern, with β-lactamase-producing bacteria undermining the effectiveness of β-lactam antibiotics [1,2]. Among Enterobacterales, extended-spectrum β-lactamases (ESBLs) have drawn significant attention due to their ability to hydrolyze third-generation cephalosporins (3GC), thereby limiting the scope of critical antibiotics used in clinical settings [3]. β-Lactamases vary among *Citrobacter* species: *Citrobacter freundii* produces a class C β-lactamase (AmpC); whereas *Citrobacter amalonaticus*, *Citrobacter gillenii*, *Citrobacter koseri*, and *Citrobacter sedlakii* produce class A β-lactamases, CdiA, CKO-1, and Sed-1, respectively (Table 1) [4–7].

Table 1. Chromosomal β-lactamases produced by the *Citrobacter* genus.

Species	β-Lactamase (Class)	Regulator Gene	Expression
C. freundii complex			
C. braakii	AmpC (C)	AmpR	Inducible
C. freundii	AmpC (C)	AmpR	Inducible
C. murliniae	AmpC (C)	AmpR	Inducible
C. youngae	AmpC (C)	AmpR	Inducible
C. werkmanii	AmpC (C)	AmpR	Inducible
C. portucalensis	AmpC (C)	AmpR	Inducible
C. gillenii	GIL-1 (A)	-	Constitutive
C. sedlakii	Sed-1 (A)	SedR	Inducible
C. farmeri	Sed-1 (A)	SedR	Inducible
C. rodentium	Sed-1 (A)	SedR	Inducible
C. amalonaticus	CdiA (A)	CdiR	Inducible
C. koseri	CKO-1 (A)	-	Constitutive
C. cronae	AmpC (C)	AmpR	Inducible
C. pasteurii	AmpC (C)	AmpR	Inducible

Gram-negative *C. freundii* expresses an inducible AmpC β-lactamase, which is regulated by the LysR-type transcriptional regulator *ampR* and the cytosolic N-acetylmuramyl L-alanine amidase AmpD [8–10]. Induction of AmpC involves the conversion of *ampR* from acting as a repressor—in the absence of an inducer—to becoming an activator. Certain AmpR mutants act as constant activators, causing constitutive overproduction of AmpC, regardless of the presence of an inducer (Figure 1, Supplementary Table S1) [8]. Specifically, the Asp135Asn AmpR substitution correlates with substantially higher β-lactamase activity in several Gram-negative bacteria, including *C. freundii*, *Enterobacter cloacae* complex, and *Pseudomonas aeruginosa* (Supplementary Figure S1) [11–14]. Mutations that permanently inactivate AmpD induce and increase muropeptide content in the cytoplasm and change the conformation of AmpR, repurposing it into a transcriptional activator [15,16]. Constitutive production of AmpC due to mutations in *ampR* or *ampD* is linked to successful resistance to 3GC [9,15].

In 2001, *C. sedlakii*, a member of the *C. freundii* complex, was first reported to harbor a chromosomally encoded Sed-1 (Table 1) [17]. This enzyme is associated with intrinsic resistance to narrow-spectrum β-lactams such as penicillins and narrow-spectrum cephalosporins. The LysR-type transcriptional regulator *sedR* is located upstream of $bla_{\text{Sed-1}}$. Similarly to AmpC, Sed-1 expression is thought to be regulated by *sedR* and *ampD*. However, the regulatory mechanism and strong resistance exhibited by Sed-1-producing bacteria remain poorly understood owing to limited reports of *C. sedlakii* infections. This is partly due to the difficulty of identifying *C. sedlakii* isolates using standard automated bacterial identification systems [18]. Understanding the genetic and functional characteristics of Sed-1 is essential for elucidating its role in β-lactam resistance and potential clinical implications.

Here, we sought to understand the mechanism for successful resistance to 3GC by investigating the regulators affecting Sed-1 production. To this end, we focused on *C. sedlakii* NR2807, which exhibited resistance to 3GC, a characteristic of ESBL-producing bacteria, and *C. sedlakii* ATCC51115, a 3GC-sensitive reference strain. The objectives of this study were to: (1) confirm the taxonomic identity of NR2807 using various molecular methods; (2) evaluate the genetic and phenotypic features of *C. sedlakii* strains; (3) characterize genetic variations in $bla_{\text{Sed-1}}$, *sedR*, and *ampD*; and (4) investigate the regulatory role of *sedR* and *ampD* in β-lactamase expression. Additionally, we aimed to determine the functional impact of mutations in $bla_{\text{Sed-1}}$ on cephalosporin hydrolysis and substrate specificity by analyzing kinetic parameters.

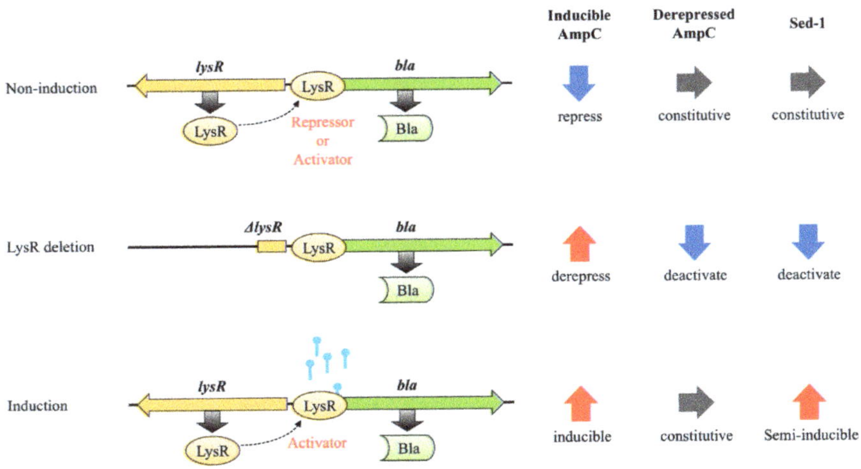

Figure 1. Schematic representation of Sed-1 and AmpC β-lactamase regulation. Inducible AmpC producers are repressed by AmpR but increase production when induced. Deletion of the *ampR* gene results in slightly higher basal expression. Derepressed AmpC producers are constitutively hyperproducing owing to mutated AmpR (Asp135Asn) with or without induction. Deletion of *ampR* results in lower expression. Sed-1 producers are semi-inducible, mirroring inducible AmpC producers. Deletion of the *sedR* gene results in reduced expression, akin to derepressed AmpC producers.

To facilitate comprehension of specialized terminology used throughout this manuscript, a glossary of technical terms has been provided in the Supplementary Materials (Supplementary Table S2).

2. Results

2.1. Identification of C. sedlakii NR2807

Microscan Walkaway showed that NR2807 was most similar to *Citrobacter farmeri* (93.15%). Matrix-assisted laser desorption/ionization-time of flight mass spectrometry (MALDI-TOF MS) identified NR2807 as *C. amalonaticus* with a similarity of 99.9%. Comparison of *C. sedlakii* ATCC51115, *C. amalonaticus* FDAARGOS_1489, *C. farmeri* FDAARGOS 1423, *C. freundii* ATCC 8090, *C. koseri* FDAARGOS_86, *Citrobacter rodentium* NBRC 105723, *Citrobacter werkmanii* FDAARGOS_1524, *Citrobacter youngae* NCTC13709, and *Citrobacter braakii* FDAARGOS 1421 yielded average nucleotide identity (ANI) values of 98.88%, 84.13%, 83.86%, 81.26%, 82.51%, 86.07%, 81.25%, 81.12%, and 81.21%, respectively. Therefore, NR2807 was identified as *C. sedlakii* and was newly registered as ST1320 (*aspC* 400, *clpX* 422, *fadD* 274, *mdh* 205, *arcA* 237, *dnaG* 196, and *lysP* 226).

2.2. Antimicrobial Susceptibility and Resistance Gene

NR2807 was resistant to cefotaxime (16 μg/mL) but was inhibited by clavulanic acid (0.5 μg/mL), showing a pattern typical of ESBL-producing strains (Table 2). *C. sedlakii* ATCC51115 was resistant to ampicillin (>256 μg/mL) but was susceptible to cefotaxime (0.5 μg/mL) and ceftazidime (2 μg/mL), which is typical for class A β-lactamase-producing strains. NR2807 showed higher MIC for β-lactams and 65.47-fold higher β-lactamase activity compared to ATCC51115. Both were susceptible to cefepime, carbapenems, levofloxacin (0.125 μg/mL), and gentamicin (1 μg/mL).

Table 2. Susceptibility and β-lactamase activity of *C. sedlakii* strains and their reconstructed transformants.

Species	Strain	Genes	AmpD	MIC (μg/mL) [a]									Relative β-Lactamase Activity (U/mg Protein) [c]		
				PIP	CTX	CTX/CLA [b]	CAZ	FEP	CMZ	CFX	ATM	IPM	Basal	Induced [d]	Induced/Basal
C. sedlakii	NR2807	$bla_{Sed\text{-}1}$, *sedR*	mutant	256	32	0.5	16	4	2	8	64	0.125	8.39	8.54	1.02
C. sedlakii	ATCC-51115	$bla_{Sed\text{-}1}$, *sedR*	wild-type	8	0.5	0.25	2	0.125	2	8	1	0.25	0.13	0.40	3.11
E. coli	pCR2807/ML4947	$bla_{Sed\text{-}1}$, *sedR*	wild-type	256	8	≤0.06	4	1	1	4	16	0.125	2.78	3.68	1.32
E. coli	pCR2807/ML4953	$bla_{Sed\text{-}1}$, *sedR*	mutant	>256	64	1	16	4	1	4	64	0.125	23.50	36.66	1.56
E. coli	pCR2807 ΔSedR/ML4947	$bla_{Sed\text{-}1}$	wild-type	64	2	≤0.06	2	0.25	1	4	8	0.125	1.44	NT [e]	-
E. coli	pCR2807 ΔSedR/ML4953	$bla_{Sed\text{-}1}$	mutant	256	4	≤0.06	2	0.5	2	4	8	0.125	3.42	NT	-
E. coli	pCR2807 ΔSedR, pAmpR 135D/ML4953	$bla_{Sed\text{-}1}$, $ampR^{wt}$	mutant	64	4	≤0.06	4	1	2	4	16	0.25	2.07	6.15	2.98
E. coli	pCR2807 ΔSedR, pAmpR 135A/ML4953	$bla_{Sed\text{-}1}$, $ampR^{mt}$	mutant	>256	64	0.125	16	4	1	4	64	0.25	20.08	23.50	1.17
E. coli	pCR51115/ML4947	$bla_{Sed\text{-}1}$, *sedR*	wild-type	128	4	≤0.06	4	1	2	4	16	0.25	2.40	3.79	1.58
E. coli	pCR51115/ML4953	$bla_{Sed\text{-}1}$, *sedR*	mutant	256	16	0.5	16	4	1	4	64	0.25	16.39	22.41	1.37
E. coli	pCR51115 ΔSedR/ML4947	$bla_{Sed\text{-}1}$	wild-type	64	2	≤0.06	2	0.25	1	4	8	0.25	1.48	NT	-
E. coli	pCR51115 ΔSedR/ML4953	$bla_{Sed\text{-}1}$	mutant	256	2	≤0.06	2	0.5	1	4	8	0.125	3.35	NT	-
E. coli	pCR51115 ΔSedR, pAmp R135D/ML4953	$bla_{Sed\text{-}1}$, $ampR^{wt}$	mutant	64	2	≤0.06	4	1	1	2	8	0.25	2.85	4.14	1.45
E. coli	pCR51115 ΔSedR, pAmpR 135A/ML4953	$bla_{Sed\text{-}1}$, $ampR^{mt}$	mutant	>256	32	≤0.06	16	4	1	2	32	0.25	14.85	16.71	1.13
E. coli	ML4947	-	wild-type	2	≤0.06	≤0.06	0.25	≤0.06	1	0.5	≤0.06	≤0.06	<0.01	NT	-
E. coli	ML4953	-	mutant	4	≤0.06	≤0.06	0.5	≤0.06	1	0.5	≤0.06	≤0.06	<0.01	NT	-

[a] Antibiotics: PIP, piperacillin; CTX, cefotaxime; CLA, clavulanic acid; CAZ, ceftazidime; FEP, cefepime; CMZ, cefmetazole; CFX, cefoxitin; ATM, aztreonam; IPM, imipenem. [b] MICs were determined in the presence of clavulanic acid (5 μg/mL). [c] β-Lactamase activities are the geometric mean values of three independent cultures. The standard deviations were within 10%. [d] 1/16× MIC of cefoxitin was used as the inducer. [e] NT, Not tested.

Whole-genome analysis with ResFinder revealed that NR2807 and ATCC51115 harbored $bla_{\text{Sed-1}}$ as the only resistance gene, with *sedR* located upstream of it in the opposite direction. Sed-1 (295 amino acid residues) from NR2807 differs from that of ATCC51115 by two amino acid substitutions (Asp175Gly and Arg228His); whereas SedR (286 amino acid residues) differs by a single amino acid substitution (Thr198Ala) and AmpD (187 amino acid residues) by four amino acid substitutions (Thr50Ala, Arg101His, Asp110Val, and Gln131Leu).

2.3. *Effect of Regulator Genes on β-Lactamase Expression*

To evaluate the effect of *sedR* and *ampD* on Sed-1 production, MIC and β-lactamase activity of Sed-1-producing transformants with wild-type or mutant AmpD were measured in the presence or absence of SedR. β-lactamase activity of the wild-type (pCR2807/ML4947) was 2.78 U/mg and it increased by 1.32-fold when induced by cefoxitin (Table 2); however, that of the AmpD mutant (pCR2807/ML4953) was 8.46-fold higher, reaching 23.50 U/mg. The MIC for cefotaxime was 8-fold higher in pCR2807/ML4953 than in pCR2807/ML4947. These findings indicate that Sed-1 was upregulated by SedR and AmpD. pCR2807ΔSedR/ML4947 showed a 4-fold lower MIC for cefotaxime and 1.93-fold lower β-lactamase activity compared to pCR2807/ML4947. In this case, SedR was not repressing Sed-1, but instead acted as a positive regulator both in the absence and especially in the presence of a β-lactam inducer.

To further elucidate the role of SedR in Sed-1 production, a plasmid carrying *ampR* was transformed into the *sedR*-deficient strain (pCR2807ΔSedR/ML4953) (Figure 2). Wild-type AmpR (pAmpR135D), derived from classical AmpC producers, functions as a repressor; whereas mutant AmpR (pAmpR135A) is in a derepressed (active) form. The wild-type *ampR* strain (pCR2807ΔSedR, pAmpR135D/ML4953) showed 1.66-fold lower Sed-1 activity (2.07 U/mg) compared to pCR2807ΔSedR/ML4953 (3.42 U/mg) (Table 2). In contrast, the mutant *ampR* strain (pCR2807ΔSedR, pAmpR135A/ML4953) showed 5.87-fold higher Sed-1 activity (20.08 U/mg) and a MIC for 3GC comparable to that of pCR2807/ML4953. These features were confirmed in ATCC51115 transformants. Therefore, SedR functions similarly to the AmpR mutant.

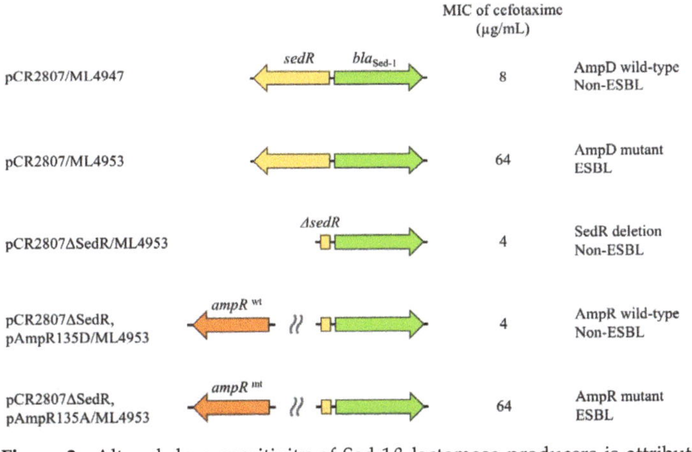

Figure 2. Altered drug sensitivity of Sed-1β-lactamase producers is attributed to mutations in regulatory genes. The AmpD mutant strain (pCR2807/ML4953) shows increased expression of Sed-1 and a higher MIC for cefotaxime (64 μg/mL), whereas *sedR* deletion (pCR2807ΔSedR/ML4953) results in decreased expression and a lower MIC (4 μg/mL). Co-expression of mutant AmpR in the *sedR*-deficient strain restores the MIC to 64 μg/mL.

2.4. Characteristics of Various Antibiotic-Resistant Mutants

To investigate the mechanism of successful antibiotic resistance by Sed-1, strains with increased MICs for antibiotics used for selection were analyzed. The antibiotic-resistant mutants appeared at a frequency of ~10^{-6}–10^{-7} for ceftazidime, cefepime, cefmetazole, aztreonam, and imipenem supplied at 2× or 4× MIC. Ten colonies of the mutants were randomly selected for each condition, and the *bla*$_{\text{Sed-1}}$, *sedR*, and *ampD* genes were sequenced and compared to those of the parent *C. sedlakii* NR2807 strain. Sequence data revealed that only *bla*$_{\text{Sed-1}}$ was altered upon ceftazidime selection. The mutant strains isolated on ceftazidime were found to possess three different sets of mutations in Sed-1 (Pro167Gln; Asp179Gly; and Ile173Met, Pro174Ala, 174_175insS) and drug sensitivity as listed in Table 3. Compared with the NR2807 parent strain, the MIC for ceftazidime was 16-fold higher, while the MIC for cefotaxime was 4- to 8-fold lower in the mutants. Strain NR4574 (Asp179Gly) also displayed a 4-fold lower MIC for aztreonam. Strain NR4575 (Ile173Met, Pro174Ala, 174_175insS) presented a 4-fold lower MIC for piperacillin. Each mutant strain was assumed to possess altered substrate specificity due to amino acid substitutions. To evaluate whether the higher MIC in ceftazidime-resistant mutants was due to mutations in Sed-1, *bla*$_{\text{Sed-1}}$, and *sedR* of strain NR4573 (Pro167Gln) were cloned into *Escherichia coli* ML4947 and compared with wild-type transformant pCR2807/ML4947. In line with the original NR4573 strain, pCR4573/ML4947 displayed higher ceftazidime MIC and lower cefotaxime MIC.

Table 3. Mutations and MICs of *C. sedlakii* NR2807 mutants.

Species	Strains	Selection [a,b]	Mutation [c]				MIC (μg/mL) [a]								
			Sed-1 [d]	SedR	AmpD	Other Mutated Genes	PIP	CTX	CTX/CLA [e]	CAZ	FEP	CMZ	CFX	ATM	IPM
C. sedlakii	NR2807	-	-	-	-	-	256	32	0.5	16	4	2	8	64	0.125
C. sedlakii	NR4573	CAZ 64	P167Q	-	-	-	128	8	0.25	256	1	2	8	128	0.125
C. sedlakii	NR4574	CAZ 64	D179G	-	-	-	128	4	0.25	>256	1	2	8	16	0.125
C. sedlakii	NR4575	CAZ 64	I173M, P174A, 174_175insS	-	-	-	64	8	0.25	>256	2	2	8	32	0.125
C. sedlakii	NR4586	FEP 8	-	-	-	*citC*, *cdsA*, *ispH*	>256	256	0.25	16	16	2	4	256	0.5
C. sedlakii	NR4062	CMZ 8	-	-	-	*rseA*	256	64	16	16	16	32	64	128	0.25
C. sedlakii	NR5701	ATM 256	-	-	-	*tsuA*, *cdsA*, *ubiD*	>256	128	0.25	16	8	1	4	256	0.5
C. sedlakii	NR4584	IPM 0.5	-	-	-	*pbpA*, *cdsA*	256	32	0.5	16	4	2	8	64	1
E. coli	pCR4573 /ML4947	-	P167Q	-	-	-	64	2	≤0.06	64	0.5	1	4	16	0.125
E. coli	pCR2807 /ML4947	-	-	-	-	-	256	8	≤0.06	4	1	4	4	16	0.125

[a] Antibiotics: PIP, piperacillin; CTX, cefotaxime; CLA, clavulanic acid; CAZ, ceftazidime; FEP, cefepime; CMZ, cefmetazole; CFX, cefoxitin; ATM, aztreonam; IPM, imipenem. [b] Numbers indicate the selected antibiotics concentration (μg/mL). [c] Abbreviations; P, Pro; Q, Gln; D, Asp; G, Gly; I, Ile; M, Met; A, Ala; ins, insertion; *citC*, [Citrate [pro-3S]-lyase] ligase; *cdsA*, phosphatidate cytidylyltransferase; *ispH*, 4-hydroxy-3-methylbut-2-enyl diphosphate reductase; *rseA*, anti-sigma-E factor; *pbpA*, penicillin-binding protein 2; *tsuA*, thiosulfate utilization transporter; *ubiD*, 3-octaprenyl-4-hydroxybenzoate carboxy-lyase. [d] Amino acid numbering corresponds to class A β-lactamase (40). [e] MICs were determined in the presence of clavulanic acid (5 μg/mL).

No mutations in *bla*$_{\text{Sed-1}}$, *sedR*, and *ampD* were found in the other antibiotic-selected mutants. Mutant strain NR4584 selected on 0.5 μg/mL imipenem showed an 8-fold higher MIC for the antibiotics used for selection. Mutant strain NR4062 selected on 8 μg/mL cefmetazole showed 16-fold higher MIC for the antibiotics used for selection,

4-fold higher MIC for cefepime, and 32-fold higher MIC for cefotaxime combined with clavulanic acid. Mutant strains NR4586 and NR5701 selected on 8 μg/mL cefepime and 256 μg/mL aztreonam, respectively, exhibited 4-fold higher MIC for the antibiotic used for selection, along with 8- and 4-fold higher MIC for cefotaxime. Genomic analysis of one representative of each antibiotic mutant strain revealed mutations in various genes (Table 3). Selection with cefepime, aztreonam, and imipenem yielded mutations in *cdsA*, a phosphatidic acid cytidyl transferase; whereas cefmetazole selection yielded mutations in *rseA*, an anti-sigma-E factor.

2.5. Kinetic Parameters of Ceftazidime-Resistant Mutants

To evaluate the substrate specificity of wild-type Sed-1 and its ceftazidime-resistant mutants, kinetic parameters were measured for *E. coli* BL21 transformants harboring wild-type $bla_{\text{Sed-1}}$ (NR2807), and ceftazidime-resistant-mutant $bla_{\text{Sed-1}}$ (NR4573, NR4574, and NR4575) (Table 4). NR2807 showed lower catalytic efficiency (k_{cat}/K_m = 1.31 s^{-1} mM^{-1}) against ceftazidime, while NR4573 (Pro167Gln), NR4574 (Asp179Gly), and NR4575 (Ile173Met, Pro174Ala, 174_175insS) showed 4.24-, 11.60-, and 30.92-fold higher hydrolytic activity (k_{cat}/K_m) than NR2807. On the one hand, K_m values were more than 17.78-fold lower in the mutants than in wild-type Sed-1, resulting in substantially higher k_{cat}/K_m. On the other hand, they showed 2.11-, 21.83-, and 1.95-fold lower hydrolytic activity against cefotaxime than NR2807, respectively.

Table 4. Kinetic parameters of ceftazidime-resistant mutants of *C. sedlakii* NR2807.

β-Lactamase and Parameter	Value for Antibiotic [a]			
	Piperacillin	Cefotaxime	Ceftazidime	Aztreonam [b,c]
Wild-type (NR2807)				
k_{cat} (s^{-1})	624 ± 31.3	230 ± 21.3	4.55 ± 0.612	9.26 ± 0.133
K_m (μM)	326 ± 29.5	264 ± 30.2	3484 ± 486	29.6 ± 2.65
k_{cat}/K_m ($s^{-1}\cdot mM^{-1}$)	1910	871	1.31	313
P167Q mutant (NR4573)				
k_{cat} (s^{-1})	163 ± 1.24	172 ± 3.12	0.649 ± 0.0542	4.42 ± 0.0496
K_m (μM)	43.0 ± 1.56	417 ± 9.19	117 ± 11.9	55.1 ± 2.77
k_{cat}/K_m ($s^{-1}\cdot mM^{-1}$)	3790	413	5.55	80.2
D179G mutant (NR4574)				
k_{cat} (s^{-1})	2.57 ± 0.0551	0.213 ± 0.0017	0.0436 ± 0.0013	NH
K_m (μM)	10.5 ± 1.22	5.34 ± 0.273	2.87 ± 0.46	NH
k_{cat}/K_m ($s^{-1}\cdot mM^{-1}$)	245	39.9	15.2	ND
I173M, P174A, 174_175insS mutant (NR4575)				
k_{cat} (s^{-1})	62.5 ± 2.13	6.06 ± 0.262	7.94 ± 0.862	NH
K_m (μM)	84.0 ± 6.55	13.6 ± 1.89	196 ± 24.1	NH
k_{cat}/K_m ($s^{-1}\cdot mM^{-1}$)	744	446	40.5	ND

[a] K_m and k_{cat} were calculated as means ± SD from three independent experiments. [b] NH, no hydrolysis was detected. [c] ND, not determined.

3. Discussion

3.1. Genomic and Clinical Significance of Sed-1 in C. Sedlakii

Sed-1 β-lactamase was reported in 2001 as a chromosomally encoded class A β-lactamase in *C. sedlakii* [17]. Few reports on *C. sedlakii* isolates followed, probably because the automated bacterial identification systems commonly used in clinical practice failed to identify this species. In this study, Gram-negative rods were identified via whole-genome ANI analysis as *C. sedlakii* strain NR2807. To the best of our knowledge, this is the first reported clinical isolation of a Sed-1 producer in Japan. According to PubMLST (https://pubmlst.org/bigsdb?db=pubmlst_cfreundii_isolates, accessed on 6 August 2025), only 13 types of STs have been attributed to *C. sedlakii*, including ST1320, which was newly registered in this study.

Whole-genome analysis showed that $bla_{Sed\text{-}1}$ was the only chromosomally encoded β-lactamase in NR2807. As in the type strain ATCC51115, it coded for the regulatory gene *sedR*. Sed-1 in NR2807 differed from its counterpart in ATCC51115 by amino acid substitutions Asp175Gly and Arg228His. These substitutions have not been related to extended changes in substrate specificity of class A β-lactamases TEM and SHV [19]. Indeed, *E. coli* transformants (pCR2807/ML4947 and pCR51115/ML4947) harboring these β-lactamases showed similar MICs.

3.2. Functional Impact of AmpD Mutation on Sed-1 Expression and Antibiotic Resistance

The MIC of NR2807 revealed resistance to a broad range of β-lactams, including 3GC, as well as ESBL-like characteristics; whereas ATCC51115 had an overall lower MIC and no ESBL-like properties (Table 2). Sed-1 production was 65.47-fold higher in NR2807 than in ATCC51115, which may have a significant impact on antimicrobial susceptibility. To elucidate the mechanism regulating Sed-1 expression, $bla_{Sed\text{-}1}$–*sedR* was cloned into *E. coli* with or without the AmpD mutation, and the effect on MIC and enzyme production was assessed. Enzyme activity experiments revealed that pCR2807/ML4947 (AmpD wild-type) and pCR2807/ML4953 (AmpD mutant) increased Sed-1 production by 1.32- and 1.56-fold following induction. Without induction, Sed-1 output was 8.46-fold higher in pCR2807/ML4953 than in pCR2807/ML4957, and the same trend was observed for MIC, suggesting that increased Sed-1 resulted in higher MIC. Accordingly, SedR causes only partial induction of Sed-1, rather than the strong inducibility typically observed in AmpR. Furthermore, Sed-1 expression was strongly influenced by AmpD mutations (Figure 1). In AmpC-producing bacteria, AmpD inactivation leads to the accumulation of cytoplasmic muropeptides, which bind to and activate the transcriptional regulator AmpR. A similar mechanism is likely at play in *C. sedlakii*, where accumulated muropeptides interact with SedR, shifting it from an inactive to an active transcriptional state. This suggests that SedR requires muropeptide binding to function as a transcriptional activator, and that AmpD mutations indirectly enhance Sed-1 expression through this activation pathway. Therefore, the AmpD mutant mimics an induced state, functionally equivalent to the "Induction" scenario depicted in Figure 1, resulting in elevated β-lactamase activity and increased resistance to 3GC.

3.3. SedR Functions as a Constitutive Activator Similar to Mutant AmpR

The *sedR*-deficient strain (pCR2807ΔSedR/ML4947) displayed 1.93-fold lower Sed-1 production and lower MIC than the *sedR*-positive strain (pCR2807/ML4947). Deletion of *ampR* in inducible AmpC producers resulted in slightly more basal *ampC*, but no further induction of enzyme synthesis (Figure 1, Supplementary Table S1) [8,20]. In the AmpR (Asp135Ala) mutant, AmpR works as an activator regardless of the presence of an inducer, and its deletion lowers *ampC* expression [12]. Given that SedR functions similarly to the mutant AmpR, wild-type AmpR, and mutant AmpR were co-inserted in a *sedR*-deficient strain, and their effect on Sed-1 expression was analyzed. The Sed-1 producer (pCR2807/ML4953) shared the same MIC and enzyme activity as the mutant AmpR-carrying strain (pCR2807ΔSedR, pAmpR135D/ML4953) (Figure 2). Transcriptional regulators of the LysR family are known to act similarly on related genes [21,22]. This result indicates that SedR functions as an activator, akin to mutant AmpR, rather than as a repressor like wild-type AmpR (Figure 3). The amino acid in SedR corresponding to position 135 of AmpR is Asn (Supplementary Figure S1). The Asp135Asn substitution in AmpR has been shown to result in an active form in various bacterial species [8,11,13,14], further confirming the activator-like function of SedR (Supplementary Table S1). Hence, expression of $bla_{\text{Sed-1}}$ is regulated by SedR, which acts like the AmpR mutant to activate Sed-1 in the absence of an inducer or aids its production in the presence of an inducer. This mechanism appears unique to Sed-1 compared to other β-lactamases (Figure 1).

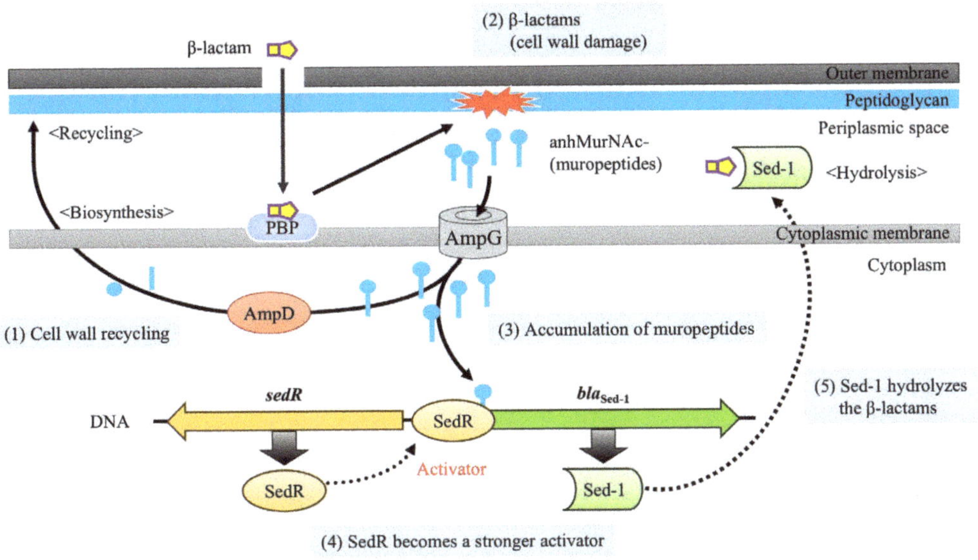

Figure 3. Schematic representation of Sed-1 β-lactamase regulation. (1) In the basal state, cell wall degradation products (muropeptides) enter the cytoplasm via AmpG and are processed by AmpD for biosynthesis or recycling. (2) When the cell wall is damaged by β-lactams, (3) muropeptides accumulate in the cytoplasm. (4) Upon interacting with muropeptides, SedR is converted into a stronger activator in a semi-inducible manner, promoting further production of Sed-1. (5) Sed-1 then hydrolyzes β-lactams in the periplasmic space.

ATCC51115 clone strains exhibited similar Sed-1 activity and MIC as NR2807 clone strains, indicating a similar expression mechanism influenced by SedR and AmpD. NR2807 has two amino acid substitutions in Sed-1 compared to ATCC51115, which may have an effect. However, increased Sed-1 production and MIC values were more likely caused by AmpD mutations than by amino acid substitutions. Therefore, NR2807 has become broadly and highly resistant due to the AmpD mutation, whereas ATCC51115 has low Sed-1 production and MIC.

3.4. Ceftazidime-Driven Mutations in Sed-1 Expand Substrate Specificity

To further elucidate the mechanism of β-lactamase expression in NR2807, experiments were conducted to generate a drug-resistant mutant strain. Similar experiments with AmpC producers, such as *C. freundii* and *E. cloacae*, yielded mutations in *ampD* and *ampR* [11,15,23], but no Sed-1-producing mutants with additional mutations in *ampD* or *sedR* could be isolated. These factors could explain the different mutation rates among bacterial species, the role of SedR as an inducible high-producing type, and NR2807 as an AmpD mutant [24]. Here, only the ceftazidime-resistant mutant strain showed an amino acid substitution in Sed-1, which likely expanded substrate specificity and, in this way—rather than via upregulation—improved resistance.

Among the amino acids involved in extended substrate specificity of class A β-lactamase mutants, single substitutions were identified in NR4573 (Pro167Gln) and NR4574 (Asp179Gly) [4,19]. The increased affinity of the mutants for ceftazidime (29.78- and 1213.94-fold decrease in *Km*, respectively) may have resulted in higher catalytic efficiency and increased MIC. Strain NR4575 had two amino acid substitutions and one insertion (Ile173Met, Pro174Ala, 174_175insS) not previously known to expand substrate specificity of class A β-lactamases. All Class A β-lactamases have a conserved structural feature, the omega loop, which spans residues 161 to 179 or 164 to 179 and forms part of the active site pocket [25–27]. The above amino acid substitutions are located in the omega-loop and likely cause a conformational change, thereby increasing enzyme activity against ceftazidime. Interestingly, Sed-1 mutants were obtained in ceftazidime only. No mutations were observed in the regulatory genes of other antibiotics, although mutations were observed in *cdsA* and *rseA* [28,29]. These genes are thought to be related to the acquisition of antibiotic resistance through extended substrate specificity.

3.5. ESBL-like Features and Future Risk in Sed-1-Producing C. Sedlakii

In the first report on Sed-1 producing *C. sedlakii*, Sed-1 showed high catalytic efficiency against narrow-spectrum β-lactams, such as aminopenicillins, carboxypenicillins, and first-generation cephalosporins. Sed-1 producers were resistant to these compounds but were susceptible to 3GC and carbapenems [17]. The wild-type ATCC51115 has a low MIC for extended-spectrum cephalosporins and is not considered an ESBL (Figure 4). Instead, NR2807 exhibited ESBL-like characteristics, such as a high MIC for extended-spectrum cephalosporins. Given reports of *C. sedlakii* carrying the ESBL gene [18,30], any inquiry into resistance mechanisms by this species will require clarification of ESBLs but also regulatory genes. Furthermore, the catalytic efficiency and MIC exhibited by the ceftazidime mutants obtained in this study clearly suggested they were ESBL producers. Given a history of ESBL types in TEM- and SHV-type Class A β-lactamases, ESBL types may also emerge in Sed-1-producing *C. sedlakii* [31,32], requiring careful monitoring.

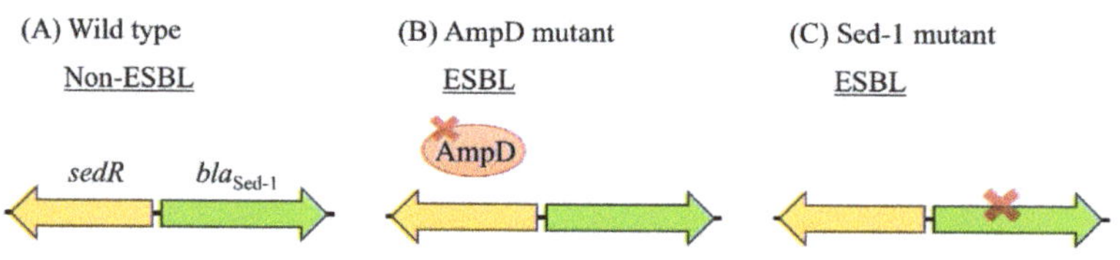

Figure 4. Schematic illustration of the evolution of Sed-1 β-lactamase activity in *C. sedlakii*. (**A**) In the wild-type strain, Sed-1 exhibits a narrow-spectrum profile and does not show ESBL properties. (**B**) In the presence of an AmpD mutation, Sed-1 expression is markedly increased, leading to ESBL-like resistance, particularly against third-generation cephalosporins (3GC). (**C**) Acquisition of Sed-1 structural mutations (e.g., in the omega-loop) can expand substrate specificity, also conferring ESBL-like properties.

3.6. Study Limitations

This study has some limitations. First, the role of SedR in Sed-1 production is not fully understood. Both strains used in the analysis were active, but there may be mutations that result in even higher output, while others may be suppressive, as in inducible AmpC producers. To elucidate the regulatory mechanism employed by SedR, the function of mutants should be analyzed. Second, several mutations were identified in *C. sedlakii* strains selected for antibiotics other than ceftazidime, but the cause of resistance to the drug remains unclear. Cloning of each candidate gene would reveal whether its mutation caused increased resistance in Sed-1-producing bacteria. Finally, this study focused on only two strains; more strains need to be collected and analyzed in detail to fully understand the mechanism of Sed-1 production. As this study reveals the possible emergence of ESBL-type Sed-1 producers, such occurrence should be carefully monitored.

4. Materials and Methods

4.1. Bacterial Strains and Antimicrobial Susceptibility Testing

Glucose-fermenting Gram-negative rod-shaped NR2807 was isolated from a blood sample of a hospitalized patient in Japan in 2018. The species was identified by MicroScan WalkAway plus (Beckman Coulter, Inc., Lane Cove West, NSW, Australia), MALDI-TOF MS (VITEK® MS; bioMérieux, North Ryde, NSW, Australia) and ANI analysis using whole-genome sequences (JSpeciesWS, Bremen, Germany) [33]. *C. sedlakii* ATCC51115 (NBRC105722) was used as the reference strain. To evaluate the effect of *sedR* and *ampD* on Sed-1 production, plasmids containing both $bla_{\text{Sed-1}}$ and *sedR* (pCR2807) or $bla_{\text{Sed-1}}$ and truncated *sedR* (pCR2807ΔSedR) were constructed using the Zero Blunt TOPO PCR Cloning Kit (Thermo Fisher Scientific, Waltham, MA, USA) with primers Sed-1down (5′-TGTCTGCGCAGGGTTCTGTTC-3′) and SedRdown (5′-TGGTACGCTGATCCCCGAAC-3′), and Sed-1down and SedRUpR (5′-CGGCGGGATGCGACAGT-3′), respectively. The plasmids were transformed into *E. coli* ML4947 (AmpD wild-type) and ML4953 (AmpD mutant) [34]. In vitro ceftazidime-resistant mutant strains were isolated from NR2807. The bacterial strains and plasmids used in this study are listed in Table 5. To evaluate the effect of LysR mutation on antimicrobial susceptibility and β-lactamase expression, *ampR* clone plasmids were transformed into *E. coli* harboring $bla_{\text{Sed-1}}$ and truncated *sedR*. The *ampR* clone plasmids pAmpR135D and pAmpR135A used in this study were previously constructed from *C. freundii* (Supplementary Figure S1) [12].

Table 5. Bacterial strains and plasmids used in this study.

Bacterial Strains or Plasmids	Characteristics [a]
Strains	
NR2807	Clinical isolate of *C. sedlakii* from Japan
ATCC51115	Reference strain of *C. sedlakii* purchased from NITE Biological Resource Center
NR4573	Ceftazidime-resistant mutant of *C. sedlakii* NR2807 with Sed-1 mutation (P167Q)
NR4574	Ceftazidime-resistant mutant of *C. sedlakii* NR2807 with Sed-1 mutation (D179G)
NR4575	Ceftazidime-resistant mutant of *C. sedlakii* NR2807 with Sed-1 mutation (I173M, P174A, 174_175insS)
ML4947	*E. coli* (F^- *galK2 galT22 hsdR hsdM lacY1 metB1 relA supE44* Rif ^{r}), cloning host with AmpD wild type
ML4953	*E. coli* (F^- *galK2 galT22 hsdR hsdM lacY1 metB1 relA supE44* Rif ^{r} *ampD9*), cloning host with AmpD mutant
TOP10	*E. coli* (F^- *mcrA* Δ(*mrr-hsdRMS-mcrBC*) φ80*lacZ*ΔM15 Δ*lacX74 nupG recA1 araD139* Δ(*ara leu*)7697 *galE*15 *galK*16 *rpsL*(*StrR*) *endA1 nupG*, cloning host for analyzing the Sed-1 β-lactamase production

Table 5. *Cont.*

Bacterial Strains or Plasmids	Characteristics [a]
BL21(DE3)	*E. coli* (F^- *ompT hsd*SB(rB^- mB^-) *gal dcm* (DE3)), cloning host for analyzing enzyme kinetic
Plasmids	
pCR2807	pCR-Blunt II-TOPO containing $bla_{\text{sed-1}}$ and *sedR* from NR2807 amplified using Sed-1down and SedRdown
pCR2807ΔSedR	pCR-Blunt II-TOPO containing $bla_{\text{sed-1}}$ from NR2807 amplified using Sed-1down and SedRUpR
pCR51115	pCR-Blunt II-TOPO containing $bla_{\text{sed-1}}$ and *sedR* from pCR51115 amplified using Sed-1down and SedRdown
pCR51115ΔSedR	pCR-Blunt II-TOPO containing $bla_{\text{sed-1}}$ from pCR51115 amplified using Sed-1down and SedRUpR
pCR4573	pCR-Blunt II-TOPO containing $bla_{\text{sed-1}}$ and *sedR* from NR4573 amplified using Sed-1down and SedRdown
pAmpR135D	pMW219 containing *ampR* fragment of wild type (AmpR135Asp) obtained from [12]
pAmpR135A	pMW219 containing *ampR* fragment of mutant (AmpR135Ala) obtained from [12]
pCR-Blunt II-TOPO	Cloning vector purchased from Thermo Fisher Scientific, Km ^{r} Zeo ^{r}
pMW219	Cloning vector purchased from Nippon Gene, Km ^{r}
pET-28a (+)	Protein expression vector purchased from Novagen, Km ^{r}

[a] Rif ^{r}, rifampin resistant; Km ^{r}, Kanamycin resistant; Zeo ^{r}, Zeocin resistant.

Antibiotic susceptibility was determined using the agar dilution method according to Clinical and Laboratory Standards Institute guidelines [35].

4.2. Whole-Genome Sequencing and Analysis

Genomic DNA of NR2807 was extracted using QIAGEN Genomic-tip 500/G. Whole genomes were sequenced using MiSeq (Illumina Inc., San Diego, CA, USA) and MinION (Oxford Nanopore Technologies, Oxford, UK), after which they were subjected to hybrid de novo assembly using Unicycler v0.5.0 [36]. Species were identified based on ANI between strain NR2807 (GenBank accession no. BAAHNL01000000.1), *C. sedlakii* ATCC51115 (GenBank accession no. GCA_000759835.1), *C. amalonaticus* FDAARGOS_1489 (GenBank accession no. GCF_020099335.1), *C. farmeri* FDAARGOS 1423 (GenBank accession no. GCA_019048065.1), *C. freundii* ATCC 8090 (GenBank accession no. CP049015.1), *C. koseri* FDAARGOS_86 (GenBank accession no. GCA_ 000783445.2), *C. rodentium* NBRC 105723 (GenBank accession no. GCA_021278985.1), *C. werkmanii* FDAARGOS_1524 (GenBank accession no. GCA_ 020341495.1), *C. youngae* NCTC13709 (GenBank accession no. GCA_ 900638065.1), and *C. braakii* FDAARGOS 1421 (GenBank accession no. GCA_019048805.1). Earlier studies recommended an ANI of 95–96% as a species demarcation cut-off [37,38].

Antimicrobial resistance genes in the genome sequence were identified by searching the ResFinder database (http://genepi.food.dtu.dk/resfinder (accessed on 6 August 2025)) using thresholds of 90% identity and 60% minimum length [39]. Amino acid numbering of Sed-1 corresponded to the class A β-lactamase [40]. Multilocus sequence typing of *Citrobacter* spp. isolates was performed as previously described [41]. Sequence types were assigned by the PubMLST database (https://pubmlst.org/organisms/citrobacter-spp (accessed on 6 August 2025)).

4.3. Measurement of β-Lactamase Activity and Induction Assays

β-Lactamase activity was measured by a colorimetric assay as previously described [34,42]. Briefly, bacterial strains were cultured in Mueller-Hinton broth to mid-logarithmic phase with shaking at 37 °C, and the protein contents in crude extracts were suspended in 50 mM phosphate buffer (pH 7.0). Enzyme activity was determined at 30 °C using a spectrophotometer (UV-1900i Plus; Shimadzu, Kyoto, Japan) with cephalothin (262 nm) as substrate. One unit of β-lactamase activity was equivalent to the amount of β-lactamase that hydrolyzed 1 μmol of β-lactam in 1 min at 30 °C. Protein concentration was measured using the Bradford assay [43]. All experiments were repeated three times, and enzymatic activity was determined as U/mg of protein. For the induction assay, bacterial strains were cultured in Mueller-Hinton broth to mid-logarithmic phase and subjected to 1/16 × MIC for cefoxitin as the inducer for 1 h [11].

4.4. Selection of Antibiotic-Resistant Strains and Detection of Sequence Changes

Antibiotic-resistant mutants were obtained by plating ~10^9 colony-forming units (CFU)/mL of late-logarithmic-phase NR2807 grown in Luria–Bertani (LB) broth on LB agar plates containing ceftazidime, cefotaxime, cefepime, cefmetazole, aztreonam, or imipenem at 2× or 4× MIC, as previously described [44]. The mutation frequency was determined by dividing the colony density in CFU/mL on LB agar plates containing the antibiotic by the total colony density in CFU/mL. DNA sequences of $bla_{\text{sed-1}}$, *sedR*, and *ampD* of selected mutants were determined by Sanger sequencing. The primers used are listed in Supplementary Table S3 [17]. Nucleotides and amino acids of selected mutants were compared to those of NR2807. For strains without mutations in these genes, whole-genome sequencing using MiSeq and annotation using DFAST were performed as above. The obtained genome sequences were compared with the sequence of NR2807 to search for mutations.

4.5. Measurement of Kinetic Parameters in NR2807 and Mutants

Amplicons of NR2807 and mutants (NR4573, NR4574, and NR4575) between primers Sed-1F-atg (5′-CTTAAAGAACGGTTTCGCCAGAC-3′) and Sed-1R+BamHI (5′-ATATGGATCCTTACTTTCCTTCCGTCAC-3′) were generated using Q5 Hot Start High-Fidelity 2× Master Mix (New England Biolabs, Ipswich, MA, USA). The amplicons were purified using the High Pure PCR Cleanup Micro Kit (Roche) and digested with *Bam*HI. The resulting fragments underwent kination by T4 Polynucleotide Kinase (TaKaRa Bio, San Jose, CA, USA) and were ligated with a previously described pET28a vector (Merck Millipore, Burlington, MA, USA) [45] using Ligation High Ver. 2 (Toyobo, Osaka, Japan). The resulting plasmids were transformed into One Shot TOP10 Chemically Competent *E. coli* (Thermo Fisher Scientific). Then, the plasmids were isolated from the transformants using the High Pure Plasmid Isolation Kit (Roche, Basel, Switzerland) and transformed into *E. coli* BL21(DE3) (Thermo Fisher Scientific) by electroporation. The resulting strains were inoculated into LB broth supplemented with 30 μg/mL kanamycin and incubated at 37 °C. Isopropyl thio-β-D-galactoside (final concentration 0.02 mM) was added when the optical density at 600 nm reached 0.5, and the culture was further incubated for 20 h at 20 °C. Cells were resuspended in buffer A (20 mM Tris/HCl, pH 7.5) and sonicated. The cell-free extract was applied to a Macro-Prep High S Support column (Bio-Rad Laboratories, South Granville, NSW, Australia) and washed with buffer A. The enzymes were eluted with buffer A containing 0.3 M NaCl. Fractions containing Sed-1 or mutant enzymes were pooled, dialyzed against buffer A, and injected into a CM-Toyopearl 650S column (Tosoh, Tokyo, Japan) equilibrated with buffer A. The enzymes were eluted with a linear gradient of NaCl (0–0.25 M) in the same buffer. Enzyme activity was determined with a spectrophotometer (V-730 BIO; JASCO, Hachioji, Japan) at 30 °C in 20 mM HEPES buffer pH7.0. The wavelengths and extinction coefficients used in this study were the same as described previously [45,46]. Protein concentration was determined by Bradford assay [43]. The enzyme was diluted with assay buffer containing 20 μg/mL bovine serum albumin to prevent denaturation. The values of the kinetic parameters (K_m and k_{cat}) were obtained by a double-reciprocal (Lineweaver-Burk) plot of initial steady-state velocities at different substrate concentrations.

4.6. Nucleotide Sequence Accession Numbers

Whole-genome DNA sequences of NR2807 were deposited in the GenBank database under accession no. BAAHNL01000000.1

5. Conclusions

In conclusion, the mechanism of Sed-1 production in *C. sedlakii* appears to differ from that of other β-lactamase-producing bacteria. SedR promotes Sed-1 expression even when not induced and partially promotes it when induced. In particular, the AmpD mutant led to strong expression of Sed-1 and, consequently, elevated resistance. The possibility of ESBL-type emergence in Sed-1 mutant strains has been noted and should be carefully monitored.

Supplementary Materials: The following supporting information can be downloaded at: https://www.mdpi.com/article/10.3390/antibiotics14080823/s1, Figure S1: Amino acid sequence alignment of the N-terminal region of AmpR and SedR (between amino acids 1 and 140); Table S1: Characteristics of inducible AmpC, derepressed AmpC, and Sed-1 producers; Table S2: Glossary of technical terms; Table S3: Primers used in the present study.

Author Contributions: Conceptualization, R.N. and A.N.; methodology, M.W., R.N., K.Y., A.N. and Y.S.; validation, M.W., R.N., K.Y., A.N., Y.S., K.S., S.N. and K.N.; investigation, M.W. and K.Y.; resources, S.N., K.E., K.N. and H.Y.; data curation, M.W., R.N., K.S., K.E. and H.Y.; writing—original draft preparation, M.W. and R.N.; writing—review and editing, M.W., R.N., K.Y., A.N., Y.S., K.S., S.N., K.E., K.N. and H.Y.; visualization, M.W., R.N. and K.Y.; supervision, H.Y.; project administration, H.Y.; funding acquisition, H.Y. All authors have read and agreed to the published version of the manuscript.

Funding: This study was supported by JSPS KAKENHI (Grant No. 22K08606). The funding agency had no role in the study design, data collection, analysis and interpretation, writing of the manuscript, and in the decision to submit the article for publication.

Institutional Review Board Statement: Not applicable.

Informed Consent Statement: Not applicable.

Data Availability Statement: Data will be made available on request.

Acknowledgments: The authors thank Tomoko Asada and Noriko Shirae from Nara Medical University for their excellent technical assistance. We are also particularly grateful to Yuki Yamada from Iwate Medical University Hospital for her continuous support and constructive suggestions.

Conflicts of Interest: The authors declare no conflicts of interest.

Abbreviations

The following abbreviations are used in this manuscript:

3GC	Third-generation cephalosporins
CFU	Colony-forming units
ESBL	Extended-spectrum β-lactamase
LB	Luria–Bertani

References

1. Tacconelli, E.; Carrara, E.; Savoldi, A.; Harbarth, S.; Mendelson, M.; Monnet, D.L.; Pulcini, C.; Kahlmeter, G.; Kluytmans, J.; Carmeli, Y.; et al. Discovery, Research, and Development of New Antibiotics: The WHO Priority List of Antibiotic-Resistant Bacteria and Tuberculosis. *Lancet Infect Dis.* **2018**, *18*, 318–327. [CrossRef]
2. Melchiorri, D.; Rocke, T.; Alm, R.A.; Cameron, A.M.; Gigante, V. Addressing Urgent Priorities in Antibiotic Development: Insights from WHO 2023 Antibacterial Clinical Pipeline Analyses. *Lancet Microbe* **2025**, *6*, 100992. [CrossRef] [PubMed]
3. Husna, A.; Rahman, M.M.; Badruzzaman, A.T.M.; Sikder, M.H.; Islam, M.R.; Rahman, M.T.; Alam, J.; Ashour, H.M. Extended-Spectrum Beta-Lactamases (Esbl): Challenges and Opportunities. *Biomedicines* **2023**, *11*, 2937. [CrossRef]
4. Philippon, A.; Slama, P.; Deny, P.; Labia, R. A Structure-Based Classification of Class a Beta-Lactamases, a Broadly Diverse Family of Enzymes. *Clin. Microbiol. Rev.* **2016**, *29*, 29–57. [CrossRef]
5. Naas, T.; Aubert, D.; Ozcan, A.; Nordmann, P. Chromosome-Encoded Narrow-Spectrum Ambler Class a Beta-Lactamase Gil-1 from *Citrobacter gillenii*. *Antimicrob. Agents Chemother.* **2007**, *51*, 1365–1372. [CrossRef] [PubMed]
6. Ortiz de la Rosa, J.M.; Bouvier, M.; Poirel, L.; Greub, G.; Blanc, D.; Nordmann, P. Cross-Reaction of Naturally-Produced Beta-Lactamases from *Citrobacter farmeri* and *Citrobacter amalonaticus* with Immunological Detection of Ctx-M Enzymes. *Diagn. Microbiol. Infect. Dis.* **2022**, *104*, 115760. [CrossRef]
7. Underwood, S.; Avison, M.B. *Citrobacter koseri* and *Citrobacter amalonaticus* Isolates Carry Highly Divergent Beta-Lactamase Genes Despite Having High Levels of Biochemical Similarity and 16s rRNA Sequence Homology. *J. Antimicrob. Chemother.* **2004**, *53*, 1076–1080. [CrossRef]
8. Bartowsky, E.; Normark, S. Purification and Mutant Analysis of *Citrobacter freundii* Ampr, the Regulator for Chromosomal Ampc Beta-Lactamase. *Mol. Microbiol.* **1991**, *5*, 1715–1725. [CrossRef]
9. Guerin, F.; Isnard, C.; Cattoir, V.; Giard, J.C. Complex Regulation Pathways of Ampc-Mediated B-Lactam Resistance in *Enterobacter cloacae* Complex. *Antimicrob. Agents Chemother.* **2015**, *59*, 7753–7761. [CrossRef] [PubMed]
10. Avison, M.B.; Underwood, S.; Okazaki, A.; Walsh, T.R.; Bennett, P.M. Analysis of Ampc Beta-Lactamase Expression and Sequence in Biochemically Atypical Ceftazidime-Resistant Enterobacteriaceae from Paediatric Patients. *J. Antimicrob. Chemother.* **2004**, *53*, 584–591. [CrossRef]
11. Kuga, A.; Okamoto, R.; Inoue, M. *Ampr* Gene Mutations That Greatly Increase Class C B-Lactamase Activity in *Enterobacter cloacae*. *Antimicrob. Agents Chemother.* **2000**, *44*, 561–567. [CrossRef]
12. Nakano, R.; Nakano, A.; Yano, H.; Okamoto, R. Role of Ampr in the High Expression of the Plasmid-Encoded Ampc Beta-Lactamase Cfe-1. *mSphere* **2017**, *2*, e00192-17. [CrossRef] [PubMed]
13. Caille, O.; Zincke, D.; Merighi, M.; Balasubramanian, D.; Kumari, H.; Kong, K.F.; Silva-Herzog, E.; Narasimhan, G.; Schneper, L.; Lory, S.; et al. Structural and Functional Characterization of *Pseudomonas aeruginosa* Global Regulator Ampr. *J. Bacteriol.* **2014**, *196*, 3890–3902. [CrossRef] [PubMed]
14. Juan, C.; Macia, M.D.; Gutierrez, O.; Vidal, C.; Perez, J.L.; Oliver, A. Molecular Mechanisms of Beta-Lactam Resistance Mediated by Ampc Hyperproduction in *Pseudomonas aeruginosa* Clinical Strains. *Antimicrob. Agents Chemother.* **2005**, *49*, 4733–4738. [CrossRef]

15. Kaneko, K.; Okamoto, R.; Nakano, R.; Kawakami, S.; Inoue, M. Gene Mutations Responsible for Overexpression of Ampc Β-Lactamase in Some Clinical Isolates of *Enterobacter cloacae*. *J. Clin. Microbiol.* **2005**, *43*, 2955–2958. [CrossRef]
16. Balcewich, M.D.; Reeve, T.M.; Orlikow, E.A.; Donald, L.J.; Vocadlo, D.J.; Mark, B.L. Crystal Structure of the Ampr Effector Binding Domain Provides Insight into the Molecular Regulation of Inducible Ampc Beta-Lactamase. *J. Mol. Biol.* **2010**, *400*, 998–1010. [CrossRef]
17. Petrella, S.; Clermont, D.; Casin, I.; Jarlier, V.; Sougakoff, W. Novel Class a Beta-Lactamase Sed-1 from *Citrobacter sedlakii*: Genetic Diversity of Beta-Lactamases within the *Citrobacter* Genus. *Antimicrob. Agents Chemother.* **2001**, *45*, 2287–2298. [CrossRef]
18. Tasnim, Y.; Stanley, C.; Rahman, M.K.; Awosile, B. *Bla* Sed-1 Beta-Lactamase-Producing *Citrobacter Sedlakii* Isolated from Horses and Genomic Comparison with Human-Derived Isolates. *J. Appl. Microbiol.* **2024**, *135*, lxae278. [CrossRef]
19. Matagne, A.; Lamotte-Brasseur, J.; Frere, J.M. Catalytic Properties of Class a Beta-Lactamases: Efficiency and Diversity. *Biochem. J.* **1998**, *330 Pt 2*, 581–598. [CrossRef]
20. Bennett, P.M.; Chopra, I. Molecular Basis of Beta-Lactamase Induction in Bacteria. *Antimicrob. Agents Chemother.* **1993**, *37*, 153–158. [CrossRef] [PubMed]
21. Maddocks, S.E.; Oyston, P.C.F. Structure and Function of the Lysr-Type Transcriptional Regulator (Lttr) Family Proteins. *Microbiology (Reading)* **2008**, *154 Pt 12*, 3609–3623. [CrossRef] [PubMed]
22. Kong, K.F.; Jayawardena, S.R.; Indulkar, S.D.; Del Puerto, A.; Koh, C.L.; Hoiby, N.; Mathee, K. *Pseudomonas aeruginosa* Ampr Is a Global Transcriptional Factor that Regulates Expression of Ampc and Poxb Beta-Lactamases, Proteases, Quorum Sensing, and Other Virulence Factors. *Antimicrob. Agents Chemother.* **2005**, *49*, 4567–4575. [CrossRef]
23. Stapleton, P.; Shannon, K.; Phillips, I. DNA Sequence Differences of *Ampd* Mutants of *Citrobacter freundii*. *Antimicrob. Agents Chemother.* **1995**, *39*, 2494–2498. [CrossRef]
24. Kohlmann, R.; Bahr, T.; Gatermann, S.G. Species-Specific Mutation Rates for Ampc Derepression in Enterobacterales with Chromosomally Encoded Inducible Ampc Beta-Lactamase. *J. Antimicrob. Chemother.* **2018**, *73*, 1530–1536. [CrossRef]
25. Arpin, C.; Labia, R.; Andre, C.; Frigo, C.; El Harrif, Z.; Quentin, C. Shv-16, a Beta-Lactamase with a Pentapeptide Duplication in the Omega Loop. *Antimicrob. Agents Chemother.* **2001**, *45*, 2480–2485. [CrossRef]
26. Hayes, F.; Hallet, B.; Cao, Y. Insertion Mutagenesis as a Tool in the Modification of Protein Function. Extended Substrate Specificity Conferred by Pentapeptide Insertions in the Omega-Loop of Tem-1 Beta-Lactamase. *J. Biol. Chem.* **1997**, *272*, 28833–28836. [CrossRef]
27. Vakulenko, S.B.; Taibi-Tronche, P.; Toth, M.; Massova, I.; Lerner, S.A.; Mobashery, S. Effects on Substrate Profile by Mutational Substitutions at Positions 164 and 179 of the Class a Tem (Puc19) Beta-Lactamase from *Escherichia coli*. *J. Biol. Chem.* **1999**, *274*, 23052–23060. [CrossRef] [PubMed]
28. Laubacher, M.E.; Ades, S.E. The Rcs Phosphorelay Is a Cell Envelope Stress Response Activated by Peptidoglycan Stress and Contributes to Intrinsic Antibiotic Resistance. *J. Bacteriol.* **2008**, *190*, 2065–2074. [CrossRef] [PubMed]
29. Sutterlin, H.A.; Zhang, S.; Silhavy, T.J. Accumulation of Phosphatidic Acid Increases Vancomycin Resistance in *Escherichia coli*. *J. Bacteriol.* **2014**, *196*, 3214–3220. [CrossRef]
30. Negrete-Gonzalez, C.; Turrubiartes-Martinez, E.; Briano-Macias, M.; Noyola, D.; Perez-Gonzalez, L.F.; Gonzalez-Amaro, R.; Nino-Moreno, P. Plasmid Carrying $Bla_{\text{ctx-M-15}}$, $Bla_{\text{per-1}}$, and $Bla_{\text{tem-1}}$ Genes in *Citrobacter* spp. from Regional Hospital in Mexico. *Infect. Dis.* **2022**, *15*, 11786337211065750. [CrossRef]
31. Paterson, D.L.; Bonomo, R.A. Extended-Spectrum Beta-Lactamases: A Clinical Update. *Clin. Microbiol. Rev.* **2005**, *18*, 657–686. [CrossRef]
32. Liakopoulos, A.; Mevius, D.; Ceccarelli, D. A Review of Shv Extended-Spectrum Beta-Lactamases: Neglected yet Ubiquitous. *Front. Microbiol.* **2016**, *7*, 1374. [CrossRef]
33. Richter, M.; Rossello-Mora, R.; Oliver Glockner, F.; Peplies, J. Jspeciesws: A Web Server for Prokaryotic Species Circumscription Based on Pairwise Genome Comparison. *Bioinformatics* **2016**, *32*, 929–931. [CrossRef]
34. Nakano, R.; Okamoto, R.; Nakano, Y.; Kaneko, K.; Okitsu, N.; Hosaka, Y.; Inoue, M. Cfe-1, a Novel Plasmid-Encoded Ampc Β-Lactamase with an *Ampr* Gene Originating from *Citrobacter freundii*. *Antimicrob. Agents Chemother.* **2004**, *48*, 1151–1158. [CrossRef] [PubMed]
35. CLSI. *Performance Standards for Antimicrobial Susceptibility Testing*, 35th ed; Clsi Supplement M100; Clinical and Laboratory Standards Institute: Wayne, PA, USA, 2025.
36. Wick, R.R.; Judd, L.M.; Gorrie, C.L.; Holt, K.E. Unicycler: Resolving Bacterial Genome Assemblies from Short and Long Sequencing Reads. *PLoS Comput. Biol.* **2017**, *13*, e1005595. [CrossRef] [PubMed]
37. Goris, J.; Konstantinidis, K.T.; Klappenbach, J.A.; Coenye, T.; Vandamme, P.; Tiedje, J.M. DNA-DNA Hybridization Values and Their Relationship to Whole-Genome Sequence Similarities. *Int. J. Syst. Evol. Microbiol.* **2007**, *57 Pt 1*, 81–91. [CrossRef]

38. Chun, J.; Rainey, F.A. Integrating Genomics into the Taxonomy and Systematics of the Bacteria and Archaea. *Int. J. Syst. Evol. Microbiol.* **2014**, *64 Pt 2*, 316–324. [CrossRef]
39. Bortolaia, V.; Kaas, R.S.; Ruppe, E.; Roberts, M.C.; Schwarz, S.; Cattoir, V.; Philippon, A.; Allesoe, R.L.; Rebelo, A.R.; Florensa, A.F.; et al. Resfinder 4.0 for Predictions of Phenotypes from Genotypes. *J. Antimicrob. Chemother.* **2020**, *75*, 3491–3500. [CrossRef] [PubMed]
40. Ambler, R.P.; Coulson, A.F.; Frere, J.M.; Ghuysen, J.M.; Joris, B.; Forsman, M.; Levesque, R.C.; Tiraby, G.; Waley, S.G. A Standard Numbering Scheme for the Class a Beta-Lactamases. *Biochem. J.* **1991**, *276 Pt 1*, 269–270. [CrossRef]
41. Bai, L.; Xia, S.; Lan, R.; Liu, L.; Ye, C.; Wang, Y.; Jin, D.; Cui, Z.; Jing, H.; Xiong, Y.; et al. Isolation and Characterization of Cytotoxic, Aggregative *Citrobacter freundii*. *PLoS ONE* **2012**, *7*, e33054. [CrossRef]
42. Waley, S.G. A Spectrophotometric Assay of Beta-Lactamase Action on Penicillins. *Biochem. J.* **1974**, *139*, 789–790. [CrossRef]
43. Bradford, M.M. A Rapid and Sensitive Method for the Quantitation of Microgram Quantities of Protein Utilizing the Principle of Protein-Dye Binding. *Anal. Biochem.* **1976**, *72*, 248–254. [CrossRef]
44. Nakano, R.; Yamada, Y.; Nakano, A.; Suzuki, Y.; Saito, K.; Sakata, R.; Ogawa, M.; Narita, K.; Kuga, A.; Suwabe, A.; et al. The Role of *Nmcr*, *Ampr*, and *Ampd* in the Regulation of the Class a Carbapenemase Nmca in *Enterobacter ludwigii*. *Front. Microbiol.* **2021**, *12*, 794134. [CrossRef] [PubMed]
45. Yamamoto, K.; Tanaka, H.; Kurisu, G.; Nakano, R.; Yano, H.; Sakai, H. Structural Insights into the Substrate Specificity of Imp-6 and Imp-1 Metallo-B-Lactamases. *J. Biochem.* **2022**, *173*, 21–30. [CrossRef] [PubMed]
46. Laraki, N.; Franceschini, N.; Rossolini, G.M.; Santucci, P.; Meunier, C.; de Pauw, E.; Amicosante, G.; Frere, J.M.; Galleni, M. Biochemical Characterization of the *Pseudomonas aeruginosa* 101/1477 Metallo-B-Lactamase Imp-1 Produced by *Escherichia coli*. *Antimicrob. Agents Chemother.* **1999**, *43*, 902–906. [CrossRef] [PubMed]

antibiotics

MDPI

Article

Investigation of WQ-3810, a Fluoroquinolone with a High Potential Against Fluoroquinolone-Resistant *Mycobacterium avium*

Sasini Jayaweera [1], Pondpan Suwanthada [1], David Atomanyi Barnes [1], Charlotte Poussier [1], Tomoyasu Nishimura [2], Naoki Hasegawa [3], Yukiko Nishiuchi [4,5], Jeewan Thapa [1], Stephen V. Gordon [6,7], Hyun Kim [8], Chie Nakajima [1,7,*] and Yasuhiko Suzuki [1,7,*]

1 Division of Bioresources, Hokkaido University International Institute for Zoonosis Control, Sapporo 001-0020, Japan; sasini@czc.hokudai.ac.jp (S.J.); dbarnes@czc.hokudai.ac.jp (D.A.B.)
2 Keio University Health Center, Yokohama 223-8521, Japan
3 Department of Infectious Diseases, Keio University School of Medicine, Tokyo 160-8582, Japan
4 Toneyama Institute for Tuberculosis Research, Osaka City University Medical School, Toyonaka 560-8552, Japan
5 Microbial Genomics and Ecology, The IDEC Institute, Hiroshima University, Higashi-Hiroshima 739-8530, Japan
6 School of Veterinary Medicine, University College Dublin, D04 V1W8 Dublin, Ireland; stephen.gordon@ucd.ie
7 Institute for Vaccine Research and Development (IVReD), Hokkaido University, Sapporo 001-0021, Japan
8 Department of Bacteriology II, Japan Institute for Health Security (JIHS), Tokyo 208-0011, Japan
* Correspondence: cnakajim@czc.hokudai.ac.jp (C.N.); suzuki@czc.hokudai.ac.jp (Y.S.); Tel.: +81-011-706-9505 (C.N.); Fax: +81-0-706-7310 (Y.S.)

Abstract: Background/Objectives: *Mycobacterium avium*, a member of *Mycobacterium avium* complex (MAC), is an emerging opportunistic pathogen causing MAC-pulmonary disease (PD). Fluoroquinolones (FQs), along with ethambutol (EMB) and rifampicin, are recommended for macrolide-resistant MAC-PD; however, FQ-resistant *M. avium* have been reported worldwide. WQ-3810 is an FQ with high potency against FQ-resistant pathogens; however, its activity against *M. avium* has not yet been studied. **Methods**: In this study, we conducted a DNA supercoiling inhibitory assay to evaluate the inhibitory effect of WQ-3810 on recombinant wild-type (WT) and four mutant DNA gyrases of *M. avium* and compared the IC_{50}s of WQ-3810 with those of ciprofloxacin (CIP), levofloxacin (LVX), and moxifloxacin (MXF). In addition, we examined WQ-3810's antimicrobial activity against 11 *M. avium* clinical isolates, including FQ-resistant isolates, with that of other FQs. Furthermore, we assessed the synergistic action of WQ-3810 with the combination of either EMB or isoniazid (INH). **Results**: In a DNA supercoiling inhibitory assay, WQ-3810 showed 1.8 to 13.7-fold higher efficacy than LVX and CIP. In the MIC assay, WQ-3810 showed 4 to 8-fold, 2 to 16-fold, and 2 to 4-fold higher antimicrobial activity against FQ-resistant isolates than CIP, LVX, and MXF, respectively. The combination of WQ-3810 and INH exhibited a synergistic relationship. **Conclusions**: The overall characteristics of WQ-3810 demonstrated greater effectiveness than three other FQs, suggesting that it is a promising option for treating FQ-resistant *M. avium* infections.

Keywords: Fluoroquinolone resistance; *Mycobacterium avium*; DNA gyrase; supercoiling inhibitory assay; minimum inhibitory concentration; checkerboard assay

1. Introduction

M. avium is a slow-growing, acid-fast, opportunistic intracellular pathogen belonging to the *Mycobacterium avium* complex (MAC) and nontuberculosis mycobacteria (NTM) [1–3]. *M. avium* is widely distributed and can be found in both natural and domestic settings, such as soil, water, and plants, as well as in showerheads and plumbing systems [2,4,5]. The most prevalent clinical manifestation of MAC infection is pulmonary disease (MAC-PD), especially found in immunocompromised individuals, such as those with HIV, and immunocompetent patients with chronic pulmonary diseases such as chronic obstructive pulmonary disease (COPD), cystic fibrosis, and previous pulmonary tuberculosis [1,5]. Previous studies indicate that the prevalence of MAC-PD is rising, with increasing cases being documented in North America, Oceania, Europe, Japan, South Korea, and elsewhere [3,6,7]. The American Thoracic Society guidelines suggest a macrolide-based multidrug regimen for the treatment of MAC-PD until a negative culture is achieved, followed by an additional year of therapy [8–10]. However, previous studies have shown that 20–40% of patients do not respond to the combination of three drugs: rifampicin, ethambutol, and macrolide [3,8–11]. For individuals whose MAC infection is resistant to first-line drugs or who are intolerant of them, a daily dosage of 400 mg of moxifloxacin (MXF) is recommended [8,11,12]. Moreover, a previous study indicated that fluoroquinolones (FQs) were prescribed for the treatment of MAC-PD in 45% of patients in European countries and 71% of patients in Japan [13]. Hence, the introduction of a novel drug with significant efficacy for treating FQ-resistant MAC is essential.

FQ, a synthetic antibiotic, has a broad-spectrum antibacterial activity that inhibits the activity of DNA gyrase [14]. DNA gyrase is a type II DNA topoisomerase, composed of two proteins; subunit A (GyrA) and subunit B (GyrB), that play a crucial role in bacterial DNA transcription and replication by adding negative supercoiling to DNA [15,16]. DNA gyrase covalently binds to double-stranded DNA and breaks it, unwinding the kinks and resealing the DNA. Quinolones bind to the cleaved ends, stabilizing the gyrase-DNA cleavage complex and preventing DNA resealing, thereby killing the bacteria [15]. Previous studies indicate that amino acids at positions 91 and 95 in the quinolone resistance-determining region (QRDR) of GyrA are essential for FQ binding, and amino acid substitutions at those positions lead to FQ-resistant *M. avium* [17–19]. In Japan, highly FQ-resistant *M. avium* clinical isolates with GyrA-Asp95Gly and GyrA-Asp95Tyr substitutions have been reported [17].

WQ-3810 is a newly developed FQ which has high lipophilicity, membrane permeability, good oral absorption, and low adverse effects [20]. It shows high efficacy against multi-drug resistant and FQ-resistant Gram-negative bacteria such as *Escherichia coli*, *Acinetobacter baumannii*, and Gram-positive bacteria such as *Streptococcus pneumonia* and methicillin-resistant *Staphylococcus aureus* [21]. According to recent studies, WQ-3810 showed high antimicrobial activity and high enzyme inhibitory activity against *Campylobacter jejuni*, *M. tuberculosis*, and *M. leprae* [22–24]. However, the efficacy of WQ-3810 against FQ-resistant *M. avium* has not been evaluated. Therefore, this study aims to evaluate the inhibitory effect of WQ-3810 on DNA gyrase against FQ-resistant *M. avium* using recombinant *M. avium* wild-type (WT) and mutant DNA gyrases, as well as the antimicrobial activity and the synergistic relationship of WQ-3810 with cell wall inhibitors on clinical isolates of *M. avium*.

2. Results

2.1. Inhibitory Effects of WQ-3810 Against M. avium WT and Mutant DNA Gyrases

The gel electrophoresis patterns shown in Figure 1 demonstrate a concentration-dependent reduction in the intensity of the supercoiled DNA band, indicating a dose-dependent inhibition of DNA gyrase activity by WQ-3810.

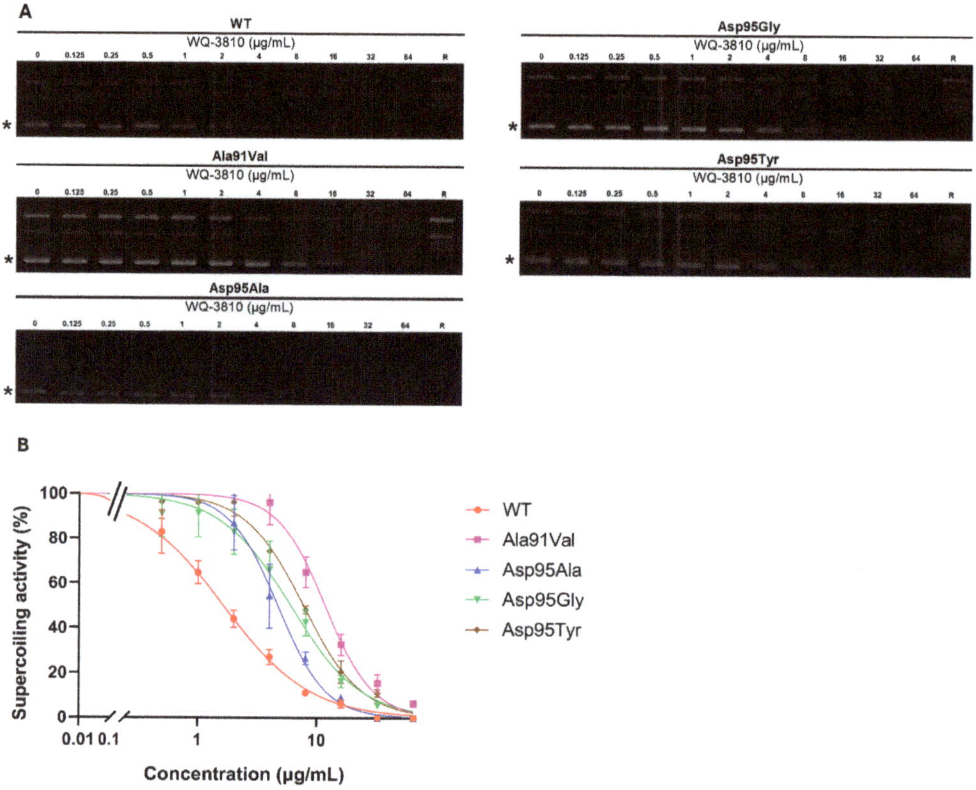

Figure 1. Inhibition of DNA supercoiling activity of *M. avium* DNA gyrase by WQ-3810. (**A**) Representative gel electrophoresis: Gel electrophoresis results demonstrate the separation of supercoiled DNA from relaxed DNA. Supercoiled DNA bands are indicated by asterisks (*). These figures are randomly selected from a triplicate assay. (**B**) Quantification of supercoiled DNA: The levels of supercoiled DNA produced in the presence of WQ-3810 were quantified.

The concentrations of WQ-3810 that inhibits 50% of DNA gyrase activity (IC_{50}) for four mutant DNA gyrases were higher than that for WT DNA gyrase (Table 1). The GyrA-Ala91Val mutant exhibited the highest IC_{50} of 9.9 μg/mL, while GyrA-Asp95Ala had the lowest IC_{50} at 4.5 μg/mL. The IC_{50}s of the GyrA-Asp95Gly mutant (7.0 μg/mL) and of the GyrA-Asp95Tyr mutant (7.8 μg/mL) demonstrated comparable resistance.

Table 1. IC_{50}s (μg/mL) of fluoroquinolones against WT and mutant DNA gyrases.

	WT	Ala91Val	Asp95Ala	Asp95Gly	Asp95Tyr	Reference
WQ-3810	1.7 ± 0.1	9.9 ± 0.6	4.5 ± 1.1	7.0 ± 0.5	7.8 ± 0.6	This study
CIP	3 ± 0.8	10.2 ± 0.3	33.4 ± 7.0	53 ± 4.7	106.8 ± 4.0	Previous study [17]
MXF	0.9 ± 0.1	9.5 ± 0.2	4.2 ± 0.3	11.3 ± 3.5	13.5 ± 0.4	Previous study [17]
LVX	3.7 ± 0.4	5 ± 1.5	26.6 ± 5.9	23.3 ± 2.4	86.1 ± 6.0	Previous study [17]

Then, we compared the inhibition results of WQ-3810 with those of other FQs, namely ciprofloxacin (CIP), levofloxacin (LVX), and moxifloxacin (MXF) (Tables 1 and 2 and Figure 2) [17]. For GyrA-WT, MXF and WQ-3810 showed comparable low IC_{50}s of 0.9 μg/mL and 1.7 μg/mL, respectively, while CIP and LVX showed higher IC_{50}s ($p < 0.05$)

(Table 1, Figure 3). CIP demonstrated the highest IC_{50}s for all mutants. For GyrA-Asp95 mutants, IC_{50}s of CIP and LVX were significantly higher than those of MXF and WQ-3810 ($p < 0.001$). Although the difference between MXF and WQ-3810 was not significant, IC_{50}s of WQ-3810 for GyrA-Asp95Gly and GyrA-Asp95Tyr (7.0 μg/mL and 7.8 μg/mL, respectively) were lower than those of MXF (11.3 μg/mL and 13.5 μg/mL, respectively).

Table 2. Structure of fluoroquinolones.

Fluoroquinolones	N_1	R_6	R_7	R_8
CIP	Cyclopropyl	F	Piperazine	H
LVX	Bridge N_1-C_8	F	4-Methylpiperazinyl	Bridge N_1-C_8
MXF	Cyclopropyl	F	Piperridino-pyrrolidinyl	CH_3O
WQ-3810	5-Amino-2,4-difluoropyridine-2-yl	F	3-Isopropylaminoazetizine-1-yl	CH_3

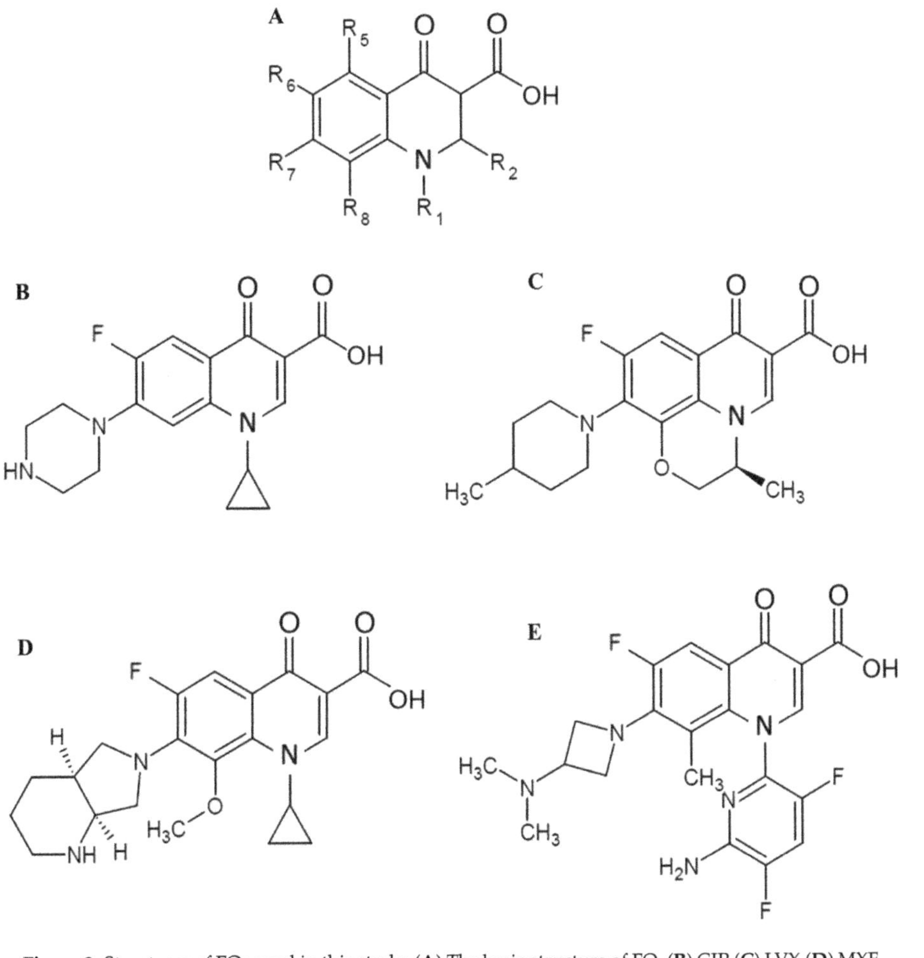

Figure 2. Structures of FQs used in this study. (**A**) The basic structure of FQ. (**B**) CIP (**C**) LVX (**D**) MXF (**E**) WQ-3810.

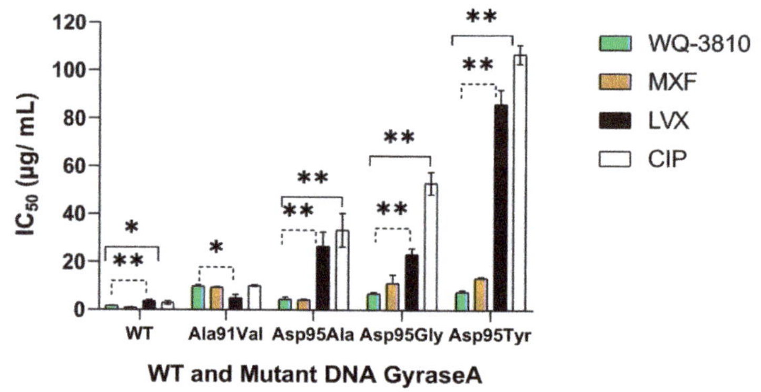

Figure 3. FQ activities against *M. avium* DNA gyrase. The amount of DNA supercoiled by DNA gyrase was measured using agarose gel electrophoresis in triplicate, and IC_{50} was calculated. The normality of the data was tested using the Shapiro–Wilk test. One-way ANOVA was used to compare the differences between the FQs. The least significant difference (LSD) test was used to identify the groups whose means were statistically significantly different. (*, $p < 0.05$, **, $p < 0.001$).

2.2. Minimum Inhibitory Concentration (MIC) of WQ-3810

Table 3 shows that WT isolates had significantly lower MICs for all four FQs, ranging from 0.125 to 4 μg/mL, compared to mutant isolates that had MICs from 4 to 128 μg/mL. The MIC ranges for WQ-3810 and MXF were 0.125–1 μg/mL and 0.25–1 μg/mL, respectively, in WT isolates. Furthermore, the study found higher MICs for GyrA-Asp95Gly and GyrA-Asp95Tyr in LVX and CIP than in MXF and WQ-3810. Notably, the MICs of MXF, ranging from 8 to 16 μg/mL for GyrA-Asp95Gly and GyrA-Asp95Tyr, respectively, were found to be higher than those of WQ-3810, which ranged from 4 to 8 μg/mL.

Table 3. Drug susceptibility of clinical isolates.

Sample ID	GyrA Mutation	MIC (μg/mL) of:						MIC (μg/mL) with 1 μg/mL EMB		MIC (μg/mL) with 1 μg/mL INH	
		LVX	CIP	MXF	WQ-3810	EMB	INH	MXF	WQ-3810	MXF	WQ-3810
Koav 3	Asp95Gly	64	32	8	4	16	4	8 (×1)	4 (×1)	4 (×0.5)	2 (×0.5)
Koav 11	Asp95Tyr	8	16	16	4	16	8	8 (×0.5)	4 (×1)	8 (×0.5)	4 (×1)
Koav 13	Asp95Gly	128	64	16	8	32	32	16 (×1)	4 (×0.5)	8 (×0.5)	4 (×0.5)
Cl-A-2	Asp95Tyr	64	32	8	4	16	4–8	4 (×0.5)	1 (×0.25)	8 (×1)	1(×0.25)
Koav 1	None (WT)	<0.5	0.5	0.25	0.125	2	8	<0.01 (×0.04)	0.01 (×0.08)	0.03 (×0.12)	0.03 (×0.24)
Koav 2	None (WT)	2	1	1	0.5	16	128	ND	ND	ND	ND
Koav 15	None (WT)	4	2	1	1	16	32	1 (×1)	1 (×1)	1(×1)	1 (×1)
Koav 19	None (WT)	2	2	0.5	0.5	16	8–16	0.5 (×1)	0.25 (×0.5)	0.13 (×0.25)	0.13 (×0.5)
Koav 20	None (WT)	<0.5	0.5	0.25	0.125	2	16	0.13–0.25 (×0.5–1)	0.13 (×1)	0.25 (×1)	0.13 (×1)
Koav 21	None (WT)	4	1	1	1	32	16	1 (×1)	1 (×1)	0.5 (×0.5)	1 (×1)
Koav 25	None (WT)	4	2	1	1	32	8	1 (×1)	0.5 (×0.5)	0.5 (×0.5)	0.5 (×0.5)
Reference		Previous study [17]	This study	This study	This study	This study	This study				

Concentration ranges as follows: LVX (0.5–128 μg/mL), CIP (0.5–128 μg/mL), MXF (0.125–32 μg/mL), WQ-3810 (0.125–32 μg/mL), EMB (0.0625–128 μg/mL), INH (0.0625–128 μg/mL), ND: not determined.

2.3. Effects of Cell Wall Synthesis Inhibitors on the MIC of WQ-3810

We used the combination of 1 µg/mL of EMB and INH with MXF and WQ-3810 given in Table 3 to determine the effects of cell wall synthesis inhibitors on the MIC of MXF and WQ-3810. MXF and WQ-3810 yielded comparable or lower MIC results when used together with either 1 µg/mL EMB or 1 µg/mL INH for WT. WQ-3810 in combination with 1 µg/mL EMB or 1 µg/mL INH showed lower MICs than MXF for the mutants, as did WQ-3810 and MXF alone (Table 3).

Subsequently, we calculated the mean fraction inhibitory concentration index (FICIm) shown in Table 4, Tables S1 and S2 of WQ-3810 and MXF with the combination of either EMB or INH. The combination of INH with WQ-3810 and MXF showed a synergistic relationship, with mean FICIms of $\leq$0.5, whereas most of the mean FICIms of WQ-3810 and MXF with the combination of EMB were between 0.5 and 1, indicating additive interactions. The mean FICIms of WQ-3810 and MXF with INH ranged from 0.19 to 0.71, and those of WQ-3810 and MXF with EMB ranged from 0.41 to 0.91.

Table 4. Mean FICIm of WQ-3810 and MXF with cell wall inhibitors.

Sample ID	Gyr A Mutation	Mean FICIm of WQ-3810		Mean FICIm of MXF	
		with INH	with EMB	with INH	with EMB
Koav 3	Asp95Gly	0.71	0.75	0.66	0.52
Koav 11	Asp95Tyr	0.46	0.83	0.41	0.52
Koav 13	Asp95Gly	0.31	0.83	0.25	0.6
Cl-A-2	Asp95Tyr	0.53	0.66	0.40	0.41
Koav 1	None (WT)	0.45	0.62	0.27	0.5
Koav 2	None (WT)	0.25	0.91	0.31	0.71
Koav 15	None (WT)	0.33	0.71	0.38	0.66
Koav 19	None (WT)	0.45	0.86	0.19	0.71
Koav 20	None (WT)	0.25	0.66	0.41	0.81
Koav 21	None (WT)	0.41	0.75	0.31	0.55
Koav 25	None (WT)	0.39	0.75	0.49	0.75

Concentration ranges as follows: WQ-3810 (0.125–32 µg/mL), MXF (0.125–32 µg/mL), EMB (0.0625–128 µg/mL), and INH (0.0625–128 µg/mL). Each experiment was conducted in triplicate to confirm the reproducibility. Synergistic activity was evaluated by calculating the fraction inhibitory concentration index (FICI) and taking the cube root. The mean minimum FICI (FICIm) $\leq$ 0.5 is regarded as synergistic.

3. Discussion

Novel FQs are essential for treating MAC-PD, as cases of the disease are steadily increasing worldwide [6,7]. Furthermore, reports of FQ-resistant *M. avium* strains have been documented [17,25]. WQ-3810 is an FQ with structural modifications at the N_1, R_7, and R_8 positions, which confer considerable efficacy against a range of FQ-resistant bacteria, including *M. tuberculosis* and *M. leprae* [22–24].

This study sought to evaluate the efficacy of WQ-3810 against *M. avium* through a supercoiling inhibitory assay, an MIC assay, and a checkerboard assay. We used recombinant DNA gyrases, including WT, and those with Ala91Val, Asp95Ala, Asp95Gly, and Asp95Tyr substitutions in GyrA, for the supercoiling inhibitory assay, given the significant association of these substitutions with FQ resistance and their impact on resistance mechanisms [17,18].

The outcomes of the supercoiling inhibitory assay indicated that WQ-3810 exhibited greater potency compared to LVX and CIP in inhibiting the activity of both WT and mutant

DNA gyrases. Notably, the IC_{50} of CIP and LVX against GyrA-WT were 1.7–2.1 times greater than those of WQ-3810. Additionally, IC_{50} of LVX and CIP against GyrA-Asp95Ala, GyrA-Asp95Gly, and GyrA-Asp95Tyr were 5.9–7.4-fold, 3.3–7.5-fold, and 11–13.6-fold higher than that of WQ-3810, respectively. MXF exhibited a 0.5-fold reduction of IC_{50} in inhibitory effect against GyrA-WT in comparison to WQ-3810. However, MXF demonstrated a 1.6-fold and 1.7-fold higher IC_{50} for GyrA-Asp95Gly and GyrA-Asp95Tyr, respectively, than that of WQ-3810, indicating that WQ-3810 serves as a more efficacious FQ concerning the amino acid substitutions at position 95.

The observed results may be elucidated by the distinct properties at positions N_1, R_7, and R_8 within WQ-3810, MXF, CIP, and LVX (Figure 2 and Table 2), alongside the mode of action of FQs, which interact with the DNA-DNA gyrase complex to inhibit its function, ultimately culminating in bacterial cell death [26]. The carbonyl substituents at C_3 and C_4 of the quinolone ring utilize a noncatalytic Mg^{2+} ion and four water molecules to establish a water-metal ion bridge (Figure 4) [27,28]. In *E. coli*, two of the water molecules interact with the GyrA residues Ser83 and Asp87, while R_7 interacts with GyrB [26–28]. Amino acid position 83 in *E. coli* corresponds to position 91 in *M. avium* and 90 *in M. tuberculosis* (Supplementary Figure S1). Since the amino acid sequence of the GyrA QRDR of *M. avium* is identical to that of *M. tuberculosis* and *M. leprae*, we referred to the results of molecular docking simulations using *M. tuberculosis* gyrase [22]. *M. avium* forms a partial water metal ion interaction with Asp95 due to the substitution of serine (Ser83) with alanine (Ala91) (Figure 4, Supplementary Figure S1) [28]. Consequently, *M avium* gyrase possesses a single amino acid for bridge anchoring, thereby resulting in weaker DNA gyrase-FQ interactions compared to other bacterial species, which can form two interactive bridges. Moreover, FQ-resistance may be exacerbated by substituting the bulky, hydrophobic valine with Ala91, as valine impairs the optimal interaction of water molecules with Mg^{2+}, resulting in an IC_{50} of FQs against GyrA-Ala91Val that was 1.3-fold to 10.5-fold higher than that of the GyrA-WT (Table 1). Furthermore, the substitution of Asp95, an acidic amino acid residue, with a larger hydrophobic tyrosine blocks the binding site of water molecules with Mg^2, resulting in Asp95Tyr exhibiting greater resistance to all FQs. Unique substituents introduced at positions N_1 and R_7 of the FQs can hydrophobically interact in binding to DNA gyrase, increasing the binding affinity to GyrB. Moreover, modifying the quinolones at positions N_1, R_3, R_6, R_7, and R_8 has been shown to overcome resistance and increase the drug's effectiveness against resistant bacteria by enhancing pharmacokinetics, reducing toxicity, and improving activity [14,15].

Considering the structural properties of the substituent at N_1, R_7, and R_8 positions of the quinolone ring, we hypothesized that WQ-3810 has the potential for multiple interactions with GyrB in the FQ binding pocket, thereby enhancing efficacy (Figure 4). Specifically, the 5-amino-2,4-difluoropyridyl group at position N_1 is significantly distorted out of the quinolone core, and this outlying conformation allows it to interact with the sites that are inaccessible to conventional FQs [21]. In fact, our previous docking simulation analysis using the modified *M. tuberculosis* gyrase FQ-binding pocket showed that WQ-3810 formed a hydrogen bond between D461 of GyrB and the amino group on its N_1 pyridyl ring, while MXF did not have any specific binding sites (Figure 4) [22]. The bulky N_1 residue also forms van der Waals interactions with the GyrB pocket, while the azetidinyl group at the R_7 position engages in both hydrophobic interactions and hydrogen bonding. Additionally, the methyl group at the R_8 position enhances the overall stability, which may contribute to WQ-3810 higher binding affinity for DNA bases, leading to a lower IC_{50} [28]. Although the piperazine group at the R_7 position of CIP interacts with GyrB through hydrogen bonds [23], these interactions are weaker than those formed by the substituents in WQ-3810, resulting in a higher IC_{50} value for CIP. Even though MXF and CIP have a similar cyclopropyl group at the N_1 position, MXF's efficacy is comparable to that of WQ-3810 due to its azabicyclic group at the R_7 position and methoxy group at the R_8 position, which forms additional hydrogen bonds and van der Waals interactions with GyrB [23]. However, WQ-3810 demonstrates higher potency against the amino acid substituents at position 95 in GyrA, suggesting a higher binding affinity for DNA gyrase than MXF, possibly due to the hydrogen bond between the D461 of GyrB and the amino group on the N_1 substituent of WQ-3810 (Figure 4).

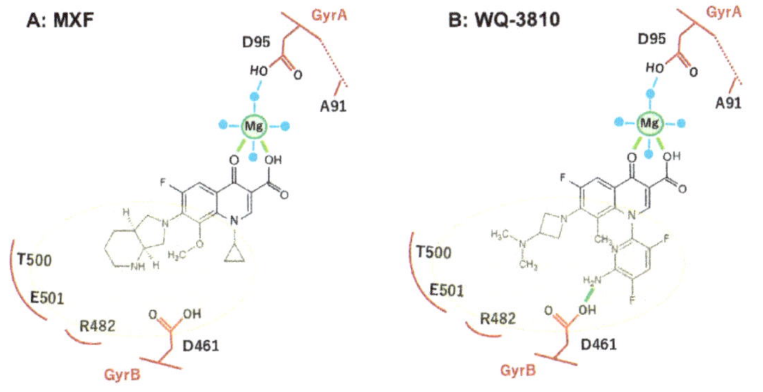

Figure 4. Predicted binding mode of MXF (**A**) and WQ-3810 (**B**) in the quinolone binding pocket formed by GyrA and GyrB. Mg: magnesium ion, blue dot: water molecule, blue line: water/Mg ion bridge, yellow-green line: hydrogen bond, pale-yellow circle: van der Waals interaction area. Amino acid numbers in GyrB are for the *M. tuberculosis* position.

We then performed a MIC assay on 11 clinical isolates and found that WQ-3810 exhibited a lower MIC than LVX and CIP, with reductions of 4 to 16-fold and 1 to 8-fold, respectively. Notably, WQ-3810's MIC of 0.125 to 1 μg/mL was comparable to MXF's MIC of 0.25 to 1 μg/mL against WT strains, and WQ-3810's MIC of 4 to 8 μg/mL was half that of MXF's MIC of 8 to 16 μg/mL for the mutant strains, demonstrating its superiority against mutant strains. The results of the supercoiling inhibitory assay indicated that mutant gyrases exhibited the lowest IC_{50}s at position 95, further supporting the superiority of WQ-3810.

Since MXF-containing regimens have improved the treatment outcomes for MAC-PD [29], MXF has been suggested as a potential secondary drug [12,13,30,31] for macrolide-resistant MAC isolates or for patients who are unable to undergo macrolide therapy. The proposed breakpoints were ≤1 μg/mL, 2 μg/mL, and >4 μg/mL, indicating susceptible, intermediate resistant, and resistant, respectively, [12]. WQ-3810 demonstrated superior antimicrobial activity compared to MXF, suggesting this novel agent may be an effective replacement for MXF for MAC-PD.

Before reaching the target protein, i.e., DNA gyrase, FQs need to pass through the bacterial cell wall and adequately accumulate inside the bacterial cell to exhibit strong antimicrobial activity. However, the mycobacterial cell wall consists of a thick, lipid-rich mycolic acid, which accounts for 60% of the mycobacterial mass in *M. tuberculosis*, making it highly hydrophobic [32]. Therefore, many researchers have attempted to develop lipophilic FQs to enhance cell wall penetration [33]. A previous study indicated that the bulky azetidinyl group at the R_7 position of WQ-3810 increases the drug's lipophilicity by introducing an alkyl group into 7-(3-aminoazetidin-1-yl) FQ [20]. Results from this study suggest that WQ-3810 exhibits increased permeability, facilitating easy penetration due to the high lipophilic substituent at the C_7 and C_8 positions of the compound. Our findings show that the MIC of WQ-3810 was half that of MXF for Asp95Gly and Asp95Tyr, demonstrating superior antimicrobial activity. In contrast, another study used recombinant *M. tuberculosis* biovar. *bovis* Bacille Calmette–Guerin strains found that WQ-3810 displayed weaker antimycobacterial activity compared to MXF [23]. This discrepancy may arise from differences in intrinsic drug resistance mechanisms, such as variations in efflux pump mechanisms [34,35], distinct structural characteristics of the cell wall, and different physiological traits among the species [36].

We furthermore used the cell wall synthesis inhibitors INH and EMB in a checkerboard assay to evaluate their interaction with WQ-3810 in *M. avium* clinical isolates. These inhibitors block the synthesis of the lipid-rich mycobacterial cell wall, increasing its permeability and facilitating the passage of FQs to the DNA gyrase [9,10]. EMB inhibits arabinosyl-transferase, disrupting arabinogalactan in the cell wall [37]. INH inhibits cell wall production by preventing the synthesis of mycolic acids [38]. MXF was selected among three drugs for the checkerboard assay to compare with WQ-3810 due to its comparable results in other assays. Our findings indicate that both WQ-3810 and MXF exhibit a synergistic relationship (mean FICIm < 0.5) when used in combination with INH, while they showed additive effects (mean FICIm between 0.5 and 1) when combined with EMB, suggesting that INH enhances the permeability of *M. avium's* cell wall to antimicrobial agents compared to EMB. However, previous studies have shown that EMB synergizes with various other drugs [23,37]. We therefore measured the MIC of WQ-3810 and MXF with the combination of 1 μg/mL of EMB and INH. For both WT and mutant strains, the MICs of both FQs were similar to or lower, when they were used alone, highlighting that WQ-3810 can be effective in combination drug therapy with other agents. Notably, when combined with either 1 μg/mL INH or EMB for mutant strains, WQ-3810 demonstrated a lower MIC than the MIC of MXF. WQ-3810's superiority was further supported by the findings from the supercoiling inhibitory assay and MIC assay, which indicated that the mutant gyrase with a substitution at position 95 exhibited the lowest IC_{50} and MIC values. Our findings suggest that WQ-3810 is a promising therapeutic option, demonstrating higher activity against mutants compared to MXF, a second-line drug currently used for macrolide-resistant MAC isolates.

4. Materials and Methods

4.1. Recombinant M. avium Subsp. Hominissuis DNA Gyrases

Recombinant *M. avium* subsp. *hominissuis* WT and mutant DNA gyrases were used. Four mutant DNA gyrases were selected that showed single amino acid substitutions compared to WT: Ala91Val, Asp95Ala, Asp95Gly, and Asp95Tyr were selected. All the DNA gyrase used in the present study was produced in our previous study [17].

4.2. Clinical Isolates of M. avium Subsp. Hominissuis

Ten clinical *M. avium* subsp. *hominissuis* isolates from Keio University Hospital, Tokyo, and one from Osaka Habikino Medical Center, Osaka, were used. The isolates included 7 WT strains, 2 strains with the Asp95Gly mutation, and 2 strains with the Asp95Tyr mutation. MIC assay and Checkerboard assay were not performed for the clinical strains with Ala91Val and Asp95Ala, as these mutations were not found in the clinical isolates [17].

4.3. Antimicrobial Compounds and Reagents

WQ 3810 was a gift from Wakunaga Pharmaceutical Co., Ltd. (Tokyo, Japan). CIP, MXF, and LVX were purchased from LKT Laboratories, Inc. (St. Paul, MN, USA). EMB was obtained from MP Biomedicals (Santa Ana, CA, USA). The chemical structures of these four FQs are shown in Figure 1 and Table 2. Kanamycin (KAN), isoniazid (INH), and Tween 80 were purchased from Fujifilm Wako Pure Chemical Co., Ltd. (Osaka, Japan). GelRed (10,000× concentration) was bought from Biotium (San Francisco, CA, USA). Relaxed pBR322 DNA was obtained from John Innes Enterprises Ltd. (Norwich, UK). Agarose gel I was obtained from Dojindo (Kumamoto, Japan). Lambda DNA-HindIII DNA marker was obtained from New England Biolabs, Inc. (Ipswich, MA, USA). 2% Ogawa medium was obtained from Serotec Co., Ltd. (Sapporo, Japan). BD Difco Middlebrook 7H9 broth and BD BBL Middlebrook OADC Enrichment were obtained from Becton, Dickinson, and Company (Franklin Lakes, NJ, USA).

4.4. FQs Inhibited DNA Gyrase Supercoiling Assay

The DNA supercoiling inhibitory assay was performed using the purified recombinant DNA gyrases as described previously (Supplementary Figure S2) [17]. In brief, 30 μL of the reaction mixture was prepared containing 1× DNA gyrase reaction buffer (35 mM of tris-HCl (pH 7.5), 24 mM of KCl, 6 mM of $MgCl_2$, 5 mM of DTT, 1.8 mM of spermidine, 6.5% glycerol (wt/vol), 0.36 mg/mL of BSA), 50 mM ATP, 1.5 nM relaxed pBR322 DNA, 7.5 nM Gyr A, 7.5 nM Gyr B and serially diluted WQ-3810. WQ-3810 was utilized at a concentration range of 0.125 μg/mL to 64 μg/mL. Then, the reaction mixture was incubated at 37 °C for 60 min, and the reaction was stopped by adding 8 μL of 5× dye (5% SDS, 25% glycerol, 0.25mg/mL bromophenol blue). Next, 10 μL of each reaction mixture and 5 μL of 500 ng/mL Lambda DNA-HindIII DNA marker (New England Biolabs) were loaded onto a 1% agarose I gel in 1× Tris-acetate-EDTA buffer for electrophoresis at 50 mA for 96 min. Subsequently, the gels were stained with 1× GelRed for 30 min, and the presence of a supercoiled DNA band was visualized under UV light. The amount of supercoiled DNA was quantified by measuring the intensity of the bands by using ImageJ 1.54d software (https://imagej.net/ij/download.html, accessed on 24 May 2023). The concentration of WQ-3810 that reduced DNA gyrase activity by 50% (IC_{50}) was calculated with the AAT Bioquest IC_{50} calculator web tool (https://www.aatbio.com/tools/ic50-calculator, accessed on 24 May 2023). All the assays were run in triplicate on the same day to eliminate experimental bias and to confirm their reproducibility.

4.5. Minimum Inhibitory Concentration (MIC) Assay

The MIC assay was performed using the microdilution method in Middlebrook 7H9 medium, supplemented with 10% oleic acid-albumin-dextrose-catalase (OADC) and 0.05% Tween 80 following the protocol recommended previously [17]. Briefly, the preserved isolates were inoculated in 2% Ogawa medium, and solid culture was sub-cultured in 4 mL of 7H9 broth and incubated at 37 °C until the optical density (OD) at 590 nm reached 0.15. Next, this culture was further diluted 40 times and used for the MIC assay. The assay was performed in a 96-well round-bottom culture plate (As One Co. Ltd., Osaka, Japan). Each well contained a 200 μL mixture of 100 μL of starting culture and 100 μL of serially diluted drug in 7H9 broth. The outer wells of the plate contained 200 μL of sterile distilled water. Each plate of the MIC assay contained two drug controls with KAN at 25 μg/mL to observe growth inhibition, two growth controls without any drugs to see bacterial growth, and two medium controls without bacteria to check for contamination of the medium. For this study, WQ-3810, MXF, EMB, and INH were used. After sealing the plate with a plastic membrane, it was placed in a container with moist cotton and incubated at 37 °C for 14 days. Each experiment was conducted in triplicate to confirm the assay's reproducibility. Bacterial growth was monitored on days 0, 1, 7, 10, and 14 by taking a picture. The MIC value, the lowest concentration of the drug at which bacterial growth was completely inhibited, was determined on day 14.

4.6. Checkerboard Assay

A checkerboard assay was conducted to evaluate the interaction between FQs (WQ-3810 and MXF) and cell wall synthesis inhibitors (INH and EMB) in clinical *M. avium* subsp. *hominissuis* isolates. Each agent's fraction inhibitory concentration index (FICI) was determined in the presence of sub-inhibitory concentrations of another agent, following the protocol previously recommended [23,39]. The assay was performed in Middlebrook 7H9 broth, supplemented with 10% OADC and 0.5% Tween 80. Bacterial cultures were prepared to an OD of 0.15 at 590 nm and diluted 40-fold in Middlebrook 7H9 broth. Then 100 µL of the diluted culture was added to each well of a sterile round-bottom microtiter plate containing serially diluted concentrations of FQs (WQ-3810 or MXF) (50 µL per well) and either INH or EMB (50 µL per well). The test plates were incubated at 37 °C for 14 days, and bacterial growth was monitored by taking photos of the plates on days 0, 1, 7, 10, and 14. Each experiment was conducted in triplicate to confirm the assay's reproducibility, and the cube root value was calculated. The FICI for each drug combination was calculated using the equation: FICI = (MIC of drug A in the presence of drug B/MIC of drug A alone) + (MIC of drug B in the presence of drug A/MIC of drug B alone). The FICIm values ≤ 0.5 were considered indicative of synergistic interaction.

4.7. Statistical Analysis

Statistical analysis was performed using RStudio version 2024.04.1 to compare the effects of four different FQs on inhibitory concentration (IC) values. A one-way analysis of variance (ANOVA) was used to identify any overall significant differences among the drugs. Furthermore, a post hoc Least Significant Difference (LSD) test was conducted to identify specific drug pairs with significant differences.

5. Conclusions

Our study assessed the effectiveness of WQ-3810, a novel FQ, against *M. avium* DNA gyrases and clinical isolates. WQ-3810 showed a stronger inhibitory effect than LVX and CIP and was comparable to or superior to MXF for both WT and mutant DNA gyrases. WQ-3810 exhibited lower MICs against clinical isolates, especially FQ-resistant isolates, than other FQs. Furthermore, WQ-3810 showed a synergistic interaction in combination with INH. Our results indicate that WQ-3810 is a promising therapeutic option, demonstrating superior efficacy compared to MXF, which is currently available as a second-line treatment for MAC-PD resistant to macrolide therapy. WQ-3810 may also be a potential treatment option for FQ-resistant *M. avium* infections.

Supplementary Materials: The following supporting information can be downloaded at https://www.mdpi.com/article/10.3390/antibiotics14070704/s1, Table S1: Interaction between WQ-3810 and cell wall inhibitors. Table S2: Interaction between MXF and cell wall inhibitors. Figure S1: Amino acid sequence alignment of DNA gyrases QRDR from various bacterial species. Figure S2: SDS-PAGE analysis of purified recombinant *M. avium* DNA gyrase subunits.

Author Contributions: Conceptualization, Y.S.; methodology, P.S., J.T., H.K., and Y.S.; validation, S.J., J.T., and Y.S.; formal analysis, S.J. and C.P.; investigation, S.J., D.A.B., H.K., and J.T.; resources, P.S., T.N., N.H., Y.N., and J.T.; writing—original draft preparation, S.J.; writing—review and editing, C.P., D.A.B., S.V.G., T.N., N.H., Y.N., C.N., and Y.S.; visualization, S.J. and H.K.; supervision, Y.S., S.V.G., and C.N.; funding acquisition, Y.S. and C.N. All authors have read and agreed to the published version of the manuscript.

Funding: This research was funded by a grant from the Ministry of Education, Culture, Sports, Science, and Technology (MEXT), Japan, for the Joint Research Program of Hokkaido University International Institute for Zoonosis Control, and in part by Japan Agency for Medical Research and Development (AMED) under Grant Number JP20wm0125008 and JP223fa627005 to Y.S. and by JSPS KAKENHI Grant Number 22K05999 to C.N.

Institutional Review Board Statement: Not applicable.

Informed Consent Statement: Not applicable.

Data Availability Statement: The original contributions presented in this study are included in the article/Supplementary Materials. Further inquiries can be directed to the corresponding authors.

Conflicts of Interest: The authors declare no conflicts of interest.

References

1. Mizzi, R.; Plain, K.M.; Whittington, R.; Timms, V.J. Global Phylogeny of *Mycobacterium avium* and Identification of Mutation Hotspots During Niche Adaptation. *Front. Microbiol.* **2022**, *13*, 892333. [CrossRef] [PubMed]
2. Boonjetsadaruhk, W.; Kaewprasert, O.; Nithichanon, A.; Ananta, P.; Chaimanee, P.; Salao, K.; Phoksawat, W.; Laohaviroj, M.; Sirichoat, A.; Fong, Y.; et al. High rate of reinfection and possible transmission of *Mycobacterium avium* complex in Northeast Thailand. *One Health* **2022**, *14*, 100374. [CrossRef] [PubMed]
3. Busatto, C.; Vianna, J.S.; Da Silva, L.V.; Ramis, I.B.; Da Silva, P.E.A. *Mycobacterium avium*: An overview. *Tuberculosis* **2019**, *114*, 127–134. [CrossRef] [PubMed]
4. Honda, J.R.; Virdi, R.; Chan, E.D. Global Environmental Nontuberculous Mycobacteria and Their Contemporaneous Man-Made and Natural Niches. *Front. Microbiol.* **2018**, *9*, 2029. [CrossRef]
5. Park, H.-E.; Kim, K.-M.; Trinh, M.P.; Yoo, J.-W.; Shin, S.J.; Shin, M.-K. Bigger problems from smaller colonies: Emergence of antibiotic-tolerant small colony variants of *Mycobacterium avium* complex in MAC-pulmonary disease patients. *Ann. Clin. Microbiol. Antimicrob.* **2024**, *23*, 25. [CrossRef]
6. Cowman, S.; Van Ingen, J.; Griffith, D.E.; Loebinger, M.R. Non-tuberculous mycobacterial pulmonary disease. *Eur. Respir. J.* **2019**, *54*, 1900250. [CrossRef]
7. Namkoong, H.; Kurashima, A.; Morimoto, K.; Hoshino, Y.; Hasegawa, N.; Ato, M.; Mitarai, S. Epidemiology of Pulmonary Nontuberculous Mycobacterial Disease, Japan1. *Emerg. Infect. Dis.* **2016**, *22*, 1116–1117. [CrossRef]
8. Griffith, D.E.; Aksamit, T.; Brown-Elliott, B.A.; Catanzaro, A.; Daley, C.; Gordin, F.; Holland, S.M.; Horsburgh, R.; Huitt, G.; Iademarco, M.F.; et al. An Official ATS/IDSA Statement: Diagnosis, Treatment, and Prevention of Nontuberculous Mycobacterial Diseases. *Am. J. Respir. Crit. Care Med.* **2007**, *175*, 367–416. [CrossRef]
9. Koh, W.-J.; Moon, S.M.; Kim, S.-Y.; Woo, M.-A.; Kim, S.; Jhun, B.W.; Park, H.Y.; Jeon, K.; Huh, H.J.; Ki, C.-S.; et al. Outcomes of *Mycobacterium avium* complex lung disease based on clinical phenotype. *Eur. Respir. J.* **2017**, *50*, 1602503. [CrossRef]
10. Jeong, B.-H.; Jeon, K.; Park, H.Y.; Kim, S.-Y.; Lee, K.S.; Huh, H.J.; Ki, C.-S.; Lee, N.Y.; Shin, S.J.; Daley, C.L.; et al. Intermittent Antibiotic Therapy for Nodular Bronchiectatic *Mycobacterium avium* Complex Lung Disease. *Am. J. Respir. Crit. Care Med.* **2015**, *191*, 96–103. [CrossRef]
11. Daley, C.L.; Iaccarino, J.M.; Lange, C.; Cambau, E.; Wallace, R.J.; Andrejak, C.; Böttger, E.C.; Brozek, J.; Griffith, D.E.; Guglielmetti, L.; et al. Treatment of nontuberculous mycobacterial pulmonary disease: An official ATS/ERS/ESCMID/IDSA clinical practice guideline. *Eur. Respir. J.* **2020**, *56*, 2000535. [CrossRef] [PubMed]
12. *M24-A2*; Susceptibility Testing of Mycobacteria, Nocardiae, and Other Aerobic Actinomycetes; Approved Standard—Second Edition. Clinical and Laboratory Standards Institute: Wayne, PA, USA, 2011.
13. Van Ingen, J.; Wagner, D.; Gallagher, J.; Morimoto, K.; Lange, C.; Haworth, C.S.; Floto, R.A.; Adjemian, J.; Prevots, D.R.; Griffith, D.E. Poor adherence to management guidelines in nontuberculous mycobacterial pulmonary diseases. *Eur. Respir. J.* **2017**, *49*, 1601855. [CrossRef] [PubMed]
14. Pham, T.D.M.; Ziora, Z.M.; Blaskovich, M.A.T. Quinolone antibiotics. *MedChemComm* **2019**, *10*, 1719–1739. [CrossRef]
15. Spencer, A.C.; Panda, S.S. DNA Gyrase as a Target for Quinolones. *Biomedicines* **2023**, *11*, 371. [CrossRef]
16. Wigley, D.B. Structure and Mechanism of DNA Gyrase. In *Nucleic Acids and Molecular Biology*; Eckstein, F., Lilley, D.M.J., Eds.; Nucleic Acids and Molecular Biology; Springer: Berlin/Heidelberg, Germany, 1995; Volume 9, pp. 165–176. ISBN 978-3-642-79490-2.
17. Thapa, J.; Chizimu, J.Y.; Kitamura, S.; Akapelwa, M.L.; Suwanthada, P.; Miura, N.; Toyting, J.; Nishimura, T.; Hasegawa, N.; Nishiuchi, Y.; et al. Characterization of DNA Gyrase Activity and Elucidation of the Impact of Amino Acid Substitution in GyrA on Fluoroquinolone Resistance in Mycobacterium avium. *Microbiol. Spectr.* **2023**, *11*, e05088-22. [CrossRef]

18. Cambau, E.; Sougakoff, W.; Jarlier, V. Amplification and nucleotide sequence of the quinolone resistance-determining region in the *gyrA* gene of mycobacteria. *FEMS Microbiol. Lett.* **1994**, *116*, 49–54. [CrossRef] [PubMed]
19. World Health Organization. *Catalogue of Mutations in Mycobacterium Tuberculosis Complex and Their Association with Drug Resistance*; World Health Organization: Geneva, Switzerland, 2021; ISBN 978-92-4-002817-3.
20. Itoh, K.; Kuramoto, Y.; Amano, H.; Kazamori, D.; Yazaki, A. Discovery of WQ-3810: Design, synthesis, and evaluation of 7-(3-alkylaminoazetidin-1-yl)fluoro-quinolones as orally active antibacterial agents. *Eur. J. Med. Chem.* **2015**, *103*, 354–360. [CrossRef]
21. Kazamori, D.; Aoi, H.; Sugimoto, K.; Ueshima, T.; Amano, H.; Itoh, K.; Kuramoto, Y.; Yazaki, A. In vitro activity of WQ-3810, a novel fluoroquinolone, against multidrug-resistant and fluoroquinolone-resistant pathogens. *Int. J. Antimicrob. Agents* **2014**, *44*, 443–449. [CrossRef]
22. Park, J.-H.; Yamaguchi, T.; Ouchi, Y.; Koide, K.; Mori, S.; Kim, H.; Mukai, T.; Nakajima, C.; Suzuki, Y. WQ-3810 inhibits DNA gyrase activity in ofloxacin-resistant *Mycobacterium leprae*. *J. Infect. Chemother.* **2020**, *26*, 335–342. [CrossRef]
23. Ouchi, Y.; Mukai, T.; Koide, K.; Yamaguchi, T.; Park, J.-H.; Kim, H.; Yokoyama, K.; Tamaru, A.; Gordon, S.V.; Nakajima, C.; et al. WQ-3810: A new fluoroquinolone with a high potential against fluoroquinolone-resistant *Mycobacterium tuberculosis*. *Tuberculosis* **2020**, *120*, 101891. [CrossRef]
24. Koide, K.; Kim, H.; Whelan, M.V.X.; Belotindos, L.P.; Tanomsridachchai, W.; Changkwanyeun, R.; Usui, M.; Ó Cróinín, T.; Thapa, J.; Nakajima, C.; et al. WQ-3810, a fluoroquinolone with difluoropyridine derivative as the R1 group exerts high potency against quinolone-resistant *Campylobacter jejuni*. *Microbiol. Spectr.* **2024**, *12*, e04322-23. [CrossRef] [PubMed]
25. Yamaba, Y.; Ito, Y.; Suzuki, K.; Kikuchi, T.; Ogawa, K.; Fujiuchi, S.; Hasegawa, N.; Kurashima, A.; Higuchi, T.; Uchiya, K.; et al. Moxifloxacin resistance and genotyping of *Mycobacterium avium* and *Mycobacterium intracellulare* isolates in Japan. *J. Infect. Chemother.* **2019**, *25*, 995–1000. [CrossRef] [PubMed]
26. Bush, N.G.; Diez-Santos, I.; Abbott, L.R.; Maxwell, A. Quinolones: Mechanism, Lethality and Their Contributions to Antibiotic Resistance. *Molecules* **2020**, *25*, 5662. [CrossRef] [PubMed]
27. Carter, H.E.; Wildman, B.; Schwanz, H.A.; Kerns, R.J.; Aldred, K.J. Role of the Water–Metal Ion Bridge in Quinolone Interactions with Escherichia coli Gyrase. *Int. J. Mol. Sci.* **2023**, *24*, 2879. [CrossRef]
28. Blower, T.R.; Williamson, B.H.; Kerns, R.J.; Berger, J.M. Crystal structure and stability of gyrase-fluoroquinolone cleaved complexes from *Mycobacterium tuberculosis*. *Proc. Natl. Acad. Sci. USA* **2016**, *113*, 1706–1713. [CrossRef]
29. Koh, W.-J.; Hong, G.; Kim, S.-Y.; Jeong, B.-H.; Park, H.Y.; Jeon, K.; Kwon, O.J.; Lee, S.-H.; Kim, C.K.; Shin, S.J. Treatment of Refractory *Mycobacterium avium* Complex Lung Disease with a Moxifloxacin-Containing Regimen. *Antimicrob. Agents Chemother.* **2013**, *57*, 2281–2285. [CrossRef]
30. Khadawardi, H.; Marras, T.K.; Mehrabi, M.; Brode, S.K. Clinical efficacy and safety of fluoroquinolone containing regimens in patients with *Mycobacterium avium* complex pulmonary disease. *Eur. Respir. J.* **2020**, *55*, 1901240. [CrossRef]
31. Griffith, D.E. Macrolide-Resistant *Mycobacterium avium* Complex: I Feel Like I've Been Here Before. *Ann. Am. Thorac. Soc.* **2016**, *13*, 1881–1882. [CrossRef]
32. Jacobo-Delgado, Y.M.; Rodríguez-Carlos, A.; Serrano, C.J.; Rivas-Santiago, B. *Mycobacterium tuberculosis* cell-wall and antimicrobial peptides: A mission impossible? *Front. Immunol.* **2023**, *14*, 1194923. [CrossRef]
33. Emami, S.; Shafiee, A.; Foroumadi, A. Quinolones: Recent Structural and Clinical Developments. *Iran. J. Pharm. Res.* **2005**, *3*, 123–136.
34. Rodrigues, L.; Sampaio, D.; Couto, I.; Machado, D.; Kern, W.V.; Amaral, L.; Viveiros, M. The role of efflux pumps in macrolide resistance in *Mycobacterium avium* complex. *Int. J. Antimicrob. Agents* **2009**, *34*, 529–533. [CrossRef] [PubMed]
35. Rossi, E.D.; Aínsa, J.A.; Riccardi, G. Role of mycobacterial efflux transporters in drug resistance: An unresolved question. *FEMS Microbiol. Rev.* **2006**, *30*, 36–52. [CrossRef] [PubMed]
36. Bannantine, J.P.; Etienne, G.; Laval, F.; Stabel, J.R.; Lemassu, A.; Daffé, M.; Bayles, D.O.; Ganneau, C.; Bonhomme, F.; Branger, M.; et al. Cell wall peptidolipids of *Mycobacterium avium*: From genetic prediction to exact structure of a nonribosomal peptide. *Mol. Microbiol.* **2017**, *105*, 525–539. [CrossRef] [PubMed]
37. Kim, H.-J.; Lee, J.S.; Kwak, N.; Cho, J.; Lee, C.-H.; Han, S.K.; Yim, J.-J. Role of ethambutol and rifampicin in the treatment of *Mycobacterium avium* complex pulmonary disease. *BMC Pulm. Med.* **2019**, *19*, 212. [CrossRef]
38. Timmins, G.S.; Deretic, V. Mechanisms of action of isoniazid. *Mol. Microbiol.* **2006**, *62*, 1220–1227. [CrossRef]
39. Odds, F.C. Synergy, antagonism, and what the chequerboard puts between them. *J. Antimicrob. Chemother.* **2003**, *52*, 1. [CrossRef]

antibiotics

MDPI

Article

Multidrug Resistance and Virulence Traits of *Salmonella enterica* Isolated from Cattle: Genotypic and Phenotypic Insights

Nada A. Fahmy [1], Sumin Karna [1], Angel Bhusal [1], Ajran Kabir [1], Erdal Erol [2] and Yosra A. Helmy [1,*]

[1] Department of Veterinary Science, Martin-Gatton College of Agriculture, Food, and Environment, University of Kentucky, Lexington, KY 40546, USA
[2] Veterinary Diagnostic Laboratory, Martin-Gatton College of Agriculture, Food, and Environment, University of Kentucky, Lexington, KY 40511, USA
* Correspondence: yosra.helmy@uky.edu

Abstract: Background/Objective: Non-typhoidal *Salmonella* is a leading cause of food-borne illness worldwide and presents a significant One Health concern due to zoonotic transmission. Although antibiotic therapy remains a standard approach for treating salmonellosis in severe cases in animals, the widespread misuse of antibiotics has contributed to the emergence of multidrug-resistant (MDR) *Salmonella* strains. This study provides insights into the genotypic and phenotypic characteristics among *Salmonella* isolates from necropsied cattle. **Methods:** A total of 1008 samples were collected from necropsied cattle. *Salmonella enterica* subspecies were identified by MALDI-TOF MS and subsequently confirmed by serotyping. The biofilm-forming ability of the isolated bacteria was assessed using a crystal violet assay. The motility of the isolates was assessed on soft agar plates. Additionally, the antimicrobial resistance genes (ARGs) and virulence genes were investigated. Antimicrobial resistance patterns were investigated against 19 antibiotics representing 9 different classes. **Results:** *Salmonella* species were isolated and identified in 27 necropsied cattle. *Salmonella* Dublin was the most prevalent serotype (29.6%). Additionally, all the isolates were biofilm producers at different levels of intensity, and 96.3% of the isolates exhibited both swarming and swimming motility. Furthermore, virulence genes, including *invA*, *hilA*, *fimA*, and *csgA*, were detected in all the isolates. The highest resistance was observed to macrolides (azithromycin and clindamycin) (100%), followed by imipenem (92.6%), and chloramphenicol (85.2%). All isolates were multidrug-resistant, with a multiple antibiotic resistance (MAR) index ranging between 0.32 and 0.74. The aminoglycoside resistance gene *aac(6′)-Ib* was detected in all the isolates (100%), whereas the distribution of other antimicrobial resistance genes (ARGs) varied among the isolates. **Conclusions:** The increasing prevalence of MDR *Salmonella* poses a significant public health risk. These resistant strains can reduce the effectiveness of standard treatments and elevate outbreak risks. Strengthening surveillance and regulating antibiotic use in livestock are essential to mitigating these threats.

Keywords: antimicrobial resistance; biofilm; MDR; motility; pathogenicity; *Salmonella*; virulence

1. Introduction

Livestock, such as cattle, sheep, and goats, serve as major reservoirs for diverse *Salmonella* serotypes and play a critical role in the transmission of infection to humans [1].

Nontyphoidal salmonellosis outbreaks are frequently associated with the consumption of contaminated meat, poultry, and processed food products [2]. According to the Centers for Disease Control and Prevention (CDC), 82.2% of Americans consume beef weekly, with 67% specifically consuming ground beef [3]. Consequently, nontyphoidal *Salmonella* is estimated to cause approximately 1.35 million infections annually, resulting in approximately 26,500 hospitalizations and 420 deaths in the USA. Thus *Salmonella* represents one of the highest economic burdens among foodborne pathogens, with total annual costs exceeding $17.1 billion [3–5]. Similarly, on a global scale, the World Health Organization (WHO) identifies *Salmonella* species among the 31 leading pathogens capable of causing both intestinal and systemic diseases in humans, contributing to an estimated 150 million illnesses and 60,000 fatalities annually worldwide [6–8]. However, the true burden of *Salmonella* infections is likely much higher, as the CDC estimates that only 1 in every 30 cases is laboratory confirmed, largely due to underdiagnosis and underreporting [9].

Salmonella species are antigenically classified according to the Kauffmann–White scheme, which differentiates over 2600 serotypes based on variations in somatic (O), flagellar (H), and capsular (K) antigens [10]. This significant diversity contributes to variation in host adaptation, pathogenic severity, and clinical outcomes [11]. *Salmonella enterica* subsp. *enterica* is the most prevalent subspecies and primarily infects warm-blooded animals [12]. The National Veterinary Services Laboratory has reported *Salmonella* Dublin as the most prevalent serotype in livestock, specifically cattle, followed by *Salmonella* Cerro and *Salmonella* Typhimurium [13]. While some serotypes are host-restricted, others can infect a wide range of animal species, contributing to their significance as zoonotic and foodborne pathogens [14]. The zoonotic transmission of *Salmonella* may occur through direct or indirect contact with infected animals or their environments, ingestion of contaminated animal-derived food or water, or via intermediate hosts, such as insect vectors [15,16]. Spillover to humans typically depends on the convergence of environmental conditions, pathogen characteristics, and host susceptibility [17,18]. In humans, high-risk groups, including infants, young children, the elderly, and immunocompromised patients, are particularly vulnerable, especially when hygiene is inadequate or when the gut microbiota is disrupted due to dysbiosis. In animals, the risk of infection is elevated by factors such as stress, co-infections, herd size, and suboptimal farm management practices [16,19].

The infection cycle of bovine salmonellosis typically begins with oral ingestion, especially in calves with immature gut microbiota or gastrointestinal stasis [20]. *Salmonella* adheres to and invades enterocytes and penetrates the intestinal lining, triggering inflammation or being engulfed by immune cells [21]. Once inside macrophages, it survives and disseminates through the lymphatic system, targeting lymphoid tissues and contributing to the development of bacteremia and systemic disease. Additionally, systemic spread can occur via pharyngeal lymphoid tissues, bypassing the intestinal tract [13,22]. Antibiotic therapy is frequently used as the initial strategy to manage infections caused by *Salmonella* spp. in both humans and animals. The extensive and often misuse of antibiotics in the livestock sector has raised significant concerns regarding the emergence and spread of antimicrobial-resistant genes (ARGs) [23]. Infections with multidrug-resistant (MDR) *Salmonella* are associated with a two-fold increase in the mortality risk compared to drug-susceptible strains [24]. The dissemination of MDR *Salmonella* has been closely linked to antibiotic misuse in food-producing animals, thereby facilitating zoonotic transmission. Approximately 70% of MDR *Salmonella* outbreaks in the U.S. have been attributed to livestock-derived foods [25]. In addition to the development of MDR, the *Salmonella* genome is encoded with a wide array of virulence factors that contribute to its pathogenicity and adaptability. *Salmonella* pathogenesis involves sequential steps critical for host colonization. The process begins with adhesion to intestinal epithelial cells via fimbriae and outer membrane proteins [26]. This is followed by invasion, mediated by *Salmonella* pathogenicity islands (SPIs) and a type 3 secretion system (T3SS), which injects effector proteins to disrupt host cell functions and facilitate bacterial entry. Upon entry into the host, *Salmonella* activates a range of virulence mechanisms, including the modulation of host signaling pathways and the inhibition of phagocytic killing to evade immune responses and establish a successful infection [26,27]. The differences in SPI gene content likely reflect the co-evolutionary dynamics and host-specific adaptations of SPIs across diverse *Salmonella* species and serotypes [28]. Similarly, virulence plasmids play a pivotal role in facilitating systemic dissemination following oral infection in animal models [29]. The pathogenicity of *Salmonella* reflects a sophisticated genomic architecture shaped by the evolutionary acquisition of pathogenicity islands and the horizontal transfer of genes and mobile genetic elements, equipping the organism with the metabolic flexibility, regulatory control, and host-specific virulence mechanisms required for successful intracellular survival and systemic infection [30–33].

Additional factors vital to *Salmonella* colonization include flagella-driven motility, fimbriae, and adhesins for epithelial attachment [34,35], as well as toxin production that disrupts host cell function [35,36]. A key component of its virulence strategy, *Salmonella* utilizes multiple secretion systems to translocate effector proteins that manipulate host processes and enhance intracellular survival. Four major systems have been identified: type 1, 3, 4, and 6 secretion systems (T1SS–T6SS), each contributing uniquely to infection and immune evasion [36]. In addition to these classical virulence factors, biofilm formation represents a critical survival strategy that significantly enhances *Salmonella* pathogenicity [37,38]. Bacterial cells growing within a biofilm adopt a markedly different mode of life compared to those in the planktonic state, with distinct patterns of gene and protein expression [39]. Within these structured communities, biofilms create protective microenvironments that support the development of both genotypic and phenotypic diversities. This dynamic nature enables bacteria to adapt rapidly to environmental stresses [40]. One of the hallmark features of biofilm-associated cells is their elevated resistance or tolerance to antimicrobial agents, which is not typically observed in their free-living cells [37].

While postharvest interventions target surface contamination, the pathogen can inhabit multiple internal organs, posing challenges for detection and control. Studies have identified *Salmonella* in the intestinal contents of clinically healthy cattle, as well as in liver and lung tissues, emphasizing the role of systemic dissemination and subclinical carriage [3]. Conventional culture methods and advanced molecular diagnostics, such as polymerase chain reaction (PCR), have been employed to assess *Salmonella* prevalence across various tissues [41,42]. These findings underscore the need for comprehensive preharvest surveillance strategies that extend beyond fecal sampling to assess infection status and reduce public health risks accurately [3,42]. In 2024, Kentucky's livestock sector, led by beef production, played a key role in sustaining the state's agriculture, with a 16.8% rise in receipts. As one of the top beef-producing states, this growth helped drive total agricultural cash receipts to nearly $8.3 billion [43]. Thus, in this study, we aimed to investigate the phenotypic and genotypic characteristics of *Salmonella* isolates from necropsied cattle in central Kentucky, with emphasis on antimicrobial resistance profiles, prevalent serotypes, biofilm formation ability, motility behavior, and the correlation between the phenotypic and the genetic traits.

2. Materials and Methods

2.1. Sample Collection, Bacterial Isolation, and Characterization

Between January 2022 and December 2023, a total of 1008 samples were collected from individual necropsied cattle submitted to the Veterinary Diagnostic Laboratory at the University of Kentucky. Sampling was conducted through random, case-based submissions from field veterinarians across various farms in central Kentucky, U.S. These cases were submitted and investigated at the request of the attending clinicians. Samples were collected as necropsied animals became available, and diagnostic evaluations were performed accordingly. These samples were collected from various organs, including the intestine, colon, lung, and liver. The animals' ages ranged between 3 days and 12 years. They were categorized as follows: neonatal (2 to 28 days), calf (29 to 90 days), juvenile (3 months to 12 months), and adult (1 year and older) [44]. Organs were seared with sterile blades prior to sample collection to minimize surface contamination. Swab samples were then obtained and inoculated onto blood agar (BA), Hektoen enteric agar, and eosin methylene blue (EMB) agar (Hardy Diagnostics, Santa Maria, CA, USA). All primary cultures were incubated at 37 °C for 24 h. Swabs from intestinal samples were enriched in selenite broth at 37 °C for 24 h and then subcultured onto Hektoen and EMB agar, followed by incubation for an additional 48 h at 37 °C. Colonies exhibiting black pigmentation on Hektoen or a pale/white appearance on EMB were selected for identification by matrix-assisted laser desorption/ionization time-of-flight mass spectrometry (MALDI-TOF MS). Identification was performed using the direct transfer method, with a minimum score threshold of 1.7 for genus-level identification and 2.0 for species-level identification, as determined by the Biotyper software (version 4.0; Bruker Scientific Corp., San Jose, CA, USA). Serotyping was conducted at the National Veterinary Services Laboratory (NVSL) following the Kauffmann–White–Le Minor Scheme. The *Salmonella* serotype was determined based on the combination of O- and H-antigen profiles obtained from the agglutination results [45]. All isolates were preserved in 25% glycerol (*v*/*v*) and stored at −80 °C for further analysis.

2.2. Biofilm Formation Assay

The ability of the *Salmonella* isolates to form biofilm was evaluated using a crystal violet staining assay, as described previously [46,47]. An overnight culture of *Salmonella* was grown in Luria–Bertani (LB) broth (BD Difco™, Franklin Lakes, NJ, USA) and diluted to a final optical density (OD_{600}) of ~0.05 (10^7 CFU/mL). Aliquots of 100 µL of the bacterial suspensions were transferred into a 96-well microtiter plate for each isolate, with sterile LB media serving as a negative control. The plates were incubated statically at 37 °C for 48 h to allow biofilm formation. After incubation, planktonic cells were removed, and the wells were washed gently with distilled water to remove any unbound cells. To quantify biofilm formation, the remaining biofilms were stained with 160 µL of 0.1% crystal violet (*w*/*v*) for 15 min. The plates were then rinsed with distilled water to remove the extra stain and were allowed to dry for 60 min at room temperature. The crystal violet bound to the biofilms was solubilized using 30% acetic acid (*v*/*v*) per well. The biofilm formation was measured at OD_{550} nm using a microplate reader (Tecan, Morrisville, NC, USA). Each isolate was tested in eight replicates, and the assay was performed twice independently to confirm biological reproducibility. The data were presented as the mean absorbance ± standard deviation (SD). The *Salmonella* strains were considered biofilm producers when the OD_{550} value was at least three times the SD of the mean absorbance of the negative control (OD_{NC}). The biofilm formation was categorized as a strong biofilm producer (SBP) ($OD_{550} \leq 4\ OD_{NC}$), moderate biofilm producer (MBP) ($2\ OD_{NC} \leq OD_{550} < 4\ OD_{NC}$), weak biofilm producer

(WBP) ($OD_{NC} \leq OD_{550} < 2\ OD_{NC}$), and non-biofilm producer ($OD_{550} \leq OD_{NC}$), as previously described [48].

2.3. Motility Assay

Swarming and swimming motility were evaluated using soft agar plates, following the protocols described before [49,50]. The *Salmonella* isolates were cultured overnight in LB broth (BD Difco™, USA). For swimming motility, 6 μL of *Salmonella* culture (OD_{600}~0.05) was inoculated into the center of a 0.25% soft agar plate supplemented with 0.5% glucose. For the swarming motility, 6 μL of the same *Salmonella* culture (OD_{600}~0.05) was spotted onto the center of a 0.5% soft agar plate, supplemented with 0.5% glucose. The plates were incubated at 37 °C, and the formation of a turbid zone was recorded at 6 h for swimming motility. In contrast, the swarming zone was measured at 12 h to evaluate the bacteria's movement on soft agar. *Proteus mirabilis* (*P. mirabilis*) ATCC 35659 and *S.* Typhimurium ATCC 14028 were used as positive controls, while *Rhodococcus equi* was included as a negative control. The isolates were considered motile when a diffuse growth zone expanded outward from the inoculation or spotting area and negative if the growth remained confined to the inoculation point [51,52]. Three independent replicates were performed for each experiment, and the assay was performed twice.

2.4. Antimicrobial Susceptibility Testing (AST)

The antimicrobial susceptibility profiles of the *Salmonella* isolates were determined using the broth microdilution method (Thermo Scientific™ Sensititre™ Vet Bovine BOPO7F Plate, Waltham, MA, USA) that contain antibiotics, including neomycin (NEO), gentamicin (GEN), ampicillin (AMP), ceftiofur (CEF), tetracycline (TET), enrofloxacin (ENR), florfenicol (FLO), and clindamycin (CLI). Additional antibiotics were not included in the panel: amikacin (AMK), amoxicillin–clavulanic acid (AMC), meropenem (MEM), imipenem (IPM), ceftazidime (CAZ), doxycycline (DOX), trimethoprim–sulfamethoxazole (SXT), levofloxacin (LVX), ciprofloxacin (CIP), chloramphenicol (CHL), and azithromycin (AZM). The concentrations used for each antibiotic, along with their abbreviations and corresponding references, are detailed in Table S1. All the *Salmonella* isolates were cultured on Columbia agar plates supplemented with 5% sheep blood (Becton Dickinson, Franklin Lakes, NJ, USA) and incubated at 37 °C for 24 h. Antimicrobial susceptibility testing (AST) was conducted, following the manufacturer's protocol. In brief, a 0.5 McFarland standard suspension was prepared in distilled water using a spectrophotometer. From this suspension, 10 μL was added to Sensititre Mueller Hinton broth. Subsequently, 50 μL of the prepared mixture was dispensed into each well of the Sensititre™ Vet Bovine BOPO7F Plate. The plate was incubated at 37 °C, and MICs were determined after 18 h using the Thermo Scientific Sensititre OptiRead Automated Fluorometric Plate Reading System. For the additional antibiotics not included in the commercial panel, AST was performed using a standard broth microdilution method. A two-fold serial dilution of the antibiotic was prepared in a sterile 96-well microtiter plate, and 50 μL of the bacterial suspension (prepared as above) was added to each well. The plates were incubated at 37 °C for 18 h, and MICs were determined using the same OptiRead™ system. The interpretive breakpoints for each antimicrobial agent were established based on CLSI M100 and CLSI VET01S guidelines [53,54]. The multiple antibiotic resistance (MAR) index was calculated as the ratio of the number of antibiotics (19) to which the strain is resistant to the total number of antibiotics tested [55]. The MAR index exceeding 0.2 typically indicates that the isolates originate from high-risk sources with frequent antibiotic exposure [56,57]. Strains were considered MDR if they were non-susceptible to at least one agent in three or more antimicrobial classes. In contrast, strains were considered as extensively drug-resistant (XDR) if they were resistant to at least one agent in all but one or two antimicrobial categories, remaining susceptible to only one or two classes of antibiotics [57,58].

2.5. Detection of Virulence Genes and Antimicrobial Resistance Genes (ARGs)

All the *Salmonella* isolates collected in this study were screened for the presence of virulence genes using conventional PCR. Genomic DNA was extracted from overnight cultures of the *Salmonella* isolates using a boiling method, as previously described [42,59,60]. Briefly, a single colony of each *Salmonella* isolate was suspended in 100 µL of molecular-grade water and subjected to heat lysis using a thermal cycler at 95 °C for 10 min, followed by immediate chilling on ice for 10 min. The lysate was then centrifuged at 5000× *g* for 10 min, and 50 µL of the supernatant was collected as the DNA template and stored at −20 °C. DNA quality and quantity were assessed using a Thermo Scientific NanoDrop spectrophotometer (Thermo Fisher, Lexington, KY, USA). Seventeen virulence genes were selected, based on their specificity to pathogenic *Salmonella* isolates, conservation across species, and functional relevance to host invasion and survival. These genes included host cell invasion and intracellular survival genes (*invA, hilA, intA, intB*), adhesion and attachment genes (*fimA, fimD*), a type 3 secretion system (T3SS) (*spiA, spi4D*), a type 1 secretion system (T1SS) (*siiA, siiC*), host immune suppressor genes (*spvC, sopB*), biofilm formation genes (*csgA, csgB*), and motility and flagellar genes (*fljB, fliC, flhD*). Likewise, fifteen ARGs, representing different antibiotic classes, were selected and screened for their presence among the *Salmonella* isolates. These included genes encoding antimicrobial resistance to β-lactamase ($bla_{TEM\text{-}1B}$, bla_{CMY}, $bla_{CTX\text{-}M}$, bla_{SHV}, $bla_{OXA\text{-}9}$), aminoglycoside (*aadB, aacA3, aac(6′)*), tetracycline (*tetB*), phenicol (*floR, catB*), sulfonamide (*sul2*), macrolides (*ermB2*), streptomycin (*str*), and colistin (*mcr1-9*). The primers for each targeted virulence and antimicrobial resistance gene are listed in Tables S2 and S3.

PCR reactions were performed in a total volume of 25 µL, containing 12.5 µL of 2X GoTaq Green Master Mix (Promega, Madison, WI, USA), 0.5 µL of each forward and reverse primer (stock solution 10 µm), 2 µL of DNA template, and nuclease-free water adjusted to the final volume of 25 µL. The thermal cycling conditions were optimized for each primer set, typically involving an initial denaturation step at 94 °C for 5 min, followed by 30–35 cycles of denaturation at 94 °C for 30 s, annealing at the appropriate temperature for 30 s, and extension at 72 °C for 1 min, with a final extension step at 72 °C for 7 min. The PCR amplicons were separated using agarose gel electrophoresis with 1.5% agarose gels stained with ethidium bromide, and then visualized using a gel documentation system (Bio-Rad, Hercules, CA, USA).

2.6. Statistical Analysis

A two-way Analysis of Variance (ANOVA) was employed to evaluate the effects of two independent variables on the categorical outcomes of the biofilm development and motility experiments. The relationship between the sample type and *Salmonella* prevalence was evaluated using Fisher's Exact Test, and descriptive statistics were used to calculate the 95% confidence interval (CI). This method is preferred over the normal approximation, especially for small sample sizes or proportions near 0 or 1. These analyses were conducted using the R Studio software (2025.05.0+496) [61]. The hierarchical clustering heatmaps were generated using the Complex Heatmap package in RStudio [62]. The Pearson's correlation coefficient was calculated to assess the relationship between the ARGs and virulence genes. The correlation matrix was illustrated using the R Studio Corr package, providing a clear graphical representation of the relationships among the variables [63]. The statistical significance was determined at a p-value of less than 0.05 for all analyses.

3. Results

3.1. Occurrence and Serotype Distribution of Salmonella

Out of 1008 necropsied cattle, 27 (2.7%) were positive for *Salmonella* isolates, with a prevalence of 59.3% (n = 16) in females, compared to 40.7% (n = 11) in males. Among the positive cases, the highest occurrence was observed in adult cattle (≥1 year), accounting for 37.0% (10/27; 95% CI: 18.8–55.3), followed by juveniles (3–12 months), with a prevalence of 33.3% (9/27; 95% CI: 15.6–51.1). Calves (29–90 days) represented 18.5% (5/27; 95% CI: 3.9–33.2), while neonates (2–28 days) had the lowest occurrence at 11.1% (3/27; 95% CI: 0.0–23.0). Most of the isolates, 77.8% (21/27), were recovered from the intestine (95% CI: 62.1–93.5). Conversely, the lowest occurrence (3.7%) was from the lungs and kidneys (1/27; 95% CI: 0.0–10.8). Notably, *S.* Dublin was the most prevalent serotype within the isolates (29.6%; 8/27; 95% CI: 12.4–46.9). This was followed by *S.* Typhimurium, *S.* Muenster (11.1% each), *S.* Thompson, *S.* Worthington, *S.* Montevideo, and *S.* Anatum (7.4%). Additionally, *S.* Cerro, *S.* Hartford, *S.* Meleagridis, *S.* Newport, and *S.* III 38:(k): z35 were detected in 3.7% of the isolates (Table 1 and Table S4).

Table 1. Occurrence of *Salmonella* isolates categorized by gender, age group, organ of isolation, and serotypes.

Group	Variable	Count (n)	Percentage (%)	95% CI (Lower–Upper)	p-Value
Gender	Female	16	59.3	40.7–77.8	0.4421
	Male	11	40.7	22.2–59.3	0.4421
Age Group	Neonatal	3	11.1	0.0–23.0	<0.05
	Calf	5	18.5	3.9–33.2	<0.05
	Juvenile	9	33.3	15.6–51.1	<0.05
	Adult	10	37	18.8–55.3	<0.05
Organ	Intestine	21	77.8	62.1–93.5	<0.0001
	Liver	2	7.4	0.0–17.3	<0.05
	Colon	2	7.4	0.0–17.3	<0.05
	Lung	1	3.7	0.0–10.8	<0.05
	Kidney	1	3.7	0.0–10.8	<0.05
Serotypes	Dublin	8	29.6	12.4–46.9	<0.05
	Typhimurium	3	11.1	0.0–23.0	<0.05
	Muenster	3	11.1	0.0–23.0	<0.05
	Montevideo	2	7.4	0.0–17.3	<0.05
	Thompson	2	7.4	0.0–17.3	<0.05
	Worthington	2	7.4	0.0–17.3	<0.05
	Anatum	2	7.4	0.0–17.3	<0.05
	Cerro	1	3.7	0.0–10.8	<0.05
	Hartford	1	3.7	0.0–10.8	<0.05
	Meleagridis	1	3.7	0.0–10.8	<0.05
	Newport	1	3.7	0.0–10.8	<0.05
	III 38:(k): z35	1	3.7	0.0–10.8	<0.05

Percentages are presented with 95% confidence intervals, calculated using the Clopper–Pearson method, based on a total of 27 *Salmonella* isolates. *Chi*-square or Fisher's exact test was used for group comparisons, depending on the expected cell counts.

3.2. Biofilm Formation Within the Salmonella Isolates

The biofilm production ability of the *Salmonella* isolates was evaluated, and varying levels of biofilm formation were observed among the tested isolates. The results showed that 48.2% (13/27) of the isolates were SBPs (95% CI: 29.30–66.99%), while 44.4% (12/27) were MBPs (95% CI: 25.70–63.19%), and 7.4% (2/27) were WBPs (95% CI: 0.00–17.29%) (Figure 1A). The mean optical density (OD_{550}) values indicated biofilm biomass ranges between 0.1 and 0.3 (Table S5). Isolates C14, C19, and C25 showed the highest OD_{550}, while C3 and C24 possessed the lowest OD_{550} (Figure 1B).

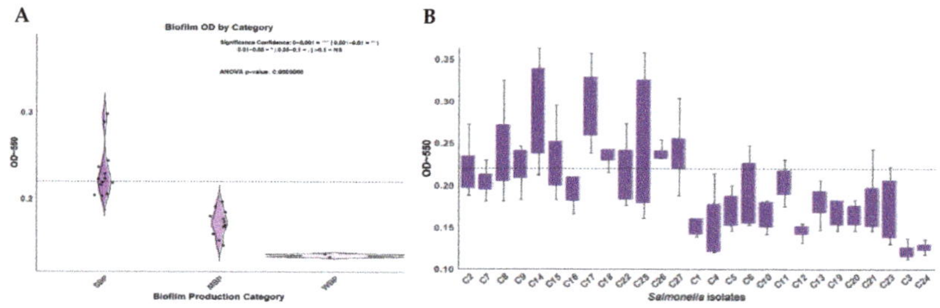

Figure 1. The quantification and categorization of biofilm formation among the bacterial isolates. (**A**) A box plot illustrating the mean biofilm biomass of the isolated bacteria using the crystal violet assay. The y-axis represents optical density (OD_{550}), and the x-axis lists the corresponding bacterial isolates. The error bars represent the standard deviation across the replicates. (**B**) A violin plot illustrating biofilm formation within the *Salmonella* isolates (OD_{550}). The samples were categorized as strong biofilm producers (SBPs), moderate biofilm producers (MBPs), and weak biofilm producers (WBPs). A horizontal dashed line represents a cutoff value.

3.3. *Swimming and Swarming Motility*

The motility of the *Salmonella* strains was assessed for surface swarming (growth on media with 0.5% agar) and swimming (growth on media with 0.25% agar). Out of the 27 isolates, 96.3% (26 out of 27) exhibited swimming, displaying either a featureless or bull's-eye morphology, with an average of 33.8 ± 0.2 mm. Additionally, 96.3% of strains demonstrated the ability to swarm along the agar surface, forming colonies that were featureless, smooth, and flat. The featureless colonies resulted from the cells spreading evenly and continuously outward from the inoculation point as a monolayer, with an average of 29.3 ± 0.2 mm (Figure S1). The highest swarming motility was observed in C2 (*S.* III 38:(k): z35) and isolates C15 and C19, which are *S.* Worthington (Figure 2). A chi-square test of independence was conducted. The results indicated no statistically significant association between the serotype and swarming motility (χ^2 = X, p = 0.134), nor between the serotype and swimming motility (χ^2 = Y, p = 0.349).

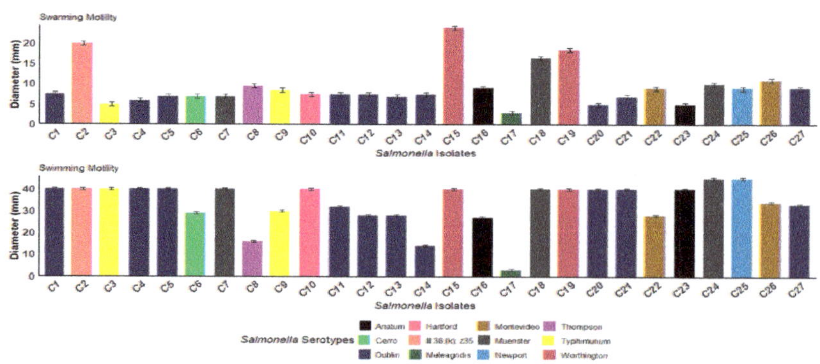

Figure 2. The swarming and swimming motility of the *Salmonella* isolates. The bar graphs represent the average motility diameters (mm) for each isolate. Each bar corresponds to a distinct *Salmonella* isolate, and the color coding indicates the different serotypes. The swarming and swimming motility assays were conducted on semi-solid agar, and the measurements were recorded after incubation. The error bars represent the standard deviations.

3.4. Distribution of Virulence Genes Among Salmonella Isolates

Molecular screening for virulence genes revealed that all the isolates (100%, $n = 27$) harbored *invA*, *hilA*, *fimA*, *csgA*, *csgB*, and *flhD*, indicating a consistent trait of invasion, motility, and biofilm formation. Additionally, 96.3% (26/27) of the isolates carried both *siiA* and *siiC* genes, which are components of TISS, while 92.6% (25/27) of the isolates harbored the *spiA* gene, a key effector within SPI-2. Moreover, genes involved in host immune modulation, such as *sopB* and *spvC*, were detected in 92.6% (25/27) and 74.1% (20/27) of the isolates, respectively. Interestingly, all isolates harbored at least one of the motility-associated genes *fljB*, *fliC*, or *flhD*, confirming their genetic capacity for flagellar-driven motility. The gene-specific prevalence is illustrated in Figure 3.

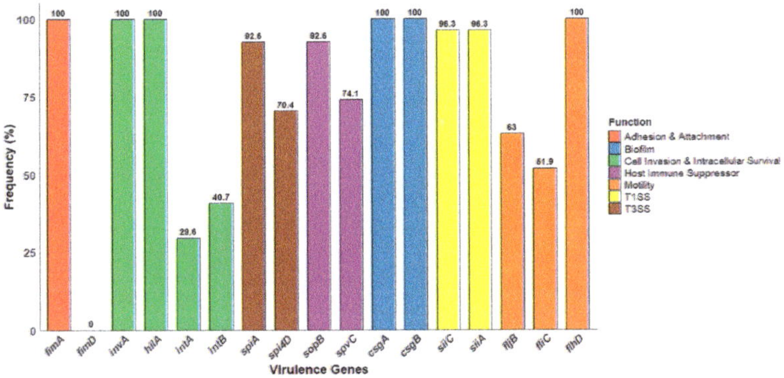

Figure 3. The frequency distribution of the virulence genes among the *Salmonella* isolates. The bar chart illustrates the percentage of isolates harboring each gene. The data reflect the prevalence of specific genes associated with pathogenicity, including those involved in adhesion and attachment, biofilm formation, host invasion, motility, a type 1 secretion system (T1SS), and a type 3 secretion system (T3SS).

3.5. Antibiotic Susceptibility Profile

The antibiotic susceptibility testing for the *Salmonella* isolates was performed using broth microdilution against 19 antibiotics that cover nine different antimicrobial classes. Our results showed that 92.6% (25 out of 27) of the isolates were resistant to carbapenem (imipenem), and 85.2% (23 out of 27) were resistant to chloramphenicol. Similarly, 51.9% of the isolates (14 out of 27) were resistant to tetracycline, and 18.5% (5 out of 27) to doxycycline. In contrast, meropenem demonstrated the highest susceptibility, with activity observed against most isolates, 96.3% (26 out of 27), followed by trimethoprim–sulfamethoxazole (88.9%; 24 out of 27) (Figure 4 and Table S6). A total of 18 distinct phenotypic antimicrobial resistance profiles were identified among the 27 *Salmonella* isolates. The most prevalent profile, observed in five isolates across various serotypes, including *S.* Dublin, *S.* Muenster, *S.* Anatum, *S.* Thompson, and *S.* Typhimurium, included resistance to NEO, GEN, IPM, CHL, AZM, and CLI. Notably, *S.* Dublin exhibited the highest diversity of resistance profiles, being associated with eight distinct profiles. Multi-drug resistance (MDR) patterns frequently involved aminoglycosides (NEO and GEN), β-lactams (AMP, CEF, and CAZ), tetracyclines (TET and DOX), phenicols (CHL), fluoroquinolones (LVX, CIP, and ENR), macrolides (AZM), and lincosamides (CLI). One isolate of *S.* Montevideo and one of *S.* Muenster shared an identical profile (NEO, GEN, AMK, IPM, TET, DOX, AZM, and CLI). In contrast, some profiles were serotype-specific, such as the extended resistance profile found only in *S.* Dublin (NEO, GEN, AMK, AMP, CEF, MEM, IPM, TET, LVX, CIP, ENR, CHL,

AZM, and CLI) (Figure 5). All the isolates were classified as MDR, as each demonstrated resistance to at least one antibiotic in three or more antimicrobial classes. Interestingly, 100% (n = 27) of the isolates were MDR to macrolides (azithromycin and clindamycin). Likely, 100% (n = 27) of the isolates were MDR to aminoglycoside (neomycin), and 29.6% (n = 8) to aminoglycoside (neomycin, gentamicin, and amikacin). The MAR index was calculated for each isolate, with values ranging from 0.32 to 0.74, indicating MDR across all the isolates. Notably, strain C14 exhibited the highest MAR index at 0.74, followed by C11 at equal to 0.68, and C5 and C12, with a MAR index equal to 0.63. These isolates are considered XDR as they were susceptible to only one or two classes of antibiotics (Table S7).

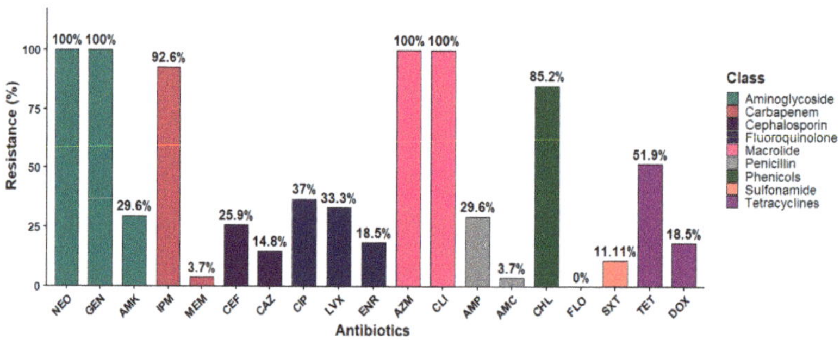

Figure 4. The antibiotic resistance frequency within the collected *Salmonella* isolates. The bar chart illustrates the percentage of isolates resistant to antibiotics within each antimicrobial class. All the isolates possessed 100% resistance to NEO, GEN, CLI, and AZM. Neomycin (NEO), gentamicin (GEN), amikacin (AMK), amoxicillin and clavulanic acid (AMC), ampicillin (AMP), meropenem (MEM), imipenem (IPM), ceftazidime (CAZ), ceftiofur (CEF), tetracycline (TET), doxycycline (DOX), trimethoprim and sulfamethoxazole (SXT), levofloxacin (LVX), ciprofloxacin (CIP), enrofloxacin (ENR), florfenicol (FLO), chloramphenicol (CHL), azithromycin (AZM), and clindamycin (CLI).

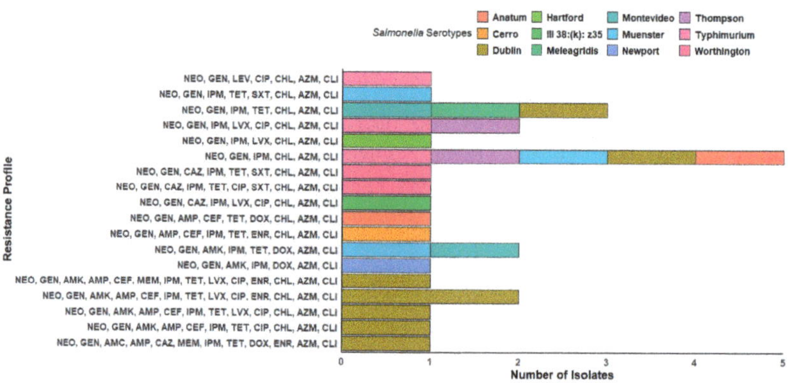

Figure 5. The distribution of antimicrobial resistance profiles among the *Salmonella* isolates, categorized by serotype. Each row represents the number of isolates sharing the same resistance profile, with the colors indicating the different *Salmonella* serotypes.

3.6. Distribution of ARGs Among Salmonella Isolates

Our *Salmonella* isolates harbored a variety of antimicrobial resistance genes (ARGs), including the aminoglycoside resistance gene *aac(6′)-Ib*, which was detected in all the isolates (100%; 27/27). The prevalence of other ARGs varied across the isolates; notably, the streptomycin resistance gene (*strA*) was detected in 74.1% (20/27) of the isolates, while macrolides (*ermB2*) were detected in 70.1% (19/27). Similarly, phenicol resistance genes (*floR* and *catB*) were identified in (59.3%, $n = 16$) and (51.9%, $n = 14$), respectively. Additionally, 59.3% (16/27) of the isolates harbor the sulfonamide resistance (*sul2*) gene. Interestingly, the β-lactamase resistant gene ($bla_{OXA\text{-}9}$) was not detected in any of the isolates. Among the screened *mcr*-1 to *mcr*-9 genes, only *mcr*-2 was detected in the *Salmonella* isolates, with a prevalence of 3.7% (1 out of 27). The frequency of each gene is illustrated in Figure 6.

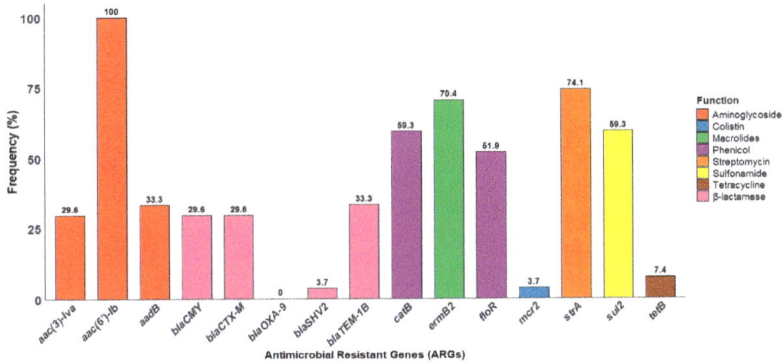

Figure 6. The frequency distribution of antimicrobial-resistant genes among the *Salmonella* isolates. The bar chart illustrates the percentage of isolates harboring each gene. All the isolates harbor aminoglycoside resistance genes *(aac(6′)-Ib)*.

3.7. Phenotypic and Genotyping Correlation

For the phenotypic data, hierarchical clustering of the *Salmonella* isolates was performed based on their phenotypic antibiotic resistance profiles. The resulting dendrogram, derived from the phenotypic resistance matrix, was used to align all virulence genes, ARGs, and phenotypic profiles. Importantly, isolates C4, C5, C11, C12, and C14 were clustered into distinct clades and classified as XDR, demonstrating resistance to multiple antibiotic classes. These isolates were all recovered from calf and juvenile animals and belonged exclusively to the *S.* Dublin serotype. Interestingly, most of the *S.* Dublin isolates demonstrated exceptionally high MAR index values (>0.5). Furthermore, all the isolates in this sub-clade carried the macrolide resistance gene (*ermB2*), sulfonamide resistance gene (*sul2*), and the streptomycin resistance gene (*strA*). Conversely, isolates C1, C2, C3, C10, and C20 were also clustered together and demonstrated lower resistance levels. Notably, all the isolates in this case were encoded by the phenicol-resistant gene (*catB*). These isolates were recovered from juvenile and calf cattle, specifically from intestinal sites, and demonstrated high swimming motility. Interestingly, despite their low resistance profiles, these isolates harbored a broad array of virulence genes. In parallel, an analysis of virulence gene distribution revealed that isolates *S.* Thompson (C8) and *S.* Muenster (C18) harbored the most virulence-associated genes (94.1%, $n = 16$) and clustered within a distinct subclade. Both were recovered from the intestine and exhibited high swarming motility and strong biofilm-forming ability. In contrast, isolates C21 (*S.* Dublin) and C23 (*S.* Anatum) possessed fewer virulence genes and clustered separately. Notably, both isolates were recovered from the colon and exhibited moderate biofilm-forming ability Figure 7.

Pearson's correlation coefficients were calculated to assess the relationships between phenotypic and genotypic profiles, with significance levels evaluated to confirm the strength of these associations. Notably, the *intA* gene (*Salmonella* integron A) exhibited a significant positive correlation with doxycycline resistance ($r = 0.73$; $p < 0.001$) and with the *tetB* gene (tetracycline resistance) ($r = 0.44$; $p < 0.05$). Additionally, *intA* was positively correlated with the *aadB* gene

($r = 0.40$; $p < 0.05$). Similarly, *intB* (*Salmonella* integron B) demonstrated a positive correlation with both *aadB* (aminoglycoside resistance gene) and *catB* (phenicol resistance gene) ($r = 0.53$; $p < 0.01$), as well as with $bla_{CTX\text{-}M}$ (β-lactam resistance gene) ($r = 0.44$; $p < 0.05$). Furthermore, significant correlations were also observed among the virulence genes. Positive correlations were observed between genes encoding components of the T3SS and host immune modulators. Specifically, *spi4D* correlated positively with both *sopB* and *spiA* ($r = 0.44$; $p < 0.05$), and *spiA* showed a significant positive correlation with *spvC* ($r = 0.48$; $p < 0.05$). Importantly, *siiA* and *siiC*, components of the T1SS, exhibited a perfect correlation ($r = 1.00$; $p < 0.001$). Additionally, correlation analysis revealed that biofilm production was positively associated with swarming motility ($r = 0.68$; $p < 0.001$), whereas a non-significant correlation was observed with swimming motility ($r = -0.30$). Additionally, a significant positive correlation was observed between the bla_{TEM} (β-lactam resistance gene) and resistance to both meropenem ($r = 0.48$; $p < 0.05$) and ampicillin ($r = 0.25$). Similarly, the bla_{CMY} (β-lactam resistance gene) showed significant positive correlations with resistance to amoxicillin–clavulanic acid ($r = 0.62$; $p < 0.001$), ampicillin ($r = 0.64$; $p < 0.001$), and ceftiofur ($r = 0.54$; $p < 0.001$). The correlation analysis of antibiotic resistance profiles revealed several significant associations, with strong positive correlations observed particularly among resistance to related antibiotic classes, such as β-lactams. For instance, resistance to ampicillin was positively correlated with resistance to amoxicillin–clavulanic acid ($r = 0.58$; $p < 0.001$) and ceftiofur ($r = 0.59$; $p < 0.001$). Moderate to strong correlations were also observed between resistance to meropenem and other β-lactam antibiotics like amoxicillin–clavulanic acid ($r = 0.62$; $p < 0.001$) and ampicillin ($r = 0.52$; $p < 0.001$), indicating possible cross-resistance or overlapping resistance pathways (Figure 8).

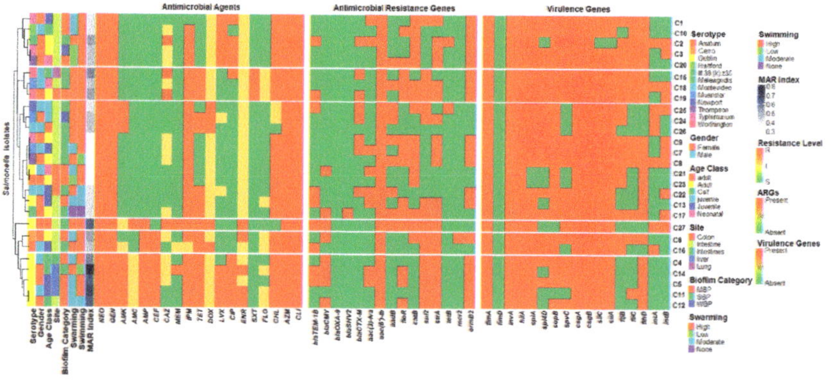

Figure 7. Hierarchical clustering was performed based on the phenotypic resistance profiles of the *Salmonella* isolates, resulting in six distinct clades, visualized as row splits with gaps. The same dendrogram structure was applied across the heatmaps, representing the presence of virulence genes, antimicrobial resistance gene (ARG) profiles, and phenotypic profiles. Phenotypic data include the MAR index, swimming, swarming, and biofilm category, age class, gender, serotype, and site of isolation for each isolate. Neomycin (NEO), gentamicin (GEN), amikacin (AMK), amoxicillin and clavulanic acid (AMC), ampicillin (AMP), meropenem (MEM), imipenem (IPM), ceftazidime (CAZ), ceftiofur (CEF), tetracycline (TET), doxycycline (DOX), trimethoprim and sulfamethoxazole (SXT), levofloxacin (LVX), ciprofloxacin (CIP), enrofloxacin (ENR), florfenicol (FLO), chloramphenicol (CHL), azithromycin (AZM), and clindamycin (CLI).

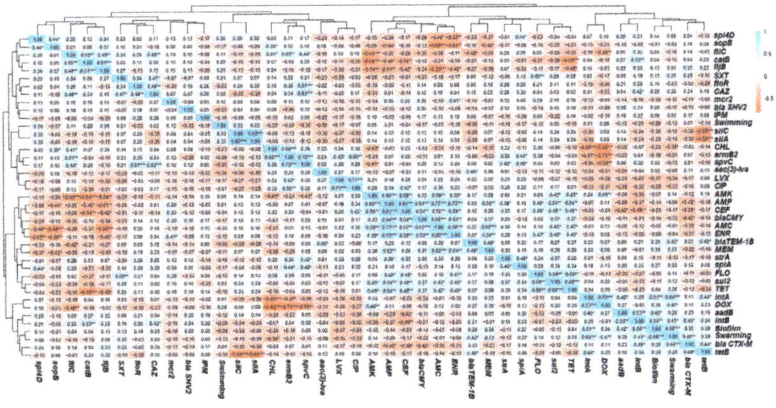

Figure 8. Correlation heatmap with a hierarchical clustering correlation coefficient (*r*) matrix between the phenotypic and genotypic patterns. The asterisks next to the correlation coefficients indicate statistical significance; *** ($p < 0.001$), ** ($p < 0.01$), * ($p < 0.05$), and no asterisk indicates $p \geq 0.05$ (not significant). Neomycin (NEO), gentamicin (GEN), amikacin (AMK), amoxicillin and clavulanic acid (AMC), ampicillin (AMP), meropenem (MEM), imipenem (IPM), ceftazidime (CAZ), ceftiofur (CEF), tetracycline (TET), doxycycline (DOX), trimethoprim and sulfamethoxazole (SXT), levofloxacin (LVX), ciprofloxacin (CIP), enrofloxacin (ENR), florfenicol (FLO), chloramphenicol (CHL), azithromycin (AZM), and clindamycin (CLI).

4. Discussion

Salmonella is recognized as one of the most prominent foodborne pathogens that is responsible for the highest economic losses globally [64]. The prevalence of *Salmonella* in cattle is a major public health concern due to the risk of zoonotic transmission and contamination through the food chain [65]. This study investigates the presence of *Salmonella* in necropsied cattle submitted to a veterinary diagnostic laboratory. It also investigates their antimicrobial resistance profiles and key virulence factors, including biofilm formation, motility, and the presence of ARGs and virulence-associated genes.

In our study, *Salmonella* was detected in 2.7% of 1008 necropsied cattle collected from January 2022 to December 2023. These findings align with earlier surveillance conducted between 2008 and 2019, during which *Salmonella* was detected in 2.5% of necropsied cattle from dairy farms across different states [66]. While this prevalence is lower than the pooled global estimate of 9% (95% CI: 7–11%) from 71 publications and 75 datasets, it falls within the range of global prevalence variations observed across continents. For instance, the pooled analysis shows that *Salmonella* prevalence in healthy cattle ranged from as low as 2% in Europe to as high as 16% in North America [67]. A Similar prevalence appears consistent across multiple animal populations, with previous reports of 1.2% in necropsied horses [45] and 1.4% among wildlife species (birds, mammals, and reptiles) [68]. In companion animals, a U.S. survey of 2965 veterinary clinical samples (2012–2014) found *Salmonella* prevalence rates of <1% in cats and 2.5% in dogs, with raw food representing a significant risk factor [69]. Recent surveillance of 2710 dairy cattle samples in Italy reported a 4.43% *Salmonella* prevalence, which is slightly higher than rates reported in other animal populations [70]. The variation in prevalence may be attributed to differences in isolation sites, as environmental and regional factors can significantly influence the presence of *Salmonella* [67]. Higher prevalence rates have often been associated with specific sampling sites, such as fecal and rectal samples [71]. In the present study, 21 out of 27 isolates (77.8%) were recovered from intestinal tissues, indicating it as the predominant site of isolation (95%

CI: 62.1–93.5, $p < 0.0001$) (Table 1). *Salmonella* infection depends on the pathogen's ability to outcompete the gut microbiota and induce inflammation, creating a niche for colonization [72]. Consistent with previous studies, the intestine was identified as the primary site of *Salmonella* isolation in both humans and animals, including cattle [67,73,74]. Although younger calves are often considered more susceptible to *Salmonella* due to their immature immune systems, limited gut microbiota, and higher susceptibility to co-infections, our study observed higher prevalence in juveniles (33.3%) and adults (37.0%). In juveniles, the dietary transition during weaning from milk to solid feed can disrupt gut microbiota stability, potentially facilitating *Salmonella* colonization [75,76]. The lower prevalence observed in calves (18.5%) and neonates (11.1%) may be partially explained by limited environmental exposure and the protective effect of passive immunity acquired through colostrum intake during the early postnatal period [77]. Additionally, prior antimicrobial treatments commonly administered during early growth stages may alter intestinal microbial balance and select for *Salmonella* colonization, including MDR strains [76].

Serotype-specific variations in the O-antigen structure enhance immune evasion and persistence, favoring survival in the nutrient-rich, permissive intestinal environment [78,79]. The most commonly reported *Salmonella enterica* serovars in cattle are *S.* Dublin, *S.* Typhimurium, *S.* Newport, *S.* Cerro, and *S.* Montevideo, with *S.* Dublin being host-adapted and frequently associated with systemic disease [67]. Among the *Salmonella* isolates recovered in this study, *S.* Dublin was the most abundant serotype, accounting for 29.6% (8/27, 95% CI: 12.4–46.9, $p < 0.05$) (Table 1), which is consistent with previous findings where *S.* Dublin was also the most frequently recovered serotype from bovines [80,81]. Additionally, *S.* Typhimurium was the second most commonly identified serotype in this study, consistent with global reports, underscoring its importance in bovine infections [82]. A recent meta-analysis estimated its prevalence at 10.1% among *Salmonella* isolates from cattle, establishing it as a key serotype in bovine salmonellosis [67]. In the National Animal Health Monitoring System (NAHMS), *S.* Typhimurium was also identified among *Salmonella* serotypes in U.S. beef cow–calf operations, though at a relatively low prevalence of 7.7% [83]. While *S.* Cerro, *S.* Newport, and *S.* Muenster are commonly reported in cattle at lower prevalence rates, this finding aligns with the distribution observed in this study [13,67]. Although earlier reports primarily highlighted *S.* Typhimurium and *S.* Dublin as the predominant serotypes isolated from cattle and beef products, more recent surveillance, including the present study, reveals a broader diversity of *Salmonella* serotypes circulating in cattle [13,16,84,85].

Salmonella spp. can produce biofilm, enhancing their survival and pathogenicity. Bacteria within biofilms exhibit increased resistance to antimicrobials, physical stresses, and host immune responses [38]. All *Salmonella* isolates exhibited biofilm production at varying levels, with 48.2% classified as strong biofilm producers. The p-value of 0.000264 is very small, indicating that the difference between at least two of the biofilm categories —SBPs, MBPs, and WBPs is statistically significant ($p < 0.01$). Notably, all the isolates harbor the curli amyloid fiber genes *csgA* and *csgB* (Figure 2), aligning with their observed biofilm-forming phenotypes. These findings are consistent with previous research demonstrating that the deletion of *csgA* impairs biofilm formation in *Salmonella* spp. [86–88]. The presence of these genes across all isolates suggests a conserved mechanism among *Salmonella* strains for establishing biofilms, which may contribute to their resilience in diverse environments and hosts [89]. Beyond biofilm formation, motility is a critical factor in *Salmonella* pathogenesis, facilitating both colonization and host invasion [90]. The widespread ability of bacteria to swarm on such media suggests their importance in surface colonization in natural environments. It is plausible that the gelatinous surfaces of animal tissues can mimic agar-like properties [91]. In our study, all isolates exhibited motility, demonstrating both swimming and swarming behavior on agar media of different concentrations, except for the non-motile *S.* Meleagridis (C17) (Figure S1). Despite harboring the flagellar genes *fliC*, *fljB*, and *flhD*, the isolate exhibited a non-motile phenotype. This discrepancy between genotype and phenotype may result from multiple factors, including gene regulation, point mutations, and epigenetic silencing [92,93]. Specifically, the presence of prophage elements that interfere with flagellar expression or assembly, as has been reported in *E. coli K-12*. The prophage regulator AppY enhances stress tolerance and biofilm formation while suppressing motility by downregulating *FlhDC*. A similar mechanism may explain the non-motile phenotype in *S.* Meleagridis (C17), despite intact flagellar genes, suggesting that prophage elements may modulate motility-related gene regulation [94–96].

Among the isolates, *S.* Worthington (C15 and C19) demonstrated the highest swarming and swimming motility. However, recent findings suggest that host factors can modulate bacterial motility. Beyond this, certain strains in our study exhibited high swarming motility but low swimming motility; this observation is in line with previous findings suggesting that bacterial strains can display diverse motility patterns influenced by environmental conditions and genetic regulation [91]. Such variability in motility may contribute to differences in biofilm architecture, consistent with the model proposed by Shrout et al., where early swarming behavior impacts the final structure of the biofilm through coordinated regulation of motility and matrix production [97,98]. Of note, the negative correlation between swimming and swarming motilities ($r = -0.18$) was not statistically significant, indicating that these behaviors are largely independent. In contrast, swarming motility, which involves coordinated surface movement, showed a significant positive correlation with biofilm formation ($r = 0.68$, $p < 0.001$) (Figure 8). In contrast, swimming motility, representing individual movement in liquid environments, appeared to play a lesser role in biofilm development. These findings support that surface-associated motility modes, rather than planktonic motility, are more closely linked to biofilm architecture and persistence [51,91,99].

Alongside biofilm formation and motility, the pathogenic success of *Salmonella* is fundamentally dependent on its ability to invade host cells, evade immune responses, and establish colonization. Cellular invasion is primarily mediated by the T3SS encoded within *Salmonella* SPI-1, which injects effector proteins into host epithelial cells [100]. All the *Salmonella* isolates in our study harbor adhesion and invasion genes (*fimA*, *hilA*, and *invA*), reflecting their conservation in association with adhesion and invasion mechanisms [101]. Additionally, they encode a broad array of effector proteins and host immune suppressors, reflecting their potential to modulate host cell signaling, evade immune detection, and enhance intracellular survival [102]. *Salmonella* is known as the *Salmonella*-containing vacuole (SCV), where it utilizes a second T3SS encoded by SPI-2 to manipulate host trafficking pathways, avoid lysosomal degradation, and promote intracellular replication [103,104]. Through the activation of SPI-1, *hilA* induces expression of effector proteins, which manipulate host cytoskeletal dynamics, promoting bacterial uptake via membrane ruffling [105]. Furthermore, 74.1% of our isolates are encoded by *spvC*, which is essential for systemic infection and functions as a phosphor–threonine lyase that disrupts host MAPK signaling, aiding in immune evasion [106]. The widespread presence of virulence factors highlights the isolates' strong potential for invasion and immune evasion.

Due to growing concerns about antibiotic resistance, we analyzed antimicrobial resistance patterns and MAR indices to assess the potential for MDR and its implications for public health. All the isolates have a MAR index exceeding 0.2, a threshold that is widely recognized as indicative of high-risk sources where antibiotics are frequently used [55]. Moreover, all the isolates were classified as multidrug-resistant (MDR), as they were non-susceptible to antibiotics from more than three different classes [58]. Among our isolates, we observed the highest susceptibility to meropenem (96.3%) and trimethoprim–sulfamethoxazole (88.9%). These drugs are *strictly restricted* in veterinary practice. Conversely, we found higher resistance to aminoglycosides (neomycin and gentamycin) and macrolides (azithromycin and clindamycin) (100%), followed by carbapenem (imipenem) (92.6%) and chloramphenicol (85.2%) (Figure 4). The higher resistance pattern to chloramphenicol aligns with surveillance data from a study assessing antibiotic profiles in bovine *Salmonella* isolates across the U.S. from 2013 to 2022 [107]. Our results support that intensive antibiotic use is a critical driver of high MAR indices. Previous studies revealed that regions with less regulated antibiotic practices frequently report similar resistance patterns, underscoring the direct impact of antimicrobial usage on resistance development [2]. In response to these findings, the FDA's new rule (GFI 263) will require that farmers and livestock owners obtain a prescription from a licensed veterinarian [108]. This observation may reflect the therapeutic use of these antibiotics for the treatment of *Salmonella* infections among adult cattle. Notably, *S.* Dublin strains in this study exhibited higher resistance profiles, consistent with previous reports. In the U.S., *S.* Dublin has emerged as one of the most MDR serotypes, often associated with resistance to critically important antimicrobial agents [109,110].

To validate the phenotypic resistance patterns, the presence of corresponding ARGs was assessed through targeted PCR analysis. We tested for the presence of key ARGs associated with resistance to β-lactams, aminoglycosides, macrolides, tetracyclines, sulfonamides, and phenicols, providing insights into the genetic basis of the observed resistance patterns. We identified resistance genes for tetracycline (*tetB*) and sulfonamides (sul2), indicating a broad resistance profile among the isolates [85]. Moreover, our isolates harbored several ARGs, including genes resistant to beta-lactamases, including $bla_{TEM\text{-}1B}$, bla_{CMY}, bla_{CTXM}, and bla_{SHV}. The $bla_{TEM\text{-}1B}$ gene typically confers resistance to penicillin, such as ampicillin. It may contribute to reduced susceptibility to carbapenems, including meropenem, particularly when additional resistance mechanisms, such as porin loss, are present [111]. Although some ARGs exhibited weak or non-significant correlations with phenotypic resistance, the bla_{CMY} gene showed strong associations with β-lactam resistance, including significant correlations with amoxicillin–clavulanic acid ($r = 0.62$, $p < 0.001$), ampicillin ($r = 0.64$, $p < 0.001$), and ceftiofur ($r = 0.54$, $p < 0.001$). Similarly, the $bla_{TEM\text{-}1B}$ gene showed significant positive correlations with resistance to meropenem ($r = 0.37$; $p < 0.05$), amoxicillin–clavulanic acid ($r = 0.62$; $p < 0.001$), ampicillin ($r = 0.64$; $p < 0.001$), and ceftiofur ($r = 0.54$; $p < 0.001$) (Figure 8), highlighting its role in mediating resistance across both penicillin and cephalosporin classes. Importantly, all isolates carried the (*aac(6′)-Ib*) gene, which encodes an aminoglycoside-modifying enzyme that confers resistance to a broad spectrum of aminoglycoside antibiotics [112]. Consistent with the genotypic profile, all isolates demonstrated phenotypic resistance to both neomycin and gentamicin [113]. Similarly, 70.1% of the isolates (19 out of 27) harbored the *ermB2* gene, which encodes an rRNA methyltransferase associated with macrolide resistance [114]. Interestingly, all the isolates (100%, $n = 27$) demonstrated MDR to macrolides, specifically azithromycin and clindamycin. Although phenotypic and genotypic correlations provide valuable insights, they may not capture all underlying resistance mechanisms. Whole-genome sequencing (WGS) is necessary to identify undetected β-lactamases, efflux systems, regulatory mutations, and mobile genetic elements that contribute to resistance. Integrating WGS with phenotypic data would provide a more comprehensive understanding of resistance pathways and virulence gene dissemination in *Salmonella* isolates [115]. While this study observed statistical correlations between phenotypic resistance patterns and the presence of specific resistance genes, it is important to recognize that correlation does not imply causation. Some associations may be spurious or non-causal, potentially resulting from genetic linkage, co-selection, or the presence of shared mobile genetic elements rather than direct gene-function relationships [116,117]. This study highlights the critical need for sustained surveillance and genetic characterization of *Salmonella* in cattle. Despite the relatively low prevalence, the detection of a broad array of virulence factors and antimicrobial resistance genes, particularly those associated with biofilm formation, motility, and immune evasion, raises concerns about persistence within herds and the potential for both intraspecies and interspecies transmission. These factors may also serve as potential targets for the development of novel therapeutics and control strategies [118,119].

5. Conclusions

Our study highlights the significant concern of AMR in *Salmonella* isolates from cattle, along with their notable virulence potential and biofilm formation capabilities. *Salmonella* Dublin was the most prevalent serotype, along with eleven other serotypes. Swarming and swimming motility were observed in most of the isolates, in addition to biofilm production. Furthermore, our isolates harbored different virulence genes and ARGs. The aminoglycoside resistance gene *aac(6′)-Ib* was detected in all the isolates (100% prevalence), and all the isolates also harbored multiple virulence genes, including *invA*, *hilA*, *fimA*, *csgA*, *csgB*, and *flhD*. Strikingly, all the isolates exhibited an MDR phenotype, characterized by high MAR index values, underscoring their potential threat to both animal and public health. The emergence of MDR *Salmonella* in cattle poses a critical threat to the livestock industry and public health. Thorough investigations using whole-genome sequencing are needed to fully understand the AMR and virulence gene profiles of pathogens. Simultaneously, studying transmission dynamics is crucial for tracing the sources of *Salmonella* infections in cattle and understanding how these infections can spread to the environment. Additionally, elucidating the distribution of ARGs and virulence factors offers valuable insights for identifying potential drug targets and guiding the development of more effective antibiotic alternatives. These efforts will assist veterinary services and regulatory authorities in implementing effective biosecurity measures to prevent the further dissemination of these pathogens. These findings underscore the importance of implementing integrated One Health–based monitoring strategies to mitigate environmental dissemination and protect public and animal health.

Supplementary Materials: The following supporting information can be downloaded at: https://www.mdpi.com/article/10.3390/antibiotics14070689/s1, Table S1. Breakpoints and antibiotic concentrations for *Salmonella* spp. according to CLSI M100 and CLSI VET01S guidelines. Table S2. Oligonucleotide primer sequences and their corresponding virulence genes. Table S3. Oligonucleotide primer sequences and their corresponding Antimicrobial Resistant Genes (ARGs). Table S4. Distribution of isolates by response animal species, gender, variable age, site of isolation, and *Salmonella* serotype. Table S5. Biofilm formation of *Salmonella* isolates. Table S6. Resistance frequencies for each class. Table S7. Multidrug Resistance Profile of *Salmonella* Isolates. Figure S1 (A) swimming motility and (B) Swarming motility for C26, C19, C17, and C16 on semi-solid agar plates after 12 h. of incubation at 37 °C, in comparison to *S.* Typhimurium ATCC 14028, *P. mirabilis* ATCC 35659, and *Rhodococcus equi*. References [120–140] are cited in the supplementary materials.

Author Contributions: Conceptualization, Y.A.H.; methodology, N.A.F., S.K., A.B., A.K., E.E. and Y.A.H.; software, N.A.F. and Y.A.H.; validation, E.E. and Y.A.H.; formal analysis, N.A.F. and Y.A.H.; investigation, N.A.F., E.E. and Y.A.H.; resources, E.E. and Y.A.H.; data curation, E.E. and Y.A.H.; writing—original draft preparation, N.A.F. and Y.A.H.; writing—review and editing, N.A.F., S.K., A.B., A.K., E.E. and Y.A.H.; visualization, N.A.F. and Y.A.H.; supervision, Y.A.H.; project administration, Y.A.H.; funding acquisition, Y.A.H. All authors have read and agreed to the published version of the manuscript.

Funding: This research was supported by the Center of Biomedical Research Excellence (COBRE) for Translational Chemical Biology (CTCB, NIH P20 GM130456), the National Center for Advancing

Translational Sciences, National Institutes of Health (grant number KL2TR001996), and the University of Kentucky Igniting Research Collaborations program.

Institutional Review Board Statement: Not applicable.

Informed Consent Statement: Not applicable.

Data Availability Statement: The original contributions presented in this study are included in the article/supplementary material. Further inquiries can be directed to the corresponding author(s).

Conflicts of Interest: The authors declare no conflicts of interest.

References

1. Nhung, N.T.; Phu, D.H.; Carrique-Mas, J.J.; Padungtod, P. A review and meta-analysis of non-typhoidal Salmonella in Vietnam: Challenges to the control and antimicrobial resistance traits of a neglected zoonotic pathogen. *One Health* **2024**, *18*, 100698. [CrossRef] [PubMed]
2. Ager, E.O.; Carvalho, T.; Silva, E.M.; Ricke, S.C.; Hite, J.L. Global trends in antimicrobial resistance on organic and conventional farms. *Sci. Rep.* **2023**, *13*, 22608. [CrossRef]
3. Nickodem, C.; Arnold, A.N.; Gehring, K.B.; Gill, J.J.; Richeson, J.T.; Samuelson, K.L.; Scott, H.M.; Smith, J.K.; Taylor, T.M.; Vinasco, J. A longitudinal study on the dynamics of Salmonella enterica prevalence and serovar composition in beef cattle feces and lymph nodes and potential contributing sources from the feedlot environment. *Appl. Environ. Microbiol.* **2023**, *89*, e00033-23. [CrossRef] [PubMed]
4. Hoffmann, S.; Batz, M.B.; Morris, J.G., Jr. Annual cost of illness and quality-adjusted life year losses in the United States due to 14 foodborne pathogens. *J. Food Prot.* **2012**, *75*, 1292–1302. [CrossRef] [PubMed]
5. Huang, H.; Naushad, S. *Salmonella: Perspectives for Low-Cost Prevention, Control and Treatment*; BoD–Books on Demand: Norderstedt, Germany, 2024.
6. Ballash, G.A.; Parker, E.M.; Mollenkopf, D.F.; Wittum, T.E. The One Health dissemination of antimicrobial resistance occurs in both natural and clinical environments. *J. Am. Vet. Med. Assoc.* **2024**, *1*, 451–458. [CrossRef]
7. Collier, S.A.; Deng, L.; Adam, E.A.; Benedict, K.M.; Beshearse, E.M.; Blackstock, A.J.; Bruce, B.B.; Derado, G.; Edens, C.; Fullerton, K.E. Estimate of burden and direct healthcare cost of infectious waterborne disease in the United States. *Emerg. Infect. Dis.* **2021**, *27*, 140. [CrossRef]
8. Billah, M.M.; Rahman, M.S. Salmonella in the environment: A review on ecology, antimicrobial resistance, seafood contaminations, and human health implications. *J. Hazard. Mater. Adv.* **2024**, *13*, 100407. [CrossRef]
9. About Salmonella Infection. Available online: https://www.cdc.gov/salmonella/about/index.html (accessed on 16 June 2025).
10. Chenu, J.; Cox, J.; Pavic, A. Classification of Salmonella enterica serotypes from Australian poultry using repetitive sequence-based PCR. *J. Appl. Microbiol.* **2012**, *112*, 185–196. [CrossRef]
11. Cheng, R.A.; Eade, C.R.; Wiedmann, M. Embracing diversity: Differences in virulence mechanisms, disease severity, and host adaptations contribute to the success of nontyphoidal Salmonella as a foodborne pathogen. *Front. Microbiol.* **2019**, *10*, 1368. [CrossRef]
12. Velasquez, C.; Macklin, K.; Kumar, S.; Bailey, M.; Ebner, P.; Oliver, H.; Martin-Gonzalez, F.; Singh, M. Prevalence and antimicrobial resistance patterns of Salmonella isolated from poultry farms in southeastern United States. *Poult. Sci.* **2018**, *97*, 2144–2152. [CrossRef]
13. Holschbach, C.L.; Peek, S.F. Salmonella in dairy cattle. *Vet. Clin. N. Am. Food Anim. Pract.* **2017**, *34*, 133. [CrossRef]
14. Soyer, Y.; Orsi, R.H.; Rodriguez-Rivera, L.D.; Sun, Q.; Wiedmann, M. Genome wide evolutionary analyses reveal serotype specific patterns of positive selection in selected Salmonella serotypes. *BMC Evol. Biol.* **2009**, *9*, 264. [CrossRef] [PubMed]
15. Rahman, M.M.; Hossain, H.; Chowdhury, M.S.R.; Hossain, M.M.; Saleh, A.; Binsuwaidan, R.; Noreddin, A.; Helmy, Y.A.; El Zowalaty, M.E. Molecular characterization of multidrug-resistant and extended-spectrum β-lactamases-producing Salmonella enterica serovars enteritidis and typhimurium isolated from raw meat in retail markets. *Antibiotics* **2024**, *13*, 586. [CrossRef] [PubMed]
16. Lamichhane, B.; Mawad, A.M.; Saleh, M.; Kelley, W.G.; Harrington, P.J.; Lovestad, C.W.; Amezcua, J.; Sarhan, M.M.; El Zowalaty, M.E.; Ramadan, H. Salmonellosis: An overview of epidemiology, pathogenesis, and innovative approaches to mitigate the antimicrobial resistant infections. *Antibiotics* **2024**, *13*, 76. [CrossRef] [PubMed]
17. Launay, A.; Wu, C.-J.; Dulanto Chiang, A.; Youn, J.-H.; Khil, P.; Dekker, J. In vivo evolution of an emerging zoonotic bacterial pathogen in an immunocompromised human host. *Nat. Commun.* **2021**, *12*, 4495. [CrossRef]
18. Plowright, R.K.; Parrish, C.R.; McCallum, H.; Hudson, P.J.; Ko, A.I.; Graham, A.L.; Lloyd-Smith, J.O. Pathways to zoonotic spillover. *Nat. Rev. Microbiol.* **2017**, *15*, 502–510. [CrossRef]

19. Bentum, K.E.; Kuufire, E.; Nyarku, R.; Osei, V.; Price, S.; Bourassa, D.; Samuel, T.; Jackson, C.R.; Abebe, W. Salmonellosis in Cattle: Sources and Risk of Infection, Control, and Prevention. *Zoonotic Dis.* **2025**, *5*, 4. [CrossRef]
20. Mohler, V.; Heithoff, D.; Mahan, M.; Walker, K.; Hornitzky, M.; Shum, L.; Makin, K.; House, J. Cross-protective immunity conferred by a DNA adenine methylase deficient Salmonella enterica serovar Typhimurium vaccine in calves challenged with Salmonella serovar Newport. *Vaccine* **2008**, *26*, 1751–1758. [CrossRef]
21. Smith, B.P. *Large Animal Internal Medicine-E-Book*; Elsevier Health Sciences: Amsterdam, The Netherlands, 2014.
22. Divers, T.J.; Peek, S.F. *Rebhun's Diseases of Dairy Cattle*; Elsevier Health Sciences: Amsterdam, The Netherlands, 2007.
23. Ventola, C.L. The antibiotic resistance crisis: Part 1: Causes and threats. *Pharm. Ther.* **2015**, *40*, 277.
24. Alenazy, R. Antibiotic resistance in Salmonella: Targeting multidrug resistance by understanding efflux pumps, regulators and the inhibitors. *J. King Saud Univ. Sci.* **2022**, *34*, 102275. [CrossRef]
25. Brown, A.; Grass, J.; Richardson, L.; Nisler, A.; Bicknese, A.; Gould, L. Antimicrobial resistance in Salmonella that caused foodborne disease outbreaks: United States, 2003–2012. *Epidemiol. Infect.* **2017**, *145*, 766–774. [CrossRef] [PubMed]
26. Rehman, T.; Yin, L.; Latif, M.B.; Chen, J.; Wang, K.; Geng, Y.; Huang, X.; Abaidullah, M.; Guo, H.; Ouyang, P. Adhesive mechanism of different Salmonella fimbrial adhesins. *Microb. Pathog.* **2019**, *137*, 103748. [CrossRef] [PubMed]
27. Dong, K.; Zhu, Y.; Deng, Q.; Sun, L.; Yang, S.; Huang, K.; Cao, Y.; Li, Y.; Wu, S.; Huang, R. Salmonella pSLT-encoded effector SpvB promotes RIPK3-dependent necroptosis in intestinal epithelial cells. *Cell Death Discov.* **2022**, *8*, 44. [CrossRef]
28. Sia, C.M.; Pearson, J.S.; Howden, B.P.; Williamson, D.A.; Ingle, D.J. Salmonella pathogenicity islands in the genomic era. *Trends Microbiol.* **2025**, *33*, 752–764. [CrossRef]
29. Gulig, P.A. Virulence plasmids of Salmonella typhimurium and other salmonellae. *Microb. Pathog.* **1990**, *8*, 3–11. [CrossRef]
30. Campioni, F.; Vilela, F.P.; Cao, G.; Kastanis, G.; dos Prazeres Rodrigues, D.; Costa, R.G.; Tiba-Casas, M.R.; Yin, L.; Allard, M.; Falcão, J.P. Whole genome sequencing analyses revealed that Salmonella enterica serovar Dublin strains from Brazil belonged to two predominant clades. *Sci. Rep.* **2022**, *12*, 10555. [CrossRef]
31. Linde, J.; Szabo, I.; Tausch, S.H.; Deneke, C.; Methner, U. Clonal relation between Salmonella enterica subspecies enterica serovar Dublin strains of bovine and food origin in Germany. *Front. Vet. Sci.* **2023**, *10*, 1081611. [CrossRef]
32. Johnson, R.; Mylona, E.; Frankel, G. Typhoidal Salmonella: Distinctive virulence factors and pathogenesis. *Cell. Microbiol.* **2018**, *20*, e12939. [CrossRef]
33. Groisman, E.A.; Ochman, H. How Salmonella became a pathogen. *Trends Microbiol.* **1997**, *5*, 343–349. [CrossRef]
34. Yim, L.; Betancor, L.; Martínez, A.; Bryant, C.; Maskell, D.; Chabalgoity, J.A. Naturally occurring motility-defective mutants of Salmonella enterica serovar Enteritidis isolated preferentially from nonhuman rather than human sources. *Appl. Environ. Microbiol.* **2011**, *77*, 7740–7748. [CrossRef]
35. Marcus, S.L.; Brumell, J.H.; Pfeifer, C.G.; Finlay, B.B. Salmonella pathogenicity islands: Big virulence in small packages. *Microbes Infect.* **2000**, *2*, 145–156. [CrossRef] [PubMed]
36. Tamamura, Y.; Tanaka, K.; Uchida, I. Characterization of pertussis-like toxin from Salmonella spp. that catalyzes ADP-ribosylation of G proteins. *Sci. Rep.* **2017**, *7*, 2653. [CrossRef] [PubMed]
37. Flemming, H.-C.; Wingender, J.; Szewzyk, U.; Steinberg, P.; Rice, S.A.; Kjelleberg, S. Biofilms: An emergent form of bacterial life. *Nat. Rev. Microbiol.* **2016**, *14*, 563–575. [CrossRef]
38. Trampari, E.; Holden, E.; Wickham, G.; Ravi, A.; Martins, L.d.O.; Savva, G.; Webber, M. Exposure of Salmonella biofilms to antibiotic concentrations rapidly selects resistance with collateral tradeoffs. *NPJ Biofilms Microbiomes* **2021**, *7*, 13. [CrossRef]
39. Sauer, K.; Stoodley, P.; Goeres, D.M.; Hall-Stoodley, L.; Burmølle, M.; Stewart, P.S.; Bjarnsholt, T. The biofilm life cycle: Expanding the conceptual model of biofilm formation. *Nat. Rev. Microbiol.* **2022**, *20*, 608–620. [CrossRef]
40. Penesyan, A.; Paulsen, I.T.; Kjelleberg, S.; Gillings, M.R. Three faces of biofilms: A microbial lifestyle, a nascent multicellular organism, and an incubator for diversity. *NPJ Biofilms Microbiomes* **2021**, *7*, 80. [CrossRef]
41. Robi, D.T.; Mossie, T.; Temteme, S. A comprehensive review of the common bacterial infections in dairy calves and advanced strategies for health management. *Vet. Med. Res. Rep.* **2024**, *15*, 1–14. [CrossRef]
42. Helmy, Y.A.; Taha-Abdelaziz, K.; Hawwas, H.A.E.-H.; Ghosh, S.; AlKafaas, S.S.; Moawad, M.M.; Saied, E.M.; Kassem, I.I.; Mawad, A.M. Antimicrobial resistance and recent alternatives to antibiotics for the control of bacterial pathogens with an emphasis on foodborne pathogens. *Antibiotics* **2023**, *12*, 274. [CrossRef]
43. Kentucky Agriculture Expects Highs and Lows in 2025. Available online: https://news.ca.uky.edu/article/kentucky-agriculture-expects-highs-and-lows-2025 (accessed on 16 June 2025).
44. das Mercês Santos, A.F.; Amparo, L.F.V.; Machado, S.C.A.; Dias, T.S.; Berto, L.H.; da Costa Abreu, D.L.; de Aquino, M.H.C.; dos Prazeres Rodrigues, D.; de Almeida Pereira, V.L. Salmonella serovars associated with human salmonellosis in Brazil (2011–2020). *Res. Soc. Dev.* **2022**, *11*, e28011830533. [CrossRef]

45. Kabir, A.; Kelley, W.G.; Glover, C.; Erol, E.; Helmy, Y.A. Phenotypic and genotypic characterization of antimicrobial resistance and virulence profiles of Salmonella enterica serotypes isolated from necropsied horses in Kentucky. *Microbiol. Spectr.* **2025**, *13*, e02501–e02524. [CrossRef]
46. Obe, T.; Nannapaneni, R.; Sharma, C.S.; Kiess, A. Homologous stress adaptation, antibiotic resistance, and biofilm forming ability of Salmonella enterica serovar Heidelberg ATCC8326 on different food-contact surfaces following exposure to sublethal chlorine concentrations. *Poult. Sci.* **2018**, *97*, 951–961. [CrossRef] [PubMed]
47. Xu, Z.; Liang, Y.; Lin, S.; Chen, D.; Li, B.; Li, L.; Deng, Y. Crystal violet and XTT assays on Staphylococcus aureus biofilm quantification. *Curr. Microbiol.* **2016**, *73*, 474–482. [CrossRef]
48. Singh, A.K.; Prakash, P.; Achra, A.; Singh, G.P.; Das, A.; Singh, R.K. Standardization and classification of in vitro biofilm formation by clinical isolates of Staphylococcus aureus. *J. Glob. Infect. Dis.* **2017**, *9*, 93–101.
49. Partridge, J.D.; Harshey, R.M. Swarming motility assays in Salmonella. In *Bacterial and Archaeal Motility*; Springer: Berlin/Heidelberg, Germany, 2023; pp. 147–158.
50. Brunelle, B.W.; Bearson, B.L.; Bearson, S.M.; Casey, T.A. Multidrug-resistant Salmonella enterica serovar Typhimurium isolates are resistant to antibiotics that influence their swimming and swarming motility. *MSphere* **2017**, *2*, e00306–e00317. [CrossRef]
51. Kearns, D.B. A field guide to bacterial swarming motility. *Nat. Rev. Microbiol.* **2010**, *8*, 634–644. [CrossRef] [PubMed]
52. Tittsler, R.P.; Sandholzer, L.A. The use of semi-solid agar for the detection of bacterial motility. *J. Bacteriol.* **1936**, *31*, 575–580. [CrossRef]
53. Shryock, T.R. *Performance Standards for Antimicrobial Disk and Dilution Susceptibility Tests for Bacteria Isolated from Animals: Approved Standard*; Clinical & Laboratory Standards Institute: Berwyn, PA, USA, 2002.
54. The Clinical and Laboratory Standards Institute CLSI. *Performance Standards for Antimicrobial Susceptibility Testing*; CLSI: Wayne, PA, USA, 2023.
55. Khan, J.A.; Irfan, A.; Soni, S.; Maherchandani, S.; Soni, S.; Maherchandani, S. Antibiogram and multiple antibiotic resistance index of Salmonella enterica isolates from poultry. *J. Pure Appl. Microbiol.* **2015**, *9*, 2495–2500.
56. Mthembu, T.P.; Zishiri, O.T.; El Zowalaty, M.E. Molecular detection of multidrug-resistant Salmonella isolated from livestock production systems in South Africa. *Infect. Drug Resist.* **2019**, *12*, 3537–3548. [CrossRef] [PubMed]
57. Davis, R.; Brown, P.D. Multiple antibiotic resistance index, fitness and virulence potential in respiratory Pseudomonas aeruginosa from Jamaica. *J. Med. Microbiol.* **2016**, *65*, 261–271. [CrossRef]
58. Magiorakos, A.-P.; Srinivasan, A.; Carey, R.B.; Carmeli, Y.; Falagas, M.; Giske, C.; Harbarth, S.; Hindler, J.; Kahlmeter, G.; Olsson-Liljequist, B. Multidrug-resistant, extensively drug-resistant and pandrug-resistant bacteria: An international expert proposal for interim standard definitions for acquired resistance. *Clin. Microbiol. Infect.* **2012**, *18*, 268–281. [CrossRef] [PubMed]
59. Kashoma, I.P.; Kassem, I.I.; John, J.; Kessy, B.M.; Gebreyes, W.; Kazwala, R.R.; Rajashekara, G. Prevalence and antimicrobial resistance of Campylobacter isolated from dressed beef carcasses and raw milk in Tanzania. *Microb. Drug Resist.* **2016**, *22*, 40–52. [CrossRef] [PubMed]
60. Hailu, W.; Helmy, Y.A.; Carney-Knisely, G.; Kauffman, M.; Fraga, D.; Rajashekara, G. Prevalence and antimicrobial resistance profiles of foodborne pathogens isolated from dairy cattle and poultry manure amended farms in northeastern Ohio, the United States. *Antibiotics* **2021**, *10*, 1450. [CrossRef] [PubMed]
61. Chakraborty, H.; Hossain, A. R package to estimate intracluster correlation coefficient with confidence interval for binary data. *Comput. Methods Programs Biomed.* **2018**, *155*, 85–92. [CrossRef] [PubMed]
62. Gu, Z.; Hübschmann, D. Make interactive complex heatmaps in R. *Bioinformatics* **2022**, *38*, 1460–1462. [CrossRef]
63. Wei, T.; Simko, V.; Levy, M.; Xie, Y.; Jin, Y.; Zemla, J. Package 'corrplot'. *Statistician* **2017**, *56*, e24.
64. Eng, S.-K.; Pusparajah, P.; Ab Mutalib, N.-S.; Ser, H.-L.; Chan, K.-G.; Lee, L.-H. Salmonella: A review on pathogenesis, epidemiology and antibiotic resistance. *Front. Life Sci.* **2015**, *8*, 284–293. [CrossRef]
65. Hoffmann, S.A.; Maculloch, B.; Batz, M. *Economic Burden of Major Foodborne Illnesses Acquired in the United States*; United States Department of Agriculture: Washington, DC, USA, 2015.
66. Wilson, D.J.; Kelly, E.J.; Gucwa, S. Causes of mortality of dairy cattle diagnosed by complete necropsy. *Animals* **2022**, *12*, 3001. [CrossRef]
67. Gutema, F.D.; Agga, G.E.; Abdi, R.D.; De Zutter, L.; Duchateau, L.; Gabriël, S. Prevalence and serotype diversity of Salmonella in apparently healthy cattle: Systematic review and meta-analysis of published studies, 2000–2017. *Front. Vet. Sci.* **2019**, *6*, 102.
68. Cummings, K.J.; Siler, J.D.; Abou-Madi, N.; Goodman, L.B.; Mitchell, P.K.; Palena, L.; Childs-Sanford, S.E. Salmonella isolated from Central New York wildlife admitted to a veterinary medical teaching hospital. *J. Wildl. Dis.* **2021**, *57*, 743–748. [CrossRef]
69. Reimschuessel, R.; Grabenstein, M.; Guag, J.; Nemser, S.M.; Song, K.; Qiu, J.; Clothier, K.A.; Byrne, B.A.; Marks, S.L.; Cadmus, K. Multilaboratory survey to evaluate Salmonella prevalence in diarrheic and nondiarrheic dogs and cats in the United States between 2012 and 2014. *J. Clin. Microbiol.* **2017**, *55*, 1350–1368. [CrossRef]

70. Parolini, F.; Ventura, G.; Rosignoli, C.; Rota Nodari, S.; D'incau, M.; Marocchi, L.; Santucci, G.; Boldini, M.; Gradassi, M. Detection and phenotypic antimicrobial susceptibility of Salmonella enterica serotypes in dairy cattle farms in the Po valley, Northern Italy. *Animals* **2024**, *14*, 2043. [CrossRef] [PubMed]
71. Hanson, D.; Loneragan, G.; Brown, T.; Nisbet, D.; Hume, M.; Edrington, T. Evidence supporting vertical transmission of Salmonella in dairy cattle. *Epidemiol. Infect.* **2016**, *144*, 962–967. [CrossRef] [PubMed]
72. Gal-Mor, O.; Boyle, E.C.; Grassl, G.A. Same species, different diseases: How and why typhoidal and non-typhoidal Salmonella enterica serovars differ. *Front. Microbiol.* **2014**, *5*, 391. [CrossRef]
73. Tauxe, R.V.; Doyle, M.P.; Kuchenmüller, T.; Schlundt, J.; Stein, C. Evolving public health approaches to the global challenge of foodborne infections. *Int. J. Food Microbiol.* **2010**, *139*, S16–S28. [CrossRef]
74. Scallan, E.; Hoekstra, R.M.; Angulo, F.J.; Tauxe, R.V.; Widdowson, M.-A.; Roy, S.L.; Jones, J.L.; Griffin, P.M. Foodborne illness acquired in the United States—Major pathogens. *Emerg. Infect. Dis.* **2011**, *17*, 7. [CrossRef]
75. Cummings, K.; Warnick, L.; Alexander, K.; Cripps, C.; Gröhn, Y.; McDonough, P.; Nydam, D.; Reed, K. The incidence of salmonellosis among dairy herds in the northeastern United States. *J. Dairy Sci.* **2009**, *92*, 3766–3774. [CrossRef]
76. Fossler, C.; Wells, S.; Kaneene, J.; Ruegg, P.; Warnick, L.; Eberly, L.; Godden, S.; Halbert, L.; Campbell, A.; Bolin, C. Cattle and environmental sample-level factors associated with the presence of Salmonella in a multi-state study of conventional and organic dairy farms. *Prev. Vet. Med.* **2005**, *67*, 39–53. [CrossRef]
77. Lindberg, A.; Robertsson, J. Salmonella typhimurium infection in calves: Cell-mediated and humoral immune reactions before and after challenge with live virulent bacteria in calves given live or inactivated vaccines. *Infect. Immun.* **1983**, *41*, 751–757. [CrossRef]
78. Yin, Y.; Zhou, D. Organoid and enteroid modeling of Salmonella infection. *Front. Cell Infect. Microbiol.* **2018**, *8*, 102. [CrossRef]
79. Crawford, R.W.; Keestra, A.M.; Winter, S.E.; Xavier, M.N.; Tsolis, R.M.; Tolstikov, V.; Bäumler, A.J. Very long O-antigen chains enhance fitness during Salmonella-induced colitis by increasing bile resistance. *PLoS Pathog.* **2012**, *8*, e1002918. [CrossRef]
80. Fritz, H.M.; Pereira, R.V.; Toohey-Kurth, K.; Marshall, E.; Tucker, J.; Clothier, K.A. Salmonella enterica serovar Dublin from cattle in California from 1993–2019: Antimicrobial resistance trends of clinical relevance. *Antibiotics* **2022**, *11*, 1110. [CrossRef] [PubMed]
81. Pires, J.; Huisman, J.S.; Bonhoeffer, S.; Van Boeckel, T.P. Multidrug resistance dynamics in Salmonella in food animals in the United States: An analysis of genomes from public databases. *Microbiol. Spectr.* **2021**, *9*, e00421. [CrossRef] [PubMed]
82. Mueller-Doblies, D.; Speed, K.; Kidd, S.; Davies, R. Salmonella Typhimurium in livestock in Great Britain–trends observed over a 32-year period. *Epidemiol. Infect.* **2018**, *146*, 409–422. [CrossRef]
83. Burson, W.C. Confined Versus Conventional Cow-Calf Management Systems: Implications for Calf Health. Ph.D. Thesis, Texas Tech University, Lubbock, TX, USA, 2017.
84. Rabsch, W.; Andrews, H.L.; Kingsley, R.A.; Prager, R.; Tschäpe, H.; Adams, L.G.; Bäumler, A.J. Salmonella enterica serotype Typhimurium and its host-adapted variants. *Infect. Immun.* **2002**, *70*, 2249–2255. [CrossRef]
85. Helmy, Y.A.; Kabir, A.; Saleh, M.; Kennedy, L.A.; Burns, L.; Johnson, B. Draft genome sequence analysis of multidrug-resistant Salmonella enterica subsp. enterica serovar Mbandaka harboring colistin resistance gene mcr-9.1 isolated from foals in Kentucky, USA. *Microbiol. Resour. Announc.* **2024**, *13*, e00724–e00737. [CrossRef]
86. Perov, S.; Lidor, O.; Salinas, N.; Golan, N.; Tayeb-Fligelman, E.; Deshmukh, M.; Willbold, D.; Landau, M. Structural insights into curli CsgA cross-β fibril architecture inspire repurposing of anti-amyloid compounds as anti-biofilm agents. *PLoS Pathog.* **2019**, *15*, e1007978. [CrossRef]
87. El Hag, M.; Feng, Z.; Su, Y.; Wang, X.; Yassin, A.; Chen, S.; Peng, D.; Liu, X. Contribution of the csgA and bcsA genes to Salmonella enterica serovar Pullorum biofilm formation and virulence. *Avian Pathol.* **2017**, *46*, 541–547. [CrossRef]
88. Yan, Z.; Yin, M.; Chen, J.; Li, X. Assembly and substrate recognition of curli biogenesis system. *Nat. Commun.* **2020**, *11*, 241. [CrossRef]
89. Siddique, A.; Azim, S.; Ali, A.; Andleeb, S.; Ahsan, A.; Imran, M.; Rahman, A. Antimicrobial resistance profiling of biofilm forming non typhoidal Salmonella enterica isolates from poultry and its associated food products from Pakistan. *Antibiotics* **2021**, *10*, 785. [CrossRef]
90. Barbosa, F.d.O.; Freitas, O.C.d.; Batista, D.F.A.; Almeida, A.M.d.; Rubio, M.d.S.; Alves, L.B.R.; Vasconcelos, R.d.O.; Barrow, P.A.; Berchieri, A. Contribution of flagella and motility to gut colonisation and pathogenicity of Salmonella Enteritidis in the chicken. *Braz. J. Microbiol.* **2017**, *48*, 754–759. [CrossRef]
91. Partridge, J.D.; Harshey, R.M. Swarming: Flexible roaming plans. *J. Bacteriol.* **2013**, *195*, 909–918. [CrossRef] [PubMed]
92. Hernday, A.; Krabbe, M.; Braaten, B.; Low, D. Self-perpetuating epigenetic pili switches in bacteria. *Proc. Natl. Acad. Sci. USA* **2002**, *99*, 16470–16476. [CrossRef] [PubMed]
93. Cui, L.; Neoh, H.-M.; Iwamoto, A.; Hiramatsu, K. Coordinated phenotype switching with large-scale chromosome flip-flop inversion observed in bacteria. *Proc. Natl. Acad. Sci. USA* **2012**, *109*, E1647–E1656. [CrossRef]
94. Derdouri, N.; Ginet, N.; Denis, Y.; Ansaldi, M.; Battesti, A. The prophage-encoded transcriptional regulator AppY has pleiotropic effects on E. coli physiology. *PLoS Genet.* **2023**, *19*, e1010672. [CrossRef]

95. Tanner, J.R.; Kingsley, R.A. Evolution of Salmonella within hosts. *Trends Microbiol.* **2018**, *26*, 986–998. [CrossRef]
96. Li, C.-S.; Lo, T.-H.; Tu, T.-J.; Chueh, D.-Y.; Yao, C.-I.; Lin, C.-H.; Chen, P.; Liu, F.-T. Cytosolic galectin-4 enchains bacteria, restricts their motility, and promotes inflammasome activation in intestinal epithelial cells. *Proc. Natl. Acad. Sci. USA* **2023**, *120*, e2207091120. [CrossRef]
97. Caiazza, N.C.; Merritt, J.H.; Brothers, K.M.; O'Toole, G.A. Inverse regulation of biofilm formation and swarming motility by Pseudomonas aeruginosa PA14. *J. Bacteriol.* **2007**, *189*, 3603–3612. [CrossRef]
98. Shrout, J.D.; Chopp, D.L.; Just, C.L.; Hentzer, M.; Givskov, M.; Parsek, M.R. The impact of quorum sensing and swarming motility on Pseudomonas aeruginosa biofilm formation is nutritionally conditional. *Mol. Microbiol.* **2006**, *62*, 1264–1277. [CrossRef]
99. Kalai Chelvam, K.; Chai, L.C.; Thong, K.L. Variations in motility and biofilm formation of Salmonella enterica serovar Typhi. *Gut Pathog.* **2014**, *6*, 1264–1277. [CrossRef]
100. Azimi, T.; Zamirnasta, M.; Sani, M.A.; Soltan Dallal, M.M.; Nasser, A. Molecular mechanisms of Salmonella effector proteins: A comprehensive review. *Infect. Drug Resist.* **2020**, *13*, 11–26. [CrossRef]
101. Baxter, M.A.; Jones, B.D. The fimYZ genes regulate Salmonella enterica serovar Typhimurium invasion in addition to type 1 fimbrial expression and bacterial motility. *Infect. Immun.* **2005**, *73*, 1377–1385. [CrossRef] [PubMed]
102. Li, W.; Ren, Q.; Ni, T.; Zhao, Y.; Sang, Z.; Luo, R.; Li, Z.; Li, S. Strategies adopted by Salmonella to survive in host: A review. *Arch. Microbiol.* **2023**, *205*, 362. [CrossRef] [PubMed]
103. Hänisch, J.; Ehinger, J.; Ladwein, M.; Rohde, M.; Derivery, E.; Bosse, T.; Steffen, A.; Bumann, D.; Misselwitz, B.; Hardt, W.D. Molecular dissection of Salmonella-induced membrane ruffling versus invasion. *Cell. Microbiol.* **2010**, *12*, 84–98. [CrossRef]
104. Gorvel, J.-P.; Méresse, S. Maturation steps of the Salmonella-containing vacuole. *Microbes Infect.* **2001**, *3*, 1299–1303. [CrossRef]
105. Olekhnovich, I.N.; Kadner, R.J. Crucial roles of both flanking sequences in silencing of the hilA promoter in Salmonella enterica. *J. Mol. Biol.* **2006**, *357*, 373–386. [CrossRef]
106. Mazurkiewicz, P.; Thomas, J.; Thompson, J.A.; Liu, M.; Arbibe, L.; Sansonetti, P.; Holden, D.W. SpvC is a Salmonella effector with phosphothreonine lyase activity on host mitogen-activated protein kinases. *Mol. Microbiol.* **2008**, *67*, 1371–1383. [CrossRef]
107. Cobo-Angel, C.; Craig, M.; Osman, M.; Cummings, K.J.; Cazer, C.L. Antimicrobial use regulations are associated with increased susceptibility among bovine Salmonella isolates from a US surveillance system. *One Health* **2025**, *20*, 100983. [CrossRef]
108. Parada, J.; Carranza, A.; Alvarez, J.; Pichel, M.; Tamiozzo, P.; Busso, J.; Ambrogi, A. Spatial distribution and risk factors associated with Salmonella enterica in pigs. *Epidemiol. Infect.* **2017**, *145*, 568–574. [CrossRef]
109. Velasquez-Munoz, A.; Castro-Vargas, R.; Cullens-Nobis, F.M.; Mani, R.; Abuelo, A. Salmonella Dublin in dairy cattle. *Front. Vet. Sci.* **2024**, *10*, 1331767. [CrossRef]
110. Srednik, M.E.; Lantz, K.; Hicks, J.A.; Morningstar-Shaw, B.R.; Mackie, T.A.; Schlater, L.K. Antimicrobial resistance and genomic characterization of Salmonella Dublin isolates in cattle from the United States. *PLoS ONE* **2021**, *16*, e0249617. [CrossRef]
111. Chakraborty, A.K.; Roy, A.K. High Prevalence of metal resistant genes in Salmonella enterica MDR Plasmids correlates severe toxicities of water with higher Typhoid AMR. *arXiv* **2020**. [CrossRef]
112. Rather, P.; Munayyer, H.; Mann, P.; Hare, R.; Miller, G.; Shaw, K. Genetic analysis of bacterial acetyltransferases: Identification of amino acids determining the specificities of the aminoglycoside 6′-N-acetyltransferase Ib and IIa proteins. *J. Bacteriol.* **1992**, *174*, 3196–3203. [CrossRef] [PubMed]
113. Mansour, M.N.; Yaghi, J.; El Khoury, A.; Felten, A.; Mistou, M.-Y.; Atoui, A.; Radomski, N. Prediction of Salmonella serovars isolated from clinical and food matrices in Lebanon and genomic-based investigation focusing on Enteritidis serovar. *Int. J. Food Microbiol.* **2020**, *333*, 108831. [CrossRef]
114. Liu, M.; Zhu, K.; Li, X.; Han, Y.; Yang, C.; Liu, H.; Du, X.; Xu, X.; Yang, H.; Song, H. Genetic characterization of a Salmonella enterica serovar Typhimurium isolated from an infant with concurrent resistance to ceftriaxone, ciprofloxacin and azithromycin. *J. Glob. Antimicrob. Resist.* **2023**, *35*, 252–256. [CrossRef]
115. Bhandari, M.; Poelstra, J.W.; Kauffman, M.; Varghese, B.; Helmy, Y.A.; Scaria, J.; Rajashekara, G. Genomic diversity, antimicrobial resistance, plasmidome, and virulence profiles of Salmonella isolated from small specialty crop farms revealed by whole-genome sequencing. *Antibiotics* **2023**, *12*, 1637. [CrossRef]
116. Pingault, J.-B.; Richmond, R.; Smith, G.D. Causal inference with genetic data: Past, present, and future. *Cold Spring Harb. Perspect. Med.* **2022**, *12*, a041271. [CrossRef]
117. Hill, A.B. The environment and disease: Association or causation? *J. R. Soc. Med.* **1965**, *58*, 295–300. [CrossRef]
118. Closs Jr, G.; Bhandari, M.; Helmy, Y.A.; Kathayat, D.; Lokesh, D.; Jung, K.; Suazo, I.D.; Srivastava, V.; Deblais, L.; Rajashekara, G. The probiotic Lacticaseibacillus rhamnosus GG supplementation reduces Salmonella load and modulates growth, intestinal morphology, gut microbiota, and immune responses in chickens. *Infect. Immun.* **2025**, *93*, e00420–e00424. [CrossRef]
119. Deblais, L.; Helmy, Y.A.; Kathayat, D.; Huang, H.-c.; Miller, S.A.; Rajashekara, G. Novel imidazole and methoxybenzylamine growth inhibitors affecting Salmonella cell envelope integrity and its persistence in chickens. *Sci. Rep.* **2018**, *8*, 13381. [CrossRef]

120. Fazl, A.A.; Salehi, T.Z.; Jamshidian, M.; Amini, K.; Jangjou, A.H. Molecular detection of invA, ssaP, sseC and pipB genes in Salmonella Typhimurium isolated from human and poultry in Iran. *Afr. J. Microbiol. Res.* **2013**, *7*, 1104–1108.
121. Li, G.; Yan, C.; Xu, Y.; Feng, Y.; Wu, Q.; Lv, X.; Yang, B.; Wang, X.; Xia, X.; Elkins, C.A. Punicalagin Inhibits Salmonella Virulence Factors and Has Anti-Quorum-Sensing Potential. *Appl. Environ. Microbiol.* **2014**, *80*, 6204–6211. [CrossRef] [PubMed]
122. Bass, L.; Liebert, C.A.; Lee, M.D.; Summers, A.O.; White, D.G.; Thayer, S.G.; Maurer, J.J. Incidence and Characterization of Integrons, Genetic Elements Mediating Multiple-Drug Resistance, in Avian *Escherichia coli*. *Antimicrob. Agents Chemother.* **1999**, *43*, 2925–2929. [CrossRef] [PubMed]
123. Goldstein, C.; Lee, M.D.; Sanchez, S.; Hudson, C.; Phillips, B.; Register, B.; Grady, M.; Liebert, C.; Summers, A.O.; White, D.G.; et al. Incidence of Class 1 and 2 Integrases in Clinical and Commensal Bacteria from Livestock, Companion Animals, and Exotics. *Antimicrob. Agents Chemother.* **2001**, *45*, 723–726. [CrossRef]
124. Sever, N.K.; Akan, M. Molecular analysis of virulence genes of Salmonella Infantis isolated from chickens and turkeys. *Microb. Pathog.* **2019**, *126*, 199–204. [CrossRef]
125. Turki, Y.; Mehr, I.; Ouzari, H.; Khessairi, A.; Hassen, A. Molecular typing, antibiotic resistance, virulence gene and biofilm formation of different Salmonella enterica serotypes. *J. Gen. Appl. Microbiol.* **2014**, *60*, 123–130. [CrossRef]
126. Allen, S.E.; Boerlin, P.; Janecko, N.; Lumsden, J.S.; Barker, I.K.; Pearl, D.L.; Reid-Smith, R.J.; Jardine, C. Antimicrobial resistance in generic Escherichia coli isolates from wild small mammals living in swine farm, residential, landfill, and natural environments in southern Ontario, Canada. *Appl. Environ. Microbiol.* **2011**, *77*, 882–888. [CrossRef]
127. Lin, L.; Liao, X.; Li, C.; Abdel-Samie, M.A.; Cui, H. Inhibitory effect of cold nitrogen plasma on Salmonella Typhimurium biofilm and its application on poultry egg preservation. *LWT* **2020**, *126*, 109340. [CrossRef]
128. Xu, Y.; Abdelhamid, A.G.; Sabag-Daigle, A.; Sovic, M.G.; Ahmer, B.M.; Yousef, A.E. The Role of Egg Yolk in Modulating the Virulence of Salmonella Enterica Serovar Enteritidis. *Front. Cell. Infect. Microbiol.* **2022**, *12*, 903979. [CrossRef]
129. Hassena, A.B.; Barkallah, M.; Fendri, I.; Grosset, N.; Neila, I.B.; Gautier, M.; Gdoura, R. Real time PCR gene profiling and detection of Salmonella using a novel target: The siiA gene. *J. Microbiol. Methods* **2015**, *109*, 9–15. [CrossRef]
130. Bearson, B.L.; Bearson, S.M. The role of the QseC quorum-sensing sensor kinase in colonization and norepinephrine-enhanced motility of Salmonella enterica serovar Typhimurium. *Microb. Pathog.* **2008**, *44*, 271–278. [CrossRef]
131. Ganesan, V. Detection of Salmonella in Blood by PCR using iroB gene. *J. Clin. Diagn. Res.* **2014**, *8*, DC01-3. [CrossRef] [PubMed]
132. Shah, T.; Zhu, C.; Shah, C.; Upadhyaya, I.; Upadhyay, A. Trans-cinnamaldehyde nanoemulsion reduces Salmonella Enteritidis biofilm on steel and plastic surfaces and downregulates expression of biofilm associated genes. *Poult. Sci.* **2025**, *104*, 105086. [CrossRef] [PubMed]
133. Chen, Z.; Bai, J.; Wang, S.; Zhang, X.; Zhan, Z.; Shen, H.; Zhang, H.; Wen, J.; Gao, Y.; Liao, M.; et al. Prevalence, antimicrobial resistance, virulence genes and genetic diversity of Salmonella isolated from retail duck meat in southern China. *Microorganisms* **2020**, *8*, 444. [CrossRef]
134. Cao, Z.Z.; Xu, J.W.; Gao, M.; Li, X.S.; Zhai, Y.J.; Yu, K.; Wan, M.; Luan, X.H. Prevalence and antimicrobial resistance of Salmonella isolates from goose farms in Northeast China. *Iran. J. Vet. Res.* **2020**, *21*, 287.
135. Asgharpour, F.; Mahmoud, S.; Marashi, A.; Moulana, Z. Molecular detection of class 1, 2 and 3 integrons and some antimicrobial resistance genes in Salmonella Infantis isolates. *Iran. J. Microbiol.* **2018**, *10*, 104.
136. Chuanchuen, R.; Padungtod, P. Antimicrobial Resistance Genes in Salmonella enterica Isolates from Poultry and Swine in Thailand. *J. Vet. Med. Sci.* **2009**, *71*, 1349–1355. [CrossRef]
137. Rebelo, A.R.; Bortolaia, V.; Kjeldgaard, J.S.; Pedersen, S.K.; Leekitcharoenphon, P.; Hansen, I.M.; Guerra, B.; Malorny, B.; Borowiak, M.; Hammerl, J.A.; et al. Multiplex PCR for detection of plasmid-mediated colistin resistance determinants, mcr-1, mcr-2, mcr-3, mcr-4 and mcr-5 for surveillance purposes. *Eurosurveillance* **2018**, *23*, 17-00672-39. [CrossRef]
138. Borowiak, M.; Fischer, J.; A Hammerl, J.; Hendriksen, R.S.; Szabo, I.; Malorny, B. Identification of a novel transposon-associated phosphoethanolamine transferase gene, mcr-5, conferring colistin resistance in d-tartrate fermenting Salmonella enterica subsp. enterica serovar Paratyphi B. *J. Antimicrob. Chemother.* **2017**, *72*, 3317–3324. [CrossRef]
139. Borowiak, M.; Baumann, B.; Fischer, J.; Thomas, K.; Deneke, C.; Hammerl, J.A.; Szabo, I.; Malorny, B. Development of a Novel mcr-6 to mcr-9 Multiplex PCR and Assessment of mcr-1 to mcr-9 Occurrence in Colistin-Resistant Salmonella enterica Isolates From Environment, Feed, Animals and Food (2011–2018) in Germany. *Front. Microbiol.* **2020**, *11*, 80. [CrossRef]
140. Gan, T.; Shu, G.; Fu, H.; Yan, Q.; Zhang, W.; Tang, H.; Yin, L.; Zhao, L.; Lin, J. Antimicrobial resistance and genotyping of Staphylococcus aureus obtained from food animals in Sichuan Province, China. *BMC Vet. Res.* **2021**, *17*, 177. [CrossRef]

antibiotics

MDPI

Article

Genomic Insights of Antibiotic-Resistant *Escherichia coli* Isolated from Intensive Pig Farming in South Africa Using 'Farm-to-Fork' Approach

Shima E. Abdalla [1], Linda A. Bester [2], Akebe L. K. Abia [1,3], Mushal Allam [4,5], Arshad Ismail [5,6], Sabiha Y. Essack [1] and Daniel G. Amoako [1,7,*]

1 Antimicrobial Research Unit, College of Health Sciences, University of KwaZulu-Natal, Durban 4000, South Africa; shimaeltayeb23@gmail.com (S.E.A.); abiaakebel@ukzn.ac.za (A.L.K.A.); essacks@ukzn.ac.za (S.Y.E.)
2 Biomedical Resource Unit, College of Health Sciences, University of KwaZulu-Natal, Durban 4000, South Africa; besterl@ukzn.ac.za
3 Environmental Research Foundation, Westville 3630, South Africa
4 Department of Genetics and Genomics, College of Medicine and Health Sciences, United Arab Emirates University, Al Ain P.O. Box 15551, United Arab Emirates; mushal.allam@uaeu.ac.ae
5 Sequencing Core Facility, National Institute for Communicable Diseases, Division of the National Health Laboratory Service, Johannesburg 2193, South Africa; arshadi@nicd.ac.za
6 Department of Biochemistry and Microbiology, Faculty of Science, Engineering and Agriculture, University of Venda, Thohoyandou 0950, South Africa
7 Department of Pathobiology, University of Guelph, Guelph, ON N1G 2W1, Canada
* Correspondence: amoakod@ukzn.ac.za

Abstract: Background/Objectives: Intensive pig farming is a critical component of food security and economic activity in South Africa; however, it also presents a risk of amplifying antimicrobial resistance (AMR). This study provides genomic insights into antibiotic-resistant *Escherichia coli* (*E. coli*) circulating across the pork production chain, using a 'farm-to-fork' approach. **Methods**: A total of 417 samples were collected from various points along the production continuum, including the farm (n = 144), transport (n = 60), and abattoir (n = 213). *E. coli* isolates were identified using the Colilert-18 system, and their phenotypic resistance was tested against 20 antibiotics. Thirty-one isolates were selected for further characterization based on their resistance profiles and sampling sources, utilizing whole-genome sequencing and bioinformatic analysis. **Results**: The isolates exhibited varying resistance to critical antibiotics used in both human and animal health, including ampicillin (31/31, 100%), tetracycline (31/31, 100%), amoxicillin–clavulanate (29/31, 94%), chloramphenicol (25/31, 81%), and sulfamethoxazole–trimethoprim (10/31, 33%). Genetic analysis revealed the presence of resistance genes for β-lactams (bla_{EC}, bla_{TEM}), trimethoprim/sulfonamides (*dfrA1*, *dfrA5*, *dfrA12*, *sul2*, *sul3*), tetracyclines (*tetA*, *tetB*, *tetR*, *tet34*), aminoglycosides (*aadA*, *strA*, *aph* variants), and phenicols (*catB4*, *floR*, *cmlA1*), most of which were plasmid-borne. Virulome analysis identified 24 genes, including toxins and adhesion factors. Mobile genetic elements included 24 plasmid replicons, 43 prophages, 19 insertion sequence families, and 7 class 1 integrons. The *E. coli* isolates belonged to a diverse range of sequence types, demonstrating significant genetic variability. Further phylogenomic analysis revealed eight major clades, with isolate clustering by sequence type alongside South African environmental and clinical *E. coli* strains, regardless of their sampling source. **Conclusions**: The genetic complexity observed across the pork production continuum threatens food safety and may impact human health. These findings underscore the need for enhanced AMR monitoring in livestock systems and support the integration of AMR surveillance into food safety policy frameworks.

Keywords: genomic epidemiology; *E. coli*; resistome; mobilome; antimicrobial resistance; pig production chain; South Africa

1. Introduction

Food-producing animals, including pigs, poultry, and cattle, are major reservoirs for foodborne pathogens, many of which are associated with significant human morbidity and mortality [1,2]. Globally, foodborne infections remain a persistent threat, accounting for millions of illnesses and numerous outbreaks annually [3,4].

Escherichia coli (*E. coli*) is a commensal bacterium typically present in the gastrointestinal tract of warm-blooded animals but capable of causing a range of intestinal and extraintestinal infections when pathogenic strains are involved [5,6]. Among foodborne pathogens, *E. coli* has gained prominence due to its increasing resistance to antibiotics of human and veterinary importance [7]. Antimicrobial resistance (AMR) in *E. coli* contributes significantly to the estimated 700,000 global annual deaths attributed to resistant infections, a figure projected to reach 10 million by 2050 in the absence of effective interventions [8,9]. Resistance in *E. coli* may arise through intrinsic mechanisms (e.g., efflux pumps, outer membrane impermeability) or acquired mechanisms such as chromosomal mutations and horizontal gene transfer via mobile genetic elements (MGEs), including plasmids, transposons, integrons, and bacteriophages [10,11]. These traits enable both commensal and pathogenic *E. coli* strains to act as reservoirs and vectors for antimicrobial resistance genes (ARGs) across the One Health interface [12].

In sub-Saharan Africa, particularly South Africa, pig farming remains a vital component of livestock agriculture and food security. However, the region faces growing concerns over antibiotic use and emerging resistance within intensive animal production systems [13,14]. Recent studies across Africa have reported high prevalence of MDR *E. coli* in livestock, underscoring the urgent need for localized surveillance strategies [15–17]. Despite increasing awareness, most AMR surveillance efforts in African livestock rely on traditional typing techniques, which lack the resolution of whole-genome sequencing (WGS) for discerning resistance mechanisms, phylogenetic relationships, and transmission pathways [18–20]. WGS provides unparalleled insights into the resistome, virulome, and mobilome of bacterial pathogens and is crucial for data-driven AMR risk assessment. To address these gaps, we employed a 'farm-to-fork' approach embedded within a One Health framework to characterize antibiotic-resistant *E. coli* circulating in the pig production continuum in the KwaZulu-Natal Province, South Africa. This study integrates phenotypic, genomic, and phylogenetic data to provide a holistic understanding of AMR risks associated with pork production systems and contributes critical regional data to support evidence-based policy and intervention design.

2. Results

2.1. Isolate Characteristics and Antibiotic Susceptibility Profiling

A total of 31 multidrug-resistant (MDR) *E. coli* isolates were selected for whole-genome sequencing and characterization based on their resistance profiles and representation across the pork production continuum comprising farm, transport, and abattoir stages. Of these, 19 isolates (61.3%) were derived from the farm environment, including 13 from pig fecal samples and 6 from slurry. Four isolates (12.9%) were recovered from the transport phase, evenly split between pre- and post-transport sampling. The remaining 8 isolates (25.8%)

were obtained from abattoir-associated matrices, including cecal contents (n = 2), carcass rinsates (n = 2), carcass swabs (n = 2), and meat cuts (n = 2) (Table S1).

Phenotypic antibiotic susceptibility testing revealed that all 31 isolates (100%) exhibited resistance to ampicillin and tetracycline antibiotic classes commonly used in livestock production (Figure 1). High levels of resistance were also observed for amoxicillin–clavulanate (29/31, 93.5%) and chloramphenicol (25/31, 80.6%) (Figure 1). Moderate resistance to sulfamethoxazole–trimethoprim was observed in 10 isolates (32.3%). Low-level resistance was detected against critical last-resort agents such as tigecycline (1/31, 3.2%), imipenem (1/31, 3.2%), and meropenem (1/31, 3.2%) (Figure 1).

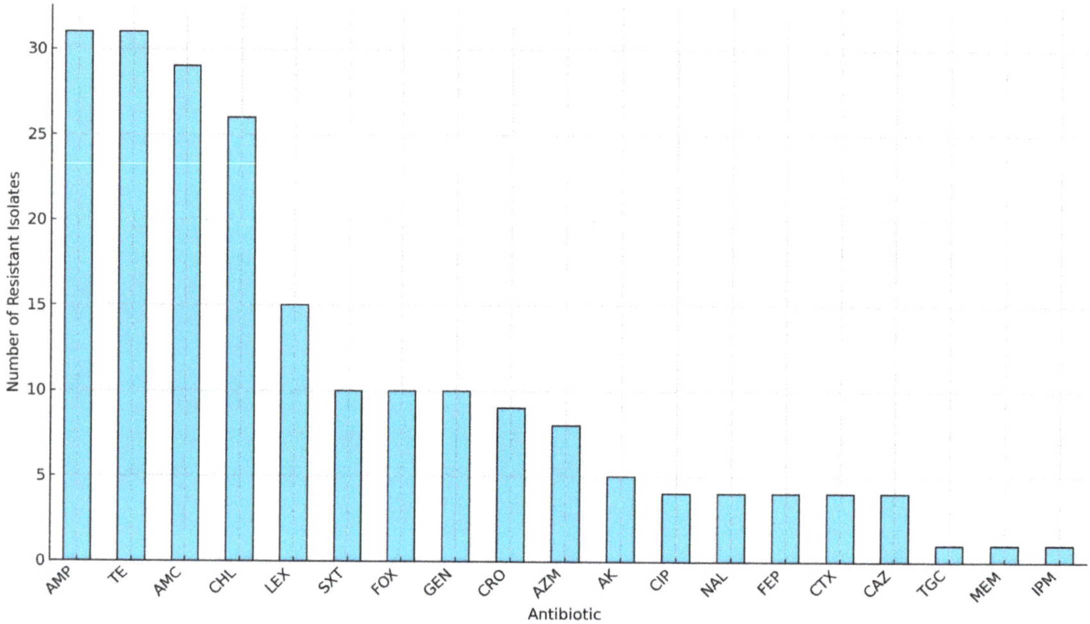

Figure 1. Overall antibiotic resistance frequencies among 31 MDR *E. coli* isolates from the farm-to-fork continuum. Key: AMP, Ampicillin; TE, Tetracycline; AMC, Amoxicillin-clavulanate; CHL, Chloramphenicol; LEX, Cephalexin; SXT, Sulfamethoxazole-trimethoprim; FOX, Cefoxitin; GEN, Gentamicin; CRO, Ceftriaxone; AZM, Azithromycin; AK, Amikacin; CIP, Ciprofloxacin; NAL, Nalidixic acid; FEP, Cefepime; CTX, Cefotaxime; CAZ, Ceftazidime; TGC, Tigecycline; MEM, Meropenem; IPM, Imipenem.

Notably, several isolates displayed identical antibiogram profiles despite originating from distinct sources along the continuum, suggesting the possible circulation or persistence of clonally related resistant strains across production stages. While phenotypic resistance data for these isolates were previously reported in trend analyses [21], the current study extends this understanding through comprehensive genomic analyses of the resistome, virulome, mobilome, and phylogenetic relationships. Figure 2 provides a schematic overview of the study workflow, including sample isolation, antimicrobial susceptibility testing, WGS, and downstream bioinformatics pipelines.

2.2. Genomic Characteristics

The genomic features of the 31 *E. coli* isolates are summarized in Supplementary Table S2. The total assembled genome sizes ranged from 4.7 to 6.1 megabases (Mb), consistent with typical *E. coli* genome architecture. The GC content varied narrowly between 50.2% and 50.8%, reflecting a conserved base composition across isolates.

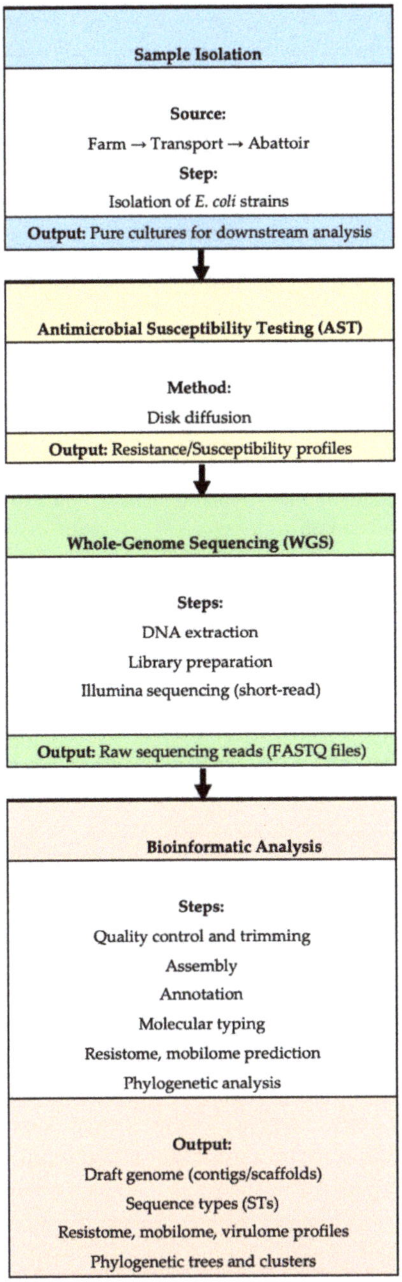

Figure 2. A workflow outlining the key steps in the study, including sample isolation, antimicrobial resistance (AMR) testing, whole-genome sequencing (WGS), and subsequent bioinformatic analysis.

Assembly quality metrics further supported the robustness of the sequencing data, with most assemblies meeting the established thresholds for downstream genomic analysis. These metrics reflect the expected diversity in assembly outcomes based on sequencing depth, strain-specific genome complexity, and repeat content. All genomes passed quality control filters and were deemed suitable for high-resolution comparative analyses of AMR, virulence, and phylogenomic structure.

2.3. Antibiotic Resistance Gene Analysis

WGS revealed a rich and diverse resistome across all *E. coli* isolates, with resistance determinants spanning major antibiotic classes across sampling sites. β-lactam resistance was the most prominent feature, characterized by the widespread presence of chromosomal and plasmid-encoded β-lactamase genes (Figure 3A and Table S1). In addition, ESBL genes bla_{TEM-1}, bla_{TEM-1B}, and $bla_{TEM-105}$ were detected. Aminoglycoside resistance was conferred by an array of modifying enzyme genes, indicating frequent acquisition of MGEs associated with this class. Resistance to tetracyclines was predominantly mediated by *tet34*, *tetA*, and *tetB*, while resistance to sulfonamides and trimethoprim was driven by *sul2*, *sul3*, *dfrA1*, *dfrA5*, and *dfrA12*. Additional resistance determinants included *catB4*, *floR*, and *cmlA1* for phenicols, and macrolide resistance genes (Figure 3B).

Of particular note, several efflux system genes were recurrently identified and are known to mediate low-level MDR across diverse antibiotic classes including macrolides, fluoroquinolones, chloramphenicol, tetracyclines, and certain aminoglycosides (Figure 3A,B). The distribution of ARGs was largely conserved across the production continuum. Farm isolates harbored a core resistome comprising *aadA*, *strA*, and *strB* in some samples, with the additional detection of *sul3*. Transport isolates exhibited comparable resistance profiles, though with slightly fewer ARGs overall. Abattoir isolates retained key resistance determinants seen upstream (Figure 3B). The phenotypes were not corroborated by ARGs in some cases, as some isolates exhibited resistance to antibiotics in the absence of known associated resistance determinants (Table S1). For example, the mechanism of β-lactam resistance in some isolates, such as WB1-1-R8, was not detected in the ResFinder and CARD databases (Table S1).

The quinolone resistance-determining regions (QRDRs) were investigated for chromosomal amino acid substitutions. Substitutions were observed in the chromosome-borne DNA gyrase (*gyrA*-S83L, D87N, E678D, S828A and *gyrB*-D185E) and topoisomerase IV (*parC*-S80I, E475D, A620V and *parE*-I136V), mediating fluoroquinolone resistance (Table 1). Similarly, amino acid substitutions in *acrA* (T104A), *acrR* (A212S, N214T), and *marB* (L27P), along with the presence of *tetA* and *tolC* genes, were identified as contributing to tigecycline resistance in the isolate resistant to this antibiotic (Table 1).

2.4. Mobilome (Plasmids, Insertion Sequences, Prophages, and Integrons) Analysis

PlasmidFinder revealed 24 different plasmid replicon types. Farm isolates carried a diverse range of plasmid replicons including Col(MG828), IncFIB(AP001918), IncFII, IncX1, and IncY—all associated with the transmission of resistance genes. In transport isolates, plasmid replicons such as ColRNAI, IncFIB(AP001918), IncFII, and IncX1 were also present. The similarity of these plasmid replicons between farm and transport isolates suggests a potential continuity of resistance mechanisms during transport. Abattoir isolates carried plasmid replicons similar to those found in the farm and transport stages, including ColRNAI, IncFIB(AP001918), IncFII, and IncX1. This consistent presence of plasmid replicons across the continuum supports the hypothesis of plasmid-mediated gene transfer throughout the pig production chain (Table S1).

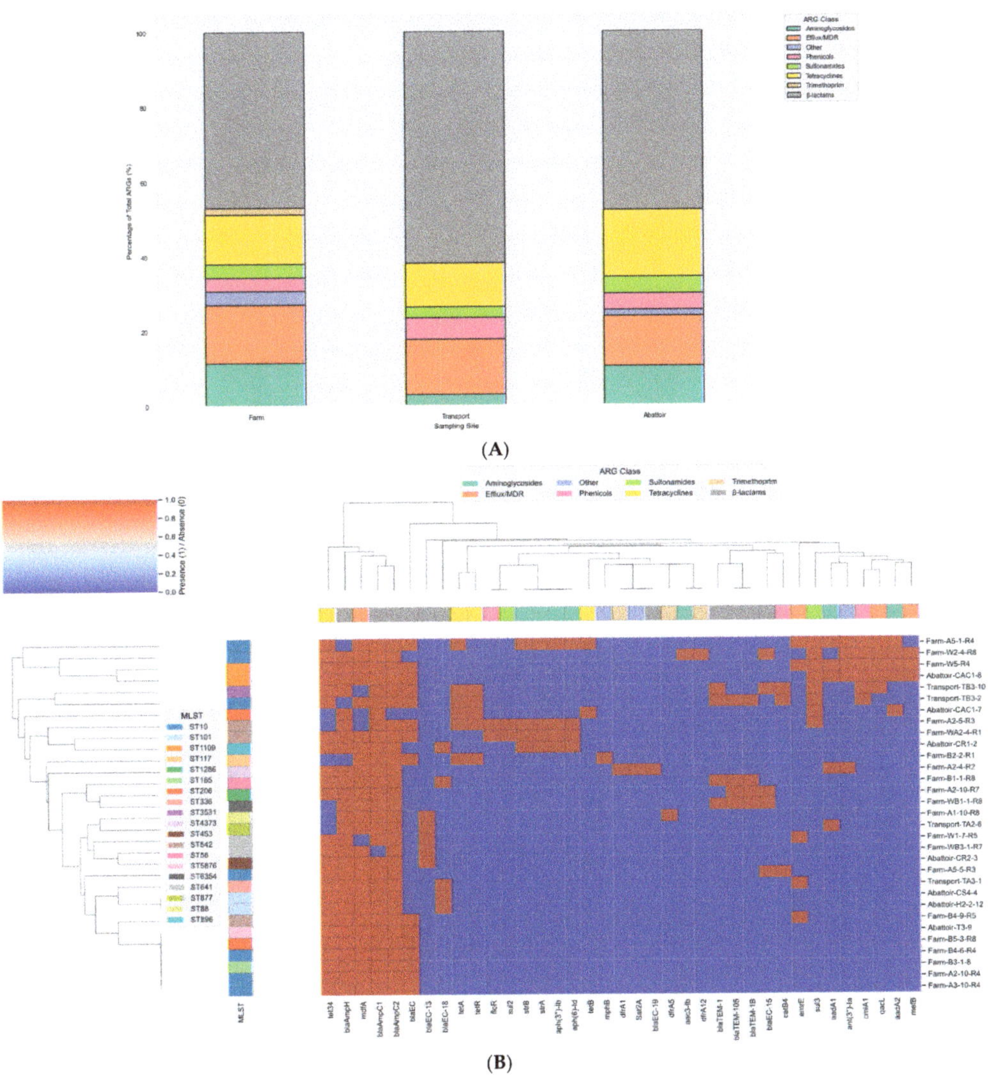

Figure 3. (**A**) Percentage distribution of ARG classes across 31 *E. coli* isolates. The stacked bar chart depicts the relative proportions of ARG classes detected in isolates obtained from farm (n = 19), transport (n = 4), and abattoir (n = 8) sources. Resistance genes were categorized into functional classes, including β-lactams, tetracyclines, aminoglycosides, sulfonamides, trimethoprim, streptothricin, and efflux/multidrug resistance (MDR) determinants. Each bar represents the percentage contribution of each ARG class to the total ARGs identified at that site, highlighting variation in resistome composition across different stages of the pork production system. (**B**) Clustermap of antimicrobial resistance gene (ARG) profiles across MDR *Escherichia coli* isolates from the pork production continuum in South Africa. Hierarchical clustering was performed based on presence/absence of ARGs, with isolates grouped according to similarity in resistome composition. Each column represents a specific ARG, while each row corresponds to an isolate labeled by its sampling source (Farm, Transport, Abattoir). The color intensity indicates gene presence (red) or absence (blue). The top color band denotes the functional class of each ARG, including β-lactams, tetracyclines, sulfonamides, aminoglycosides, trimethoprim, streptothricin, and efflux/MDR resistance determinants.

Table 1. Chromosomal mutations associated with resistance to fluoroquinolones and tigecycline.

Isolate ID	Source	Resistance Type	Genes	Amino Acid Substitutions	Associated Phenotype
A2-5-R3	Farm	Quinolone	*gyrA*, *gyrB*, *parC*, *parE*	S83L, D87N, E678D, S828A (***gyrA***); D185E (***gyrB***); E475D, A620V * (***parC***); I136V (***parE***)	Ciprofloxacin/Nalidixic acid
B4-6-R4	Farm	Quinolone	*gyrA*, *gyrB*, *parC*, *parE*	E678D, S828A (***gyrA***); D185E (***gyrB***); E475D (***parC***); I136V (***parE***)	Ciprofloxacin/Nalidixic acid
B4-9-R5	Farm	Quinolone	*gyrA*, *gyrB*, *parC*, *parE*	S83L, D87N, E678D, S828A (***gyrA***); D185E (***gyrB***); S80I, A58T *, E475D, A620V * (***parC***); I136V (***parE***)	Ciprofloxacin/Nalidixic acid
TB3-10	Transport	Quinolone	*gyrA*, *gyrB*, *parC*, *parE*	E678D, S828A (***gyrA***); D185E (***gyrB***); E475D (***parC***); I136V (***parE***)	Ciprofloxacin/Nalidixic acid
W1-7-R5	Farm	Quinolone	*gyrA*, *gyrB*, *parC*, *parE*	E678D, S828A (***gyrA***); D185E (***gyrB***); E475D (***parC***); I136V (***parE***)	Ciprofloxacin/Nalidixic acid
WA2-4-R1	Farm	Quinolone	*gyrA*, *gyrB*, *parC*, *parE*	S83L, D87N, E678D, S828A (***gyrA***); D185E (***gyrB***); A58T *, S80I, E475D, A620V * (***parC***); I136V (***parE***)	Ciprofloxacin/Nalidixic acid
B2-2-R1	Farm	Tigecycline	*acrA*, *acrR*, *marB*, *tolC*	T104A (***acrA***); A212S, N214T (***acrR***); L27P (***marB***); ***tolC*** present	Tigecycline

Note: * Novel mutations identified.

Nineteen insertion sequence (ISs) families were found in all the genomes with the most predominant being IS1595 (35.5%), followed by IS5 (32.2%) (Figure 4). In general, there was great diversity of the IS families irrespective source or sequence type, albeit there were few instances where this was not the case. For example, two *E. coli* isolates from the farm belonging to ST10 (A2-10-R4, A3-10-R4) had the same set of ISs (ISNCY, IS4 and IS1595) whilst two isolates from the abattoir belonging to ST1109 (CAC1-8, W-5-R4) also encoded the same ISs (Tn3, IS66 and IS5). The PHAge Search Tool revealed 43 intact phages, the most predominant of which was Entero_lambda found in 25 isolates, followed by Entero_P88, Salmon_Fels_2, and Shigel_SfII found in 15, 15, and 13 of the isolates, respectively (Supplementary Table S3). None of the prophages carried ARGs.

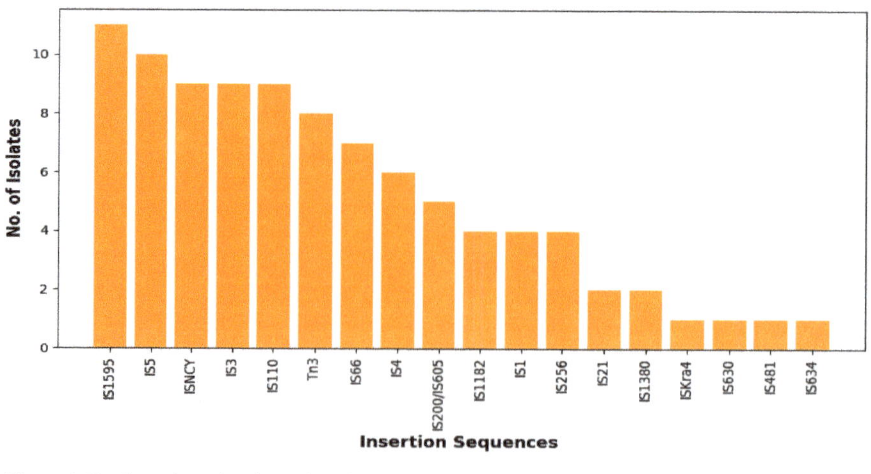

Figure 4. Total number of each predicted insertion sequence (IS) families via the ISFINDER database (https://isfinder.biotoul.fr/ accessed on 15 June 2024).

Class 1 and 2 integrons were found in the isolates, with Class 1 being the dominant (Table 2). Class 1 integrons were identified in 7 (22.6%) of the isolates found on the farm (n = 3) and abattoir (n = 4). In456 was the most frequently identified class 1 integron with resistance gene cassettes encoding resistance to aminoglycosides (n = 2) and chloramphenicol (n = 1) and bracketed by the 1S256 insertion sequences and TnAs1 transposon (Table 3). The most frequently identified gene cassette combination was *aadA1:aadA2: cmlA1*, found in two integrons, including In456 and In649 (Table 2). Similar integron types with identical gene cassettes were identified in isolates from different clonal types and sampling sites. For example, isolates from the farm A5-1-1R4 (ST10), W-5-R4 (ST1109), and abattoir CAC1-8 (ST1109) had the In456 integron type with an identical gene cassette (*aadA1:aadA2:cmlA1*). Cassette arrays did not follow clonal lineages or sources (farm and abattoir), while isolates belonging to the same STs had different gene cassettes (Table 2). The unique class 2 integron with the resistance gene cassettes (*dfrA1::aadA1::sat2*) was found in isolate A2-4-R2(ST4373) from the farm. Some ARGs were mostly co-carried on MGEs (integrons or associated with insertion sequences and/or transposons) (Table 3). The $bla_{\text{TEM-1B}}$ gene was commonly associated with *Tn3* transposon. The *Tn3* was also associated with tetracycline resistance genes in some instances. The insertion sequence (IS256) was associated with the *sul3* gene (Table 3). The resistance genes and MGEs in the *E. coli* isolates were closely related (98–100% similarity) to target sequences in the GenBank database, with most hits being plasmids with hosts from the *Enterobacterales* family (Table 3).

Table 2. Class 1 and 2 integrons, gene cassettes, and sequence types of the *E. coli* isolates.

Isolate ID	MLST [a]	Integron Class	Integron	Cassette Arrays						
				GC1	GC2 [b]	GC3	GC4	GC5	GC6	GC7
A1-10-R8	88	IntI1	In13	- [c]	*dfrA5*	-	-	-	-	-
A2-4-R2	4373	IntI2	In2-32	*dfrA1*	-	-	*aadA1*	-	-	*sat2*
A5-1-1R4	10	IntI1	In456	-	-	-	*aadA1*	*aadA2*	*cmlA1*	-
W2-4-R8	10	IntI1	In649	-	-	*dfrA12*	*aadA1*	*aadA2*	*cmlA1*	-
W-5-R4	1109	IntI1	In456	-	-	-	*aadA1*	*aadA2*	*cmlA1*	-
TB3-10	3531	IntI1	In774	-	-	-	-	-	*cmlA1*	-
CAC1-7	206	IntI1	In127	-	-	-	-	*aadA2*	-	-
CAC1-8	1109	IntI1	In456	-	-	-	*aadA1*	*aadA2*	*cmlA1*	-

Note: [a] MLST denotes multilocus sequence type; [b] GC denotes gene cassettes; and [c] denotes the missing cassette arrays due to draft genomic sequences that were fragmented during the sequencing and assembling process into different contigs.

Table 3. Mobile genetic elements associated with antibiotic resistance genes in the isolates.

Isolate ID	Contig	Synteny of Resistance Genes and MGE	Plasmid/Chromosomal Sequence with Closest Nucleotide Homology (Accession Number)
A1-10-R8	6	*::::incFII::::IS26:dfrA5:IntI1:TnAs1*	*Escherichia coli strain* 18MD05VL07 005213EC plasmid pVPS18EC0676-1, complete sequence (CP063726.1)
A2-4-R2	28	*::::emrD::::mdtL::::IntI2:dfrA1:sat2:ant(3″)-Ia::::*	No significant similarity found
A2-5-R3	602	*tet(A):tetR(A)::aph(6)-Id:aph(3″)-Ib: sul2::::IS1::::*	*Escherichia coli* strain T28R chromosome, complete genome (CP049353.1)
	14	*::::floR::IS91*	*Escherichia coli* strain AH25 chromosome, complete genome (CP055256.1)

Table 3. *Cont.*

Isolate ID	Contig	Synteny of Resistance Genes and MGE	Plasmid/Chromosomal Sequence with Closest Nucleotide Homology (Accession Number)
A2-10-R7	1	*::::TEM-1:Tn3*	*Escherichia coli* strain AMSCJX02 plasmid pAMSC5, complete sequence (CP031110.1)
A5-1-R4	60	*IS1::::aph(6)-Id:aph(3″)-Ib:ISVsa5:::: IS26:::sul3:IS256::ant(3″)-Ia:cmlA1::: IntI1::TnAs3::tetR(B):tet(B):tetC::*	*Salmonella enterica subsp. enterica serovar Indiana* strain SI67 plasmid pSI67-1, complete sequence (CP050784.1)
B1-1-R8	69	*:TEM-1:Tn3::::IS1*	*Escherichia coli* isolate MSB1_3C-sc-2280310 genome assembly, plasmid: 2 (LR890263.1)
B2-2-R1	171	*::::Tn3:::tet(A):tetR(A)::Tn3*	*Escherichia coli* O83:H1 str. NRG 857C plasmid pO83_CORR genomic sequence (CP001856.1)
CAC1-7	98	*mef(B)::::sul3::IS256:qacL:aadA1: cmlA1:ant(3″)-Ia:::intI1::TnAs1*	*Klebsiella pneumoniae* strain k9 plasmid pk9, complete sequence (CP049891.1)
CAC1-8-	79	*mef(B):::sul3:IS256::ant(3″)-Ia:cmlA1:ant(3″)-Ia:::IntI1:::TnAs1*	*Klebsiella pneumoniae* strain k9 plasmid pk9, complete sequence (CP049891.1)
W2-4-R8	127	*:::sul3:IS256::ant(3″)-Ia:cmlA1:ant(3″)-Ia::dfrA12:IntI1:TnAs3*	*Escherichia coli* strain 1919D3 plasmid p1919D3-1, complete sequence (CP046004.1)
W5R4	53	*IS26:mef(B):::sul3:IS256:qacL:aadA1: cmlA1:ant(3″)-Ia:::IntI1:TnAs1*	No significant similarity found
WA2-4-R1	145	*:aph(6)-Id:aph(3″)-Ib:sul2:::IS1::::*	*Escherichia coli* strain 13P484A chromosome, complete genome (CP019280.1)
	23	*::ISVsa3::floR::IS91*	*Escherichia coli* strain AH25 chromosome, complete genome (CP055256.1)
WB1-1-R8	59	*tetR(A):tet(A)::::TnAs1::::*	*Escherichia coli* TCJ482-1 plasmid p482-1 contig COV43U1_c1 genomic sequence (MG692709.1)
	57	*::::incFII::Tn3::TEM-1*	*Escherichia coli* strain ECOR 48 genome assembly, plasmid: RCS84_p (LT985305.1)
TB3-10	162	*:::sul3::IS256:::IntI1:::IS6*	*Escherichia coli* strain CP131_Sichuan plasmid pCP131-IncHI1, complete sequence (CP053721.1)
	376	*::tet(A):tetR(A)::TnAs1:TEM-1::*	*Klebsiella pneumoniae* strain 20130907-4 plasmid p309074-1FIIK complete sequence (MN842293.1)

2.5. Virulome in the E. coli Isolates

Virulome profiling identified 24 distinct virulence-associated genes among the 31 *E. coli* isolates, with 77.4% (24/31) carrying at least one virulence determinant (Table S4). A total of 20 isolates (64.5%) harbored multiple virulence genes. These genes were grouped into functional categories: toxins (*astA*, *vat*, *pic*, *capU*, *cba*, *cma*), adhesion factors (*lpfA*, *tsh*, *eilA*, *iha*, *stx2A*, *stx2B*, *stx2*), iron acquisition systems (*iroN*, *ireA*), immune evasion elements (*iss*, *gad*, *air*, *katP*), microcins (*mchB*, *mchC*, *mchF*, *mcmA*), and one secretion-associated gene (*aaiC*) (Figure 5). The most commonly detected virulence genes were *iss* (n = 16), *lpfA* (n = 13), *astA* (n = 12), and *gad* (n = 9). MGEs associated with these virulence genes are presented in Table 4. Farm isolates exhibited a broad virulence gene repertoire, including *gad*, *astA*, *iroN*, *iss*, and *mchF*. Transport isolates exhibited the lowest diversity in virulence gene categories compared to those from the farm and abattoir stages (Figure 5).

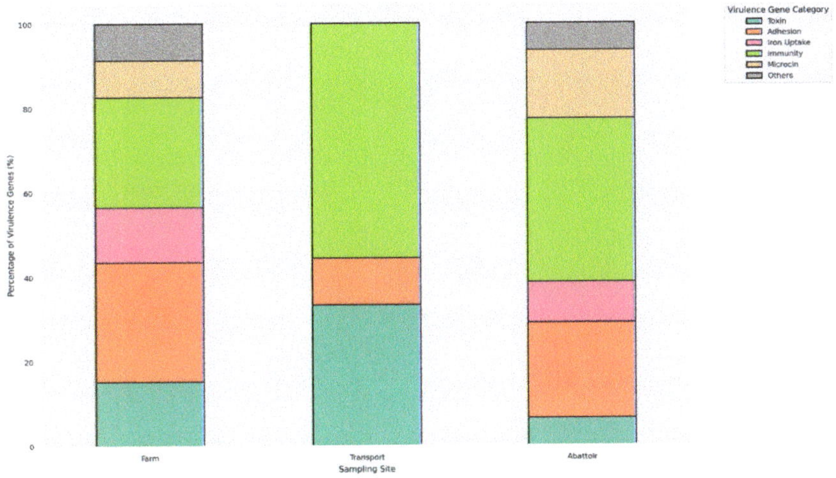

Figure 5. A plot illustrating the relative proportions of six functional virulence gene categories: toxins, adhesion factors, iron uptake systems, immune evasion mechanisms, microcins, and others, identified in isolates from farm, transport, and abattoir environments. Each bar represents the total virulence gene composition per interface, normalized to 100%, highlighting shifts in virulome across the farm continuum.

Table 4. Mobile genetic elements associated with virulence genes.

Isolate ID	Contig	Virulence and MGE	Plasmid/Chromosomal Sequence with Closest Nucleotide Homology (Accession Number)
A1-10-R8	30	*:**ireA**:IS256:IS3::*	*Escherichia coli* strain Ecol_AZ159, complete genome (CP019008.1)
A2-4-R2	28	*::::**eilA**:::IntI2*	No significant similarity found
B5-3-R8	44	*IS3 family transposase::**tsh**::IS3 family transposase:IS30-like element IS30 family transposase:IS3 family transposase*	*Escherichia coli* strain CVM N55972 plasmid pN55972-1, complete sequence (CP043759.1)
CR2-3	36	*ISAs1 family transposase:::**gadC***	*Escherichia coli* strain CVM N18EC0432 chromosome, complete genome (CP048290.1)
CS4-4	19	*::**gadC**::::IS200/IS605 family transposase*	No significant similarity found
H-1-1	29	*::**gadx**::**gadE**::::IS5-like element ISKpn26 family transposase::*	*Escherichia coli* strain HS30-1 chromosome, complete genome (CP029492.1)
H2(12)	109	*::::IS110 family transposase::ISL3 family transposase:IS30-like element IS30 family transposase::**tsh***	*Escherichia coli* strain 14EC001 chromosome, complete genome (CP024127.1)
T3-9-		*:**gad**:::ISAs1::*	No significant similarity found
TA3-1	66	*:**stb**:IS3 family transposase*	*Escherichia coli* strain RHB38-C13 plasmid pRHB38-C13_2, complete sequence (CP055625.1)
	69	*::IS1 family transposase:ISAs1 family transposase::::**gad**:*	*Escherichia coli* strain EC11 chromosome, complete genome (CP027255.1)
TB3-2		*:**gad**:::ISAs1-like element ISEc1 family transposase*	*Escherichia coli* strain LD27-1 chromosome, complete genome (CP047594.1)
TB3-10	660	*IS3 family transposase:**capU**:virK:IS30 family transposase*	*Escherichia coli* strain G4/9 chromosome, complete genome (CP060073.1)
W5-R4		***terC**::IS21 family transposase*	*Escherichia coli* PCN061, complete genome (CP006636.1)
WB1-1-R8	12	*::**gad**:::IS605 family transposase*	*Escherichia coli* strain 2 HS-C chromosome, complete genome (CP038180.1)

2.6. Multilocus Sequence Typing (MLST) and Phylogenomic Insights

In silico multilocus sequence typing identified substantial genetic diversity among the *E. coli* isolates, with representation across some distinct sequence types (STs) (Figure 3B). Farm-derived isolates included ST10, ST58, ST117, ST542, and ST4373, among others, with ST10 appearing most frequently (Table S1). Transport-stage isolates were characterized by unique sequence types such as ST877, ST336, ST3531, and ST10. Abattoir isolates showed continued diversity, comprising ST206, ST101, ST898, ST1109, and ST5876 (Figure 3B; Table S1). Whole-genome SNP-based phylogenetic analysis delineated eight major clades (Clades I–VIII), comprising isolates from this study and publicly available *E. coli* genomes from South Africa (Figure 6A). Isolates were clustered by sequence type rather than source, with study isolates co-localizing with human clinical and environmental strains. For instance, ST206 isolates formed a distinct clade (Clade XI), while ST10, ST1109, and ST101 were found in clusters containing both animal- and human-derived strains. Isolates from different provinces (KwaZulu-Natal, Gauteng, and Western Cape) were interspersed across clades, indicating no clear geographic partitioning. No clade-specific associations were observed between resistomic profiles and phylogenetic structure (Figure 6B). Resistance determinants appeared distributed across unrelated lineages, consistent with horizontal gene transfer mechanisms. The phylogenetic tree, resistome heatmap, and geographic distribution of clades are shown in Figure 6A–C.

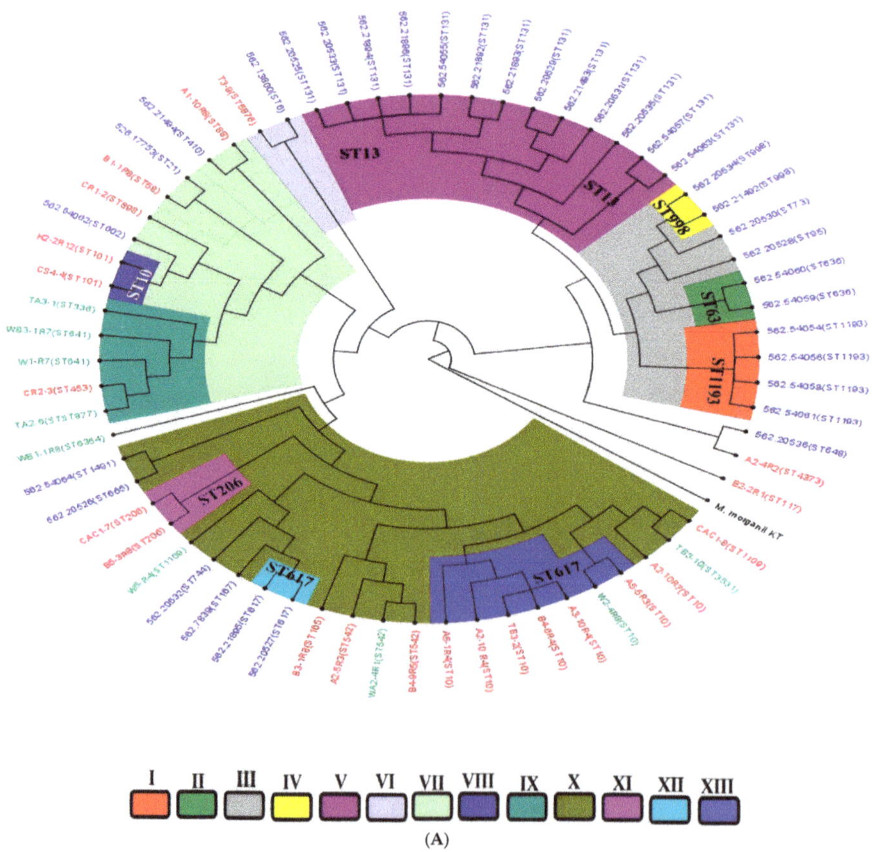

Figure 6. *Cont.*

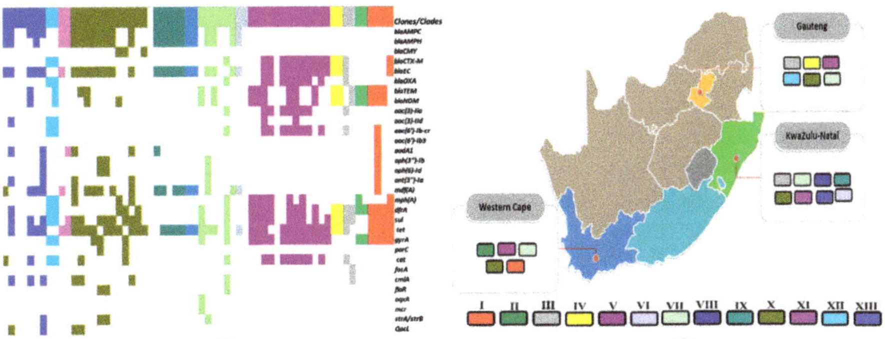

(B) (C)

Figure 6. (**A**) A phylogenetic tree depicting the relationships among *E. coli* isolates from this study alongside South African isolates from diverse One Health sources. The isolates are clustered by sequence types (STs), with tip color annotations indicating their origins: animal-source isolates (purple), human clinical isolates (blue), and environmental isolates (green). Clades are designated **I–XIII**, with distinct branch colors corresponding to each clade for ease of visualization. (**B**) A heatmap illustrating the distribution of antimicrobial resistance genes (resistome) across different clades (**I–XIII**). The color intensity represents the presence or absence of specific resistance genes within each isolate. (**C**) The geographic distribution of clades (**I–XIII**) across South African provinces, with colors matching those in the phylogenetic tree. Clade colors are consistent with those shown in the phylogenetic tree (**A**), facilitating comparison between genetic clustering and spatial distribution patterns.

3. Discussion

This study describes the genomic epidemiology of antibiotic-resistant *E. coli* isolated from an intensive pig production continuum in uMgungundlovu District, KwaZulu-Natal, South Africa, using the farm-to-fork approach (farm–transport–abattoir). Overall, the isolates showed high resistance against critically important antibiotics for humans and animals, threatening available antibiotic arsenals for treatment. The myriad of antibiotic resistance genes and mobile genetic support encoded confirms that pigs and pork could serve as potential reservoirs for antibiotic resistance. The high occurrence of virulence genes detected in the continuum also challenges food safety and is a potential risk to human health. The diversity of clones indicates the complexity of antibiotic-resistant *E. coli* within the continuum, highlighting the need for stricter sanitary conditions to prevent their transmission along the production chain. The use of antibiotics in pork production, particularly in animal feed for growth promotion and prophylaxis, has long been recognized as a major driver of antimicrobial resistance (AMR) in livestock systems [3,22]. Continuous exposure to sub-therapeutic antibiotic concentrations creates selective pressure that fosters the emergence and dissemination of resistant bacterial populations. This not only amplifies the local resistome but also facilitates horizontal gene transfer among commensal and pathogenic bacteria [23]. This study highlights the critical need for responsible antibiotic stewardship in animal husbandry and support the implementation of stricter regulations governing antibiotic use in livestock in South Africa. Such measures are essential to curb the selection of multidrug-resistant *E. coli* strains that can spread across the farm-to-fork continuum and threaten public health.

3.1. The Genetic Basis of Antibiotic Resistance and Their Mobile Genetic Support

Resistance in this study was related to several acquired antibiotic resistance genes, chromosomal mutations, and efflux genes. β-lactamases were detected in all the isolates from different sites and sources in the pork production continuum, highlighting the wide distribution of these resistant genes in healthy pigs and their products, which is associated with the high use of antimicrobials [24]. As it is normally expressed constitutively in *E. coli* [25], the plasmid-mediated *AmpC* (*bla*$_{\text{AmpC1}}$ and *bla*$_{\text{AmpC2}}$) was found in all the isolates (100%) in the current study. The hyperproduction of constitutively produced *AmpC* and plasmid-mediated *Amp*C β-lactamases that hydrolyze β-lactams antibiotics, including third- and fourth-generation cephalosporins, have been detected in pigs worldwide [26]. *AmpC*-producing bacteria were observed in clinical isolates initially, but then these bacteria were observed in companion animals as well as livestock, supporting the hypothesis that food-producing animals are the source and reservoir of these AmpC-producing bacteria [27]. The class C β-lactamase *bla*$_{\text{EC}}$ [11] were found in all the isolates. *bla*$_{EC\text{-}15}$ and *bla*$_{\text{EC-18}}$ have been reported previously in *E. coli* isolated from cattle feces in Canada [28].

ESBLs hydrolyze monobactams and broad-spectrum cephalosporins [29]. ESBL-positive *E. coli* isolated from pigs have been reported in China, Korea, and Portugal [30–32], with recent studies continuing to highlight their public health relevance. In the current study, *bla*$_{\text{TEM}}$ (*bla*$_{\text{TEM-1B}}$ and *bla*$_{\text{TEM-105}}$) was found in four isolates, three of which were farm-derived and exhibited resistance to all tested cephalosporins. These findings align with earlier reports of *bla*$_{\text{TEM}}$-harboring *E. coli* from pig farms in China, Korea, France, and more recently, from intensive farming systems in Africa, underscoring the ongoing threat posed by ESBL-producing strains in the pig production sector [31,33–35]. TEM-type derivatives, *bla*$_{\text{TEM-1B}}$ and *bla*$_{\text{TEM-2}}$, considered the most common β-lactamase found in *Enterobacterales of* both human and animal origin [31], are known to be contained within a *TniA* group of transposons or fragments from one of them [36]. In this study, all *bla*$_{\text{TEM}}$ genes were associated with *Tn3* transposase carried on plasmids (Table 3). The phenomenon of isolates carrying the same ESBL genes but showing different phenotypes, as seen in this study, has been previously documented [10]. Most ESBL-producing bacteria are multi-resistant, exhibiting additional resistance to non-cephalosporin antibiotics such as fluoroquinolones and aminoglycosides [37,38]. Therefore, the frequent use of broad-spectrum cephalosporins for prophylaxis or treatment can facilitate the emergence of MDR *E. coli.* In 2011, the Panel on Biological Hazards (BIOHAZ) highlighted the rapid emergence of ESBL/*Amp*C-producing *Enterobacterales* in food-producing animals as a major public health concern [39].

With the long-standing and extensive use of tetracycline in humans and animals, the tetracycline resistance genes, which can be horizontally transferred, have been intensively studied [10]. The *tet34* gene, which codes for an enzyme that inactivates tetracycline, predominated in this study, although to our knowledge, *tet34* has not been reported in pigs. *tetA* and *tetB* coding for energy-dependent efflux proteins, which help bacteria pump tetracycline out of the cell [10,40], were also detected. The *tetA* and *tetB* genes were adjacent to transcriptional repressor gene *tetR(A)* and *tetR(B)*, respectively. Except for one farm isolate (A2-5R3), which encoded *tetA* found on the chromosome and mediated by insertion sequence IS1, all the other *tetA* and *tetB* were conjugated with transposons and found in plasmids which may mediate their transfer [41]. More so, the presence of the *mdfA* gene belonging to the major facilitator superfamily (MFS), whose overexpression has been reported in multidrug-resistant *E. coli* strains isolated from patients [42], seems to contribute to the tetracycline resistance.

Since chloramphenicol is not widely used in veterinary medicine, some studies related their high resistance rate to the multi-resistant efflux system and/or co-selection with structurally unrelated compounds [43–45]. In this study, resistance to chloramphenicol was mainly due to gene *cmlA1* conjugated to different MGEs, such as insertion sequences (IS1, IS26, IS256), *Tn3* and class 1 integrons (In 456, In649, In774, In127), and co-carried with ARGs to aminoglycosides, macrolides, sulfonamides, trimethoprim, and tetracycline.

The *cmlA1* in this study was carried by plasmids, suggesting its possibility for horizontal transfer. The *catB* gene that encodes acetyltransferases that inactivate chloramphenicol was only detected in isolates from transport vehicles. The efflux genes *floR*, known to mediate resistance to chloramphenicol, were found on the farm only. The *floR* was associated with IS91 found in the chromosome with the closest nucleotide homology with *E. coli* strain AH25 chromosome, complete genome (CP055256.1). The presence of these three genes (*cmlA1, catB,* and *floR*) in pigs was previously reported in Australia and China [46,47]. The presence of these resistance genes in the plasmid facilitates their transmission. The co-selection of chloramphenicol resistance genes with other resistance genes belonging to other antibiotic categories may explain its high resistance rate even without selection pressure.

Previous pig studies have reported sulfamethoxazole *sul* and trimethoprim *dfrA* resistance genes [48,49]. Resistance to sulfonamides in *E. coli* frequently results from acquiring an alternative dihydropteroate synthase (DHPS) gene *sul*. Three alternative sulfonamide resistance DHPS genes [*sul1*, *sul2*, and *sul3*] in Gram-negative bacteria have been documented [50], and two genes (*sul2* and *sul3*) were reported in this study. Both *sul* and *dfrA* genes were found in co-occurrence with resistant genes belonging to other antibiotic classes. They were also found in association with different MGEs, suggesting their easy mobilization.

Genes responsible for aminoglycoside resistance were also found in this study. These genes are transmissible, encoded on conjugative plasmids, and often linked to resistance to other antimicrobials [51]. The resistance determinants (*aadA, strA, strB, aph(3″)-Ib, aph(6)-Id*) found in this study were also previously detected in *E. coli* isolated from pigs in the United States [52]. The *aadA1* and *aadA2* genes found in this study were often located in integrons (class1 and class2) carrying a gene cassette involving genes responsible for chloramphenicol and trimethoprim resistance (Table 2). The presence of integrons containing resistance gene cassettes contributes to the presence of MDR isolates [53].

Macrolides generally show modest potency against *Enterobacterales* [54]. Different macrolide resistance genes have been reported, causing resistance through various pathways. This can be through macrolide inactivation, phosphorylation (*mph A*, *mph B*), and transferable genes (*msr A, mef A or mef B*) encoding macrolide-efflux pumps [55]. The presence of the ethidium multidrug resistance protein E (*emr*E) gene which is a member of the small multidrug resistance (SMR) efflux protein family [56] in isolates phenotypically resistant to azithromycin may suggest the contribution of the gene to azithromycin resistance [57].

E. coli resistance to fluoroquinolones is primarily mediated by a specific mutation in the quinolone resistance-determining region (QRDR) within subunits constituting topoisomerase II (*gyrA* and *gyrB*) and IV (*par*C and *parE*), which are involved in DNA replication. These mutations are known to reduce the binding efficiency of fluoroquinolone drugs, thereby leading to resistance to these drugs [10]. This study found mutations in the six isolates showing phenotypic resistance to fluoroquinolones (Table 1). Mutations in *gyrA* and *parC* from *E. coli* isolated from pigs have been previously reported [57–59]. These mutations' presence enables bacteria to acquire more plasmids, which may contribute to acquiring other resistance genes. Only one farm isolate (B2-2-R1) showed phenotypic tigecycline resistance. The presence of chromosomal mutations in the genes (*acrA*, *acrR*, and *marB*) and the presence of *tet*A and *tol*C genes found in this isolate, which mediated resistance to the tigecycline, has been reported previously [60,61]. The persistence of plasmid replicons such as Col(MG828), IncFIB(AP001918), IncFII, and IncX1 across different stages of the pig production chain points to the potential role of plasmids in the dissemination of resistance genes across the continuum.

This study demonstrated a generally strong correlation between phenotypic resistance profiles and the presence of corresponding antimicrobial resistance genes (ARGs) detected through whole-genome sequencing; however, notable phenotypic–genotypic discrepancies were also observed in some isolates. These inconsistencies may be attributed to several factors, including the presence of novel or uncharacterized resistance mechanisms not captured by current databases, gene silencing or low expression levels that do not manifest phenotypically, and potential sequencing or assembly limitations that could lead to the partial or incomplete detection of ARGs [10,62]. Additionally, the phenotypic expression of resistance may be influenced by regulatory mutations, efflux pumps, or synergistic effects among multiple resistance determinants. These findings underscore the complexity of AMR expression and highlight the need for continuous updates to susceptibility testing guidelines, resistance gene databases, and improvements in bioinformatic tools to enhance genotype–phenotype correlation in AMR surveillance.

3.2. Virulome Analysis

The virulome landscape of *E. coli* isolates recovered from the pork production system revealed a concerning convergence of diversity, functional complexity, and ecological adaptation. Over three-quarters of the isolates (77.4%) encoded at least one virulence determinant, and nearly two-thirds (64.5%) harbored multiple functional genes, underscoring the latent pathogenic potential of asymptomatically carried strains within livestock systems. This polygenic virulence architecture mirrors recent findings from Australia, Mexico, and sub-Saharan Africa, where livestock-associated *E. coli* exhibits similar profiles [62,63], highlighting the global relevance of these silent zoonotic reservoirs. Functionally, virulence genes were stratified across six primary categories: adhesion (*lpfA*, *eilA*, *tsh*, *stx2A/B*), toxin production (*astA*, *vat*, *capU*), immune evasion (*iss*, *gad*, *katP*), iron acquisition (*iroN*, *ireA*), microcins (*mchB*, *mchC*, *mchF*, *mcmA*), and secretion systems (*aaiC*). The dominance of *iss*, *astA*, and *lpfA* across interfaces points to enhanced mucosal colonization and immune evasion potential hallmarks of extraintestinal pathogenic *E. coli* (ExPEC) [5]. Strikingly, a single isolate from the abattoir carried *stx2*, a defining toxin of Shiga toxin-producing *E. coli* (STEC), marking a rare but consequential event with implications for hybrid pathotype emergence at the human–animal interface.

What elevates the virulome risk landscape is not only the diversity of virulence traits but also their mobilization potential. Many genes, particularly *gad*, *tsh*, and *eilA*, were flanked by mobile genetic elements (MGEs), including insertion sequences (ISAs1-like, IS3 family), transposases, and class 1 integrons. The physical proximity of virulence loci to MGEs with some chromosomally integrated, others plasmid-borne suggests that horizontal transfer under selective pressure remains a viable evolutionary pathway [52]. This genetic plasticity aligns with the growing body of evidence that livestock ecosystems serve not only as reservoirs of AMR but also as incubators for virulence convergence. Mechanistically, the association of ExPEC-related genes with MGEs capable of mobilization raises the specter of genomic co-selection, wherein antimicrobial use may inadvertently drive the spread of virulence factors. This dual burden of virulence and resistance has been increasingly reported in clonal lineages circulating in both clinical and foodborne settings [63], reinforcing the interconnected nature of One Health risks. While the detection of *stx2* was limited to a single isolate, its presence, even in the absence of disease, reaffirms the limitations of symptom-based risk assessment in food systems. From a One Health perspective, these results underscore the silent carriage of potentially pathogenic *E. coli* within the pork production continuum. The fact that many of these virulence signatures were identified in pre-slaughter stages emphasizes the need for upstream surveillance.

Current food safety paradigms anchored in overt clinical signs may underestimate the risk posed by asymptomatic carriers of highly functional virulence machinery.

Taken together, our findings highlight the value of virulome profiling as an essential complement to AMR surveillance. The integration of functional genomics into risk assessment frameworks offers a more nuanced understanding of pathogen emergence and dissemination. As livestock systems intensify, genomic surveillance must evolve from reactive to anticipatory, enabling the early detection of high-risk strains before they reach the food chain. Future policy should prioritize molecular monitoring of virulence traits alongside resistance determinants to inform risk-based inspections, guide mitigation efforts, and prevent the silent transmission of hybrid and high-consequence *E. coli* pathotypes.

3.3. Multilocus Sequence Typing and Phylogenomic Analysis

The sequence type (ST) diversity among *E. coli* isolates across the pork production continuum revealed both ecological segregation and genetic connectivity. While distinct STs were observed within each production interface (farm, transport, and abattoir, respectively), the recurrence of lineages such as ST10, ST206, and ST101 across multiple stages suggests persistence and potential intra-system dissemination. This overlap is indicative of longitudinal transmission pathways, possibly facilitated by shared handling practices, environmental persistence, or bacterial fitness traits that enable survival across varied niches. The predominance of ST10, a member of the clonal complex CC10, is particularly noteworthy. This lineage has been frequently reported in pig-associated *E. coli* isolates from Denmark, Australia, Ireland, and Portugal [32,59], and is recognized as the dominant sequence type in swine populations across several European contexts, including Germany, Denmark, Ireland, and Spain [64]. ST10 and its related clonal complex members exhibit multidrug-resistant (MDR) phenotypes and are associated with both intestinal and extraintestinal infections in animals and humans [65]. Its detection in this study reinforces CC10′s adaptive success and zoonotic relevance and further positions it as a critical lineage at the livestock–human interface.

Beyond clonal assignment, phylogenomic reconstruction of isolates from South Africa revealed eight supported clades encompassing isolates from humans, animals, and environmental samples. Importantly, several clades showed phylogenetic continuity across these source categories, suggesting the inter-compartmental circulation of certain *E. coli* lineages. This finding is consistent with recent global reports that document the erosion of ecological boundaries in *E. coli* populations, where the same clone circulates across hosts and environments via food, water, or contact transmission. Notably, several clinical isolates (n = 10) from Gauteng and the Western Cape clustered closely with isolates from this study, indicating likely spillover from different sources. Such convergence strongly implicates the food value chain in the broader *E. coli* transmission network and highlights the limitations of treating foodborne *E. coli* as taxonomically distinct from clinical counterparts.

These data provide genomic evidence of One Health connectivity, where livestock-associated MDR *E. coli* lineages, especially those within CC10, circulate between animals, humans, and the environment. This dynamic not only challenges siloed surveillance approaches but also reinforces the urgency for integrated genomic surveillance frameworks that operate across sectors. Importantly, the phylogenetic proximity of foodborne and clinical strains highlights the potential for livestock-derived *E. coli* to acquire or co-circulate with human-adapted virulence and resistance elements, accelerating the emergence of high-risk clones. Overall, MLST and high-resolution phylogenomics expose the hidden population structure of *E. coli* circulating in the pork production ecosystem. They also reveal a mosaic of shared lineages that blur the boundaries between food safety and clinical microbiology.

This study was limited by the small number of sequenced isolates, primarily due to resource constraints. Although the dataset provided meaningful insights into the genomic diversity of *E. coli* across the farm-to-fork continuum, the limited sample size may not fully reflect the broader population structure or capture the full extent of strain transmission dynamics within and between production stages. This underscores the need for increased investment in genomic surveillance capacity in low- and middle-income settings, where high-resolution data are critical for effective antimicrobial resistance (AMR) monitoring [19,20]. Additionally, it highlights the importance of designing future studies with larger and more representative sampling frameworks spanning diverse farms, production systems, and geographic regions to generate data that can better inform national AMR control strategies.

More so, to strengthen surveillance and policy planning, future research should incorporate the whole-genome sequencing of a broader set of isolates alongside detailed epidemiological metadata. This approach would enhance the ability to detect emerging high-risk clones, identify resistance hotspots, and trace the spread of resistant bacteria across the animal–human–environment interface. Such evidence is essential for shaping data-driven policies on antimicrobial stewardship in livestock production, improving biosecurity protocols at slaughterhouses, and informing cross-sectoral One Health interventions.

4. Materials and Methods

4.1. Ethical Statement

Ethical approval was received from the Animal Research Ethics Committee (Reference: AREC/007/018) and the Biomedical Research Ethics Committee (Reference: BCA444/16) of the University of KwaZulu-Natal. The study was further approved by the South African National Department of Agriculture, Forestry, and Fisheries (Reference: 12/11/1/5).

4.2. Study Site and Sample Collection

The study was a longitudinal study conducted in the uMgungundlovu District, KwaZulu Natal, South Africa, over eighteen weeks from September 2018 to January 2019. The main sampling site was an intensive pig farm. Samples were collected from different points along the pork production system of this farm, as stipulated by the World Health Organization Advisory Group on Integrated Surveillance of Antimicrobial Resistance (WHO-AGISAR) guidelines [66]. A total of 417 samples were collected across the farm-to-fork continuum, including farm (n = 144), transport (n = 60), and abattoir (n = 213) samples, as previously described [21].

4.3. E. coli Isolation, Confirmation, and Antibiotic Susceptibility Testing

A total of 1044 *E. coli* isolates were putatively identified during enumeration using the Colilert^{TM} 18 Quanti-Tray/2000 system (IDEXX Laboratories (Pty) Ltd., Johannesburg, South Africa), followed by phenotypic confirmation on eosin methylene blue (EMB) [67]. For *E. coli* confirmation, DNA was extracted from these isolates using the boiling method and confirmed by real-time polymerase chain reaction (PCR), targeting the *uidA* gene as previously described [21]. All reactions were performed on a Quant Studio 5 (Thermo Fischer Scientific, MA, USA). *E. coli* ATCC® 25,922 was used as a positive control, while the reaction mixture with no DNA (replaced with nuclease-free water) was used as a negative template control. Isolates confirmed as *E. coli* were tested for susceptibility to 20 antibiotics using the disk diffusion method following WHO-AGISAR recommendations and CLSI 2020 guidelines, as previously published [21].

4.4. Selection of Isolates, Whole-Genome Sequencing, and Bioinformatic Analysis

4.4.1. Isolate Selection

A total of thirty-one non-duplicate *Escherichia coli* isolates were selected from a larger pool of phenotypically multidrug-resistant (MDR) strains recovered during routine antimicrobial surveillance across the pork production continuum in uMgungundlovu District, KwaZulu-Natal Province, South Africa. Isolates were obtained from diverse sampling points, including pig feces, environmental slurry, water, cecal contents, carcass rinsates, transport truck surfaces, and retail meat cuts. The sampling design targeted critical control points along the farm-to-fork continuum: farm (n = 19), transport (n = 4), and abattoir (n = 8). Isolate selection was based on the following criteria: confirmed resistance to three or more classes of antibiotics, in accordance with CLSI definitions for multidrug resistance; phenotypic diversity based on distinct antibiogram profiles; representativeness across all three sampling interfaces; and the availability of high-quality genomic DNA suitable for whole-genome sequencing (WGS). Isolates with incomplete metadata, duplicate resistance profiles, or suboptimal DNA yield or quality were excluded from the final dataset.

4.4.2. Genome Sequencing and Pre-Processing of Sequence Data

Thirty-one multidrug-resistant (MDR) isolates with diverse resistance profiles from various points along the farm-to-fork continuum were selected for whole-genome sequencing (WGS) and further characterization (Table S1). Genomic DNA (gDNA) was extracted and purified using the Gene Elute Bacterial genomic DNA kit (Sigma Aldrich, St. Louis, MI, USA) following the manufacturer's instructions. Quantification and purification were undertaken using Nano Drop8000 (Thermo Scientific, Waltham, MA, USA) at 260/280 nm wavelength, with verification by agarose gel electrophoresis. The Nextera XT DNA preparation kit was used to prepare a pair-end library, and the Illumina MiSeq machine (Illumina, San Diego, CA, USA) was used for whole-genome sequencing. FASTQC (v0.12.0) was used to assess the quality of sequenced reads, followed by trimming using Sickle (v1.33; https://github.com/najoshi/sickle accessed on 15 November 2023). The high-quality reads were then assembled de novo using SKESA (v2.4.0; https://github.com/ncbi/SKESA accessed on 19 November 2023), a k-mer-based extension tool designed for accurate genome assembly [68].

Quality control thresholds were applied as follows: a minimum read depth of 30×, genome coverage $\geq$95%, and Phred quality score $\geq$ 30. Contigs shorter than 200 bp were excluded, and assemblies with total genome sizes below 4.4 Mb or above 6.4 Mb were discarded to avoid incomplete or potentially contaminated assemblies. The resultant contiguous sequences were submitted for gene annotation to GenBank using the NCBI Prokaryotic Genome Annotation Pipeline (PGAP).

4.4.3. Molecular Typing of *E. coli* Isolates

Multilocus sequence typing (MLST) was performed in silico using the WGS data online platform MLST v2.0 (https://cge.food.dtu.dk/services/MLST/ accessed on 20 March 2024) based on the allelic profiles of the seven housekeeping genes (*adk, fumC, gyrB, icd, mdh, purA,* and *recA*) *of E. coli,* using the Achman scheme [69].

4.5. *Resistome, Mobilome, and Genetic Support Analysis*

To determine the antibiotic-resistant genes in the selected isolates, the generated contigs from the WGS data were analyzed by three different platforms to confirm and overcome the shortfalls of each platform. These platforms were the NCBI AMR finder (https://www.ncbi.nlm.nih.gov/pathogens/antimicrobial-resistance/AMRFinder/ accessed on 15 March 2024) as a high-quality, curated resistance gene database [70], the Comprehensive Antibiotic Resistance Database (CARD) (https://card.mcmaster.ca/analyze/rgi accessed on 22 May 2024) [71], and ResFinder (http://genepi.food.dtu.dk/resfinder accessed on 17 May 2024) [72]. PHAge Search Tool Enhanced Release (PHASTER; https://phaster.ca accessed on 25 May 2024) server was used to identify with default parameters, annotate, and visualize prophage sequences [73]. Insertion sequences and transposase in the genomes were predicted by blasting contigs on the ISFinder database (https://www-is.biotoul.fr/index.php accessed on 7 June 2024) [74]. Integrons were predicted by PGAP, and RAST subsystems were blasted on the INTEGRALL database to find the actual integrons [75].

4.6. *Putative Virulome Analysis*

Virulence determinants associated with *E. coli* were identified with the Virulence Finder 2.0 (https://cge.food.dtu.dk/services/VirulenceFinder/ accessed on 18 May 2024) [72].

4.7. *Phylogenomic Analyses and Metadata Insights of the E. coli Isolates*

CSI Phylogeny v1.2 (https://cge.food.dtu.dk/services/CSIPhylogeny-1.2/ accessed on 10 July 2024) was used to identify, filter, and validate SNPs from the de novo assembled contigs, and to construct a phylogenetic tree based on concatenated SNP profiles using default settings [76]. The genome of *Morganella morganii* subsp. *morganii KT* (Accession number: CP004345.1) served as the outgroup to root the tree, enabling the easy configuration of the phylogenetic distance between the strains on the branches. The phylogeny was visualized with annotations for isolate demographics, WGS in silico typing (ST), resistome, and MGE metadata to provide a comprehensive analysis of the generated phylogenomic tree.

Additionally, whole-genome sequences of *E. coli* isolates from South Africa (n = 31), along with curated sequences from the PATRIC database (https://www.patricbrc.org/ accessed on 28 June 2024) (n = 34) containing MLST, source, and geographical location data, were downloaded and analyzed alongside this study's isolates. These sequences were used for whole-genome phylogeny analysis to assess the current epidemiological and evolutionary trends (Table S5). The Figtree v1.4.4 was used to edit and visualize the phylogenetic tree. Isolates of the same clade are highlighted in the same color, whilst those of the same countries have the same label (strain name) colors. The resistome of strains with a close phylogenetic relationship to this study's isolates was searched using NCBI's Pathogen Detection database (https://www.ncbi.nlm.nih.gov/pathogens/isolates#/search/ accessed on 30 June 2024).

4.8. *Nucleotide Sequence*

All nucleotide sequences of the *E. coli* strains from this study are available in the NCBI GenBank database under BioProject ID PRJNA596168.

5. Conclusions

The genomic analysis of *E. coli* isolates from an intensive pig production system in South Africa revealed substantial genetic diversity in antimicrobial resistance genes (ARGs), virulence factors, and mobile genetic elements (MGEs) across the farm-to-fork continuum. The widespread detection of plasmid-mediated β-lactamase genes (*bla*$_{EC}$, *bla*$_{TEM}$, *bla*$_{AmpC}$), tetracycline resistance determinants (*tetA*, *tetB*, *tet34*), and integrons underscores the complexity and persistence of antimicrobial resistance in livestock environments. Importantly, the detection of virulence genes and co-localization of ARGs on transmissible elements raises concerns about the potential for zoonotic transfer and foodborne dissemination.

This study highlights the need for routine AMR surveillance integrated into food safety monitoring, particularly at critical control points like abattoirs and transport systems. Policymakers should prioritize the regulation of non-therapeutic antibiotic use in animal husbandry and support farmer education and biosecurity interventions. Future research should expand genomic surveillance across multiple farms and regions, with larger datasets and metadata to trace high-risk clones and better understand AMR transmission dynamics within a One Health framework.

Supplementary Materials: The following supporting information can be downloaded at: https://www.mdpi.com/article/10.3390/antibiotics14050446/s1, Table S1: Population, specimen source, sample type, phenotypes and genotypic characteristics of the *E. coli* isolates; Table S2: Genomic characteristics of *E. coli* isolated along the farm-to-fork continuum; Table S3: Distribution of intact prophages among the *E. coli* isolates; Table S4: Distribution of virulence genes among *E. coli* isolates; Table S5: Metadata of *E. coli* sequences from South Africa (n = 34), downloaded and analyzed alongside this study's isolates for whole-genome phylogeny analysis.

Author Contributions: Conceptualization, S.E.A., L.A.B., A.L.K.A., D.G.A. and S.Y.E.; methodology, S.E.A. and D.G.A.; formal analysis, S.E.A., D.G.A., A.L.K.A. and M.A.; investigation, S.E.A.; resources, S.Y.E., L.A.B., D.G.A. and A.I.; writing—original draft preparation, S.E.A.; writing—review and editing, S.E.A., L.A.B., A.L.K.A., M.A., A.I., D.G.A. and S.Y.E.; supervision, S.Y.E., L.A.B., A.L.K.A. and D.G.A. All authors have read and agreed to the published version of the manuscript.

Funding: The research reported in this publication was funded by the WHO Advisory Group on Integrated Surveillance of Antimicrobial Resistance (AGISAR) Research Project: "Triangulation of Antibiotic Resistance from Humans, the Food Chain and Associated Environments—A One Health Project" (Reference ID: 204517), the South African Research Chairs Initiative of the Department of Science and Technology and National Research Foundation of South Africa (Grant No. 98342), the South African Medical Research Council (SAMRC) under a Self-Initiated Research Grant and UK Medical Research Council awarded to Professor Sabiha Essack, and the College of Health Sciences, University of Kwa-Zulu Natal, South Africa. The funding sources were not involved in the study design, sample collection, data analysis, and data interpretation.

Institutional Review Board Statement: Ethical approval for this study was obtained from the Animal Research Ethics Committee (AREC/007/018, 9 February 2018) and the Biomedical Research Ethics Committee (BCA444/16, 17 March 2017) of the University of KwaZulu-Natal. Additional approval was granted by the South African National Department of Agriculture, Forestry, and Fisheries (12/11/1/5, 17 September 2018).

Informed Consent Statement: Informed consent was obtained from the farm managers and owners who participated in the study.

Data Availability Statement: The data presented in this study are openly available in National Center for Biotechnology Information [NCBI] GenBank database in the Bio-project number PRJNA596168 [https://www.ncbi.nlm.nih.gov/bioproject/?term=PRJNA596168 accessed on 24 April 2025].

Acknowledgments: We would like to acknowledge the farm managers for their cooperation and the workers for participating in this study.

Conflicts of Interest: S.Y.E. is the chairperson of the Global Respiratory Infection Partnership and a member of the Global Hygiene Council, both sponsored by an unrestricted educational grant from Reckitt, UK. All other authors declare that they have no competing interests regarding the publication of this paper.

References

1. World Health Organization (WHO). *Estimates of the Global Burden of Foodborne Diseases: Foodborne Disease Burden Epidemiology Reference Group 2007–2015*; WHO: Geneva, Switzerland, 2015.
2. Heredia, N.; Garcia, S. Animals as sources of food-borne pathogens: A review. *Anim. Nutr.* **2018**, *4*, 250–255. [CrossRef] [PubMed]
3. Odey, T.O.J.; Tanimowo, W.O.; Afolabi, K.O.; Jahid, I.K.; Reuben, R.C. Antimicrobial use and resistance in food animal production: Food safety and associated concerns in Sub-Saharan Africa. *Int. Microbiol.* **2024**, *27*, 1–23. [CrossRef]
4. World Health Organization (WHO). *Global Antimicrobial Resistance and Use Surveillance System (GLASS) Report 2023*; WHO: Geneva, Switzerland, 2023.
5. Chai, Y.; Gu, X.; Sun, Y.; Li, J. Adenylate cyclase affects the virulence of extraintestinal pathogenic *Escherichia coli* derived from sheep lungs. *Kafkas Univ. Vet. Fak. Derg.* **2024**, *30*, 63–71. [CrossRef]
6. Tadesse, D.A.; Zhao, S.; Tong, E.; Ayers, S.; Singh, A.; Bartholomew, M.J.; McDermott, P.F. Antimicrobial drug resistance in *Escherichia coli* from humans and food animals, United States, 1950–2002. *Emerg. Infect. Dis.* **2012**, *18*, 741–749. [CrossRef] [PubMed]
7. Poirel, L.; Madec, J.Y.; Lupo, A.; Schink, A.K.; Kieffer, N.; Nordmann, P.; Schwarz, S. Antimicrobial resistance in *Escherichia coli*. *Microbiol. Spectr.* **2018**, *6*, 289–316. [CrossRef]
8. O'Neill, J. *Antimicrobial Resistance: Tackling a Crisis for the Health and Wealth of Nations. The Review on Antimicrobial Resistance*; HM Government and Wellcome Trust: London, UK, 2014.
9. World Health Organization (WHO). *Ten Threats to Global Health in 2019*; WHO: Geneva, Switzerland, 2020.
10. Guo, S.; Tay, M.Y.F.; Aung, K.T.; Seow, K.L.G.; Ng, L.C.; Purbojati, R.W.; Drautz-Moses, D.I.; Schuster, S.C.; Schlundt, J. Phenotypic and genotypic characterisation of antimicrobial-resistant *Escherichia coli* isolated from ready-to-eat food in Singapore. *Food Control* **2019**, *99*, 89–97. [CrossRef]
11. Sethuvel, D.P.M.; Perumalla, S.; Anandan, S.; Michael, J.S.; Ragupathi, N.K.D.; Gajendran, R.; Walia, K.; Veeraraghavan, B. Antimicrobial resistance, virulence, and plasmid profiles among clinical isolates of *Shigella* serogroups. *Indian J. Med. Res.* **2019**, *149*, 247–253. [CrossRef]
12. Mbelle, N.M.; Feldman, C.; Osei Sekyere, J.; Maningi, N.E.; Modipane, L.; Essack, S.Y. The resistome, mobilome, virulome, and phylogenomics of multidrug-resistant *Escherichia coli* clinical isolates from Pretoria, South Africa. *Sci. Rep.* **2019**, *9*, 16457. [CrossRef]
13. Van den Honert, M.S.; Gouws, P.A.; Hoffman, L.C. Importance and implications of antibiotic resistance development in livestock and wildlife farming in South Africa: A review. *S. Afr. J. Anim. Sci.* **2018**, *48*, 401–412. [CrossRef]
14. Founou, L.L.; Amoako, D.G.; Founou, R.C.; Essack, S.Y. Antibiotic Resistance in Food Animals in Africa: A Systematic Review and Meta-Analysis. *Microb. Drug Resist.* **2018**, *24*, 648–665. [CrossRef]
15. Adenipekun, E.O.; Jackson, C.R.; Oluwadun, A.; Iwalokun, B.A.; Frye, J.G.; Barrett, J.B.; Hiott, L.M.; Woodley, T.A. Prevalence and antimicrobial resistance in *Escherichia coli* from food animals in Lagos, Nigeria. *Microb. Drug Resist.* **2015**, *21*, 358–365. [CrossRef] [PubMed]
16. Founou, L.L.; Founou, R.C.; Essack, S.Y. Antibiotic Resistance in the Food Chain: A Developing Country Perspective. *Front. Microbiol.* **2016**, *7*, 1881. [CrossRef] [PubMed]
17. Nnah, E.P.; Asante, J.; Amoako, D.G.; Abia, A.L.K.; Essack, S.Y. Antibiotic-Resistant *Escherichia coli* at One Health Interfaces in Africa: A Scoping Review. *Sci. Total Environ.* **2025**, *958*, 177580. [CrossRef]
18. Africa Pathogen Genomics Initiative (Africa-PGI). *Scaling-Up Antimicrobial Resistance Genomic Surveillance in Africa*; Africa CDC: Addis Ababa, Ethiopia, 2024.
19. Baker, K.S.; Jauneikaite, E.; Hopkins, K.L.; Lo, S.W.; Sánchez-Busó, L.; Getino, M.; Howden, B.P.; Holt, K.E.; Musila, L.A.; Hendriksen, R.S.; et al. Genomics for public health and international surveillance of antimicrobial resistance. *Lancet Microbe* **2023**, *4*, e1047–e1055. [CrossRef]
20. Kajumbula, H.M.; Amoako, D.G.; Tessema, S.K.; Aworh, M.K.; Chikuse, F.; Okeke, I.N.; Okomo, U.; Jallow, S.; Egyir, B.; Kanzi, A.M.; et al. Enhancing clinical microbiology for genomic surveillance of antimicrobial resistance implementation in Africa. *Antimicrob. Resist. Infect. Control* **2024**, *13*, 135. [CrossRef]

21. Abdalla, S.E.; Abia, A.L.K.; Amoako, D.G.; Perrett, K.; Bester, L.A.; Essack, S.Y. From farm-to-fork: *E. coli* from an intensive pig production system in South Africa shows high resistance to critically important antibiotics for human and animal use. *Antibiotics* **2021**, *10*, 18. [CrossRef]
22. Kasimanickam, V.; Kasimanickam, M.; Kasimanickam, R. Antibiotics Use in Food Animal Production: Escalation of Antimicrobial Resistance: Where Are We Now in Combating AMR? *Med. Sci.* **2021**, *9*, 14. [CrossRef]
23. Endale, H.; Mathewos, M.; Abdeta, D. Potential Causes of Spread of Antimicrobial Resistance and Preventive Measures in One Health Perspective—A Review. *Infect. Drug Resist.* **2023**, *16*, 7515–7545. [CrossRef]
24. Dohmen, W.; Bonten, M.J.; Bos, M.E.; Van Marm, S.; Scharringa, J.; Wagenaar, J.A.; Heederik, D.J. Carriage of extended-spectrum beta-lactamases in pig farmers is associated with occurrence in pigs. *Clin. Microbiol. Infect.* **2015**, *21*, 917–923. [CrossRef]
25. Santiago, G.S.; Coelho, I.S.; Bronzato, G.F.; Moreira, A.B.; Gonçalves, D.; Alencar, T.A.; Ferreira, H.N.; Castro, B.G.; Souza, M.M.S.; Coelho, S.M.O. Extended-spectrum AmpC-producing *Escherichia coli* from milk and feces in dairy farms in Brazil. *J. Dairy Sci.* **2018**, *101*, 7808–7811. [CrossRef]
26. Von Salviati, C.; Laube, H.; Guerra, B.; Roesler, U.; Friese, A. Emission of ESBL/AmpC-producing *Escherichia coli* from pig fattening farms to surrounding areas. *Vet. Microbiol.* **2015**, *175*, 77–84. [CrossRef] [PubMed]
27. Ewers, C.; Bethe, A.; Semmler, T.; Guenther, S.; Wieler, L.H. Extended-spectrum beta-lactamase-producing and AmpC-producing *Escherichia coli* from livestock and companion animals. *Clin. Microbiol. Infect.* **2012**, *18*, 646–655. [CrossRef]
28. Adator, E.H.; Walker, M.; Narvaez-Bravo, C.; Zaheer, R.; Goji, N.; Cook, S.R.; Tymensen, L.; Hannon, S.J.; Church, D.; Booker, C.W.; et al. Whole-genome sequencing differentiates presumptive extended-spectrum β-lactamase-producing *Escherichia coli*. *Microorganisms* **2020**, *8*, 448. [CrossRef]
29. Song, J.; Oh, S.S.; Kim, J.; Park, S.; Shin, J. Clinically relevant extended-spectrum β-lactamase-producing *Escherichia coli* isolates from food animals in South Korea. *Front. Microbiol.* **2020**, *11*, 604. [CrossRef]
30. Tian, G.B.; Wang, H.N.; Zou, L.K.; Tang, J.N.; Zhao, Y.W.; Ye, M.Y.; Tang, J.Y.; Zhang, Y.; Zhang, A.Y.; Yang, X. Detection of CTX-M-15, CTX-M-22, and SHV-2 extended-spectrum β-lactamases (ESBLs) in *Escherichia coli* fecal-sample isolates from pig farms in China. *Foodborne Pathog. Dis.* **2009**, *6*, 297–304. [CrossRef]
31. Rayamajhi, N.; Kang, S.G.; Lee, D.Y.; Kang, M.L.; Lee, S.I.; Park, K.Y.; Lee, H.S.; Yoo, H.S. Characterisation of TEM-, SHV-, and AmpC-type β-lactamases from cephalosporin-resistant *Enterobacteriaceae* isolated from swine. *Int. J. Food Microbiol.* **2008**, *124*, 183–187. [CrossRef]
32. Ramos, S.; Silva, N.; Dias, D.; Sousa, M.; Capelo-Martinez, J.L.; Brito, F.; Canica, M.; Igrejas, G.; Poeta, P. Clonal diversity of ESBL-producing *Escherichia coli* in pigs at slaughter level in Portugal. *Foodborne Pathog. Dis.* **2013**, *10*, 74–79. [CrossRef]
33. Li, S.; Song, W.; Zhou, Y.; Tang, Y.; Gao, Y.; Miao, Z. Spread of extended-spectrum β-lactamase-producing *Escherichia coli* from a swine farm to the receiving river. *Environ. Sci. Pollut. Res.* **2015**, *22*, 13033–13047. [CrossRef]
34. Bibbal, D.; Dupouy, V.; Ferre, J.P.; Toutain, P.L.; Fayet, O.; Prere, M.F.; Bousquet-Melou, A. Impact of three ampicillin dosage regimens on selection of ampicillin resistance in *Enterobacteriaceae* and excretion of *blaTEM* genes in swine feces. *Appl. Environ. Microbiol.* **2007**, *73*, 4785–4790. [CrossRef]
35. Strasheim, W.; Lowe, M.; Smith, A.M.; Etter, E.M.C.; Perovic, O. Whole-Genome Sequencing of Human and Porcine *Escherichia coli* Isolates on a Commercial Pig Farm in South Africa. *Antibiotics* **2024**, *13*, 543. [CrossRef]
36. Bailey, J.K.; Pinyon, J.L.; Anantham, S.; Hall, R.M. Distribution of the *blaTEM* gene and *blaTEM*-containing transposons in commensal *Escherichia coli*. *J. Antimicrob. Chemother.* **2011**, *66*, 745–751. [CrossRef] [PubMed]
37. Skaradzinska, A.; Sliwka, P.; Kuzminska-Bajor, M.; Skaradzinski, G.; Rzasa, A.; Friese, A.; Roschanski, N.; Murugaiyan, J.; Roesler, U.H. The efficacy of isolated bacteriophages from pig farms against ESBL/AmpC-producing *Escherichia coli* from pig and turkey farms. *Front. Microbiol.* **2017**, *8*, 530. [CrossRef] [PubMed]
38. Husna, A.; Rahman, M.M.; Badruzzaman, A.T.M.; Sikder, M.H.; Islam, M.R.; Rahman, M.T.; Alam, J.; Ashour, H.M. Extended-Spectrum β-Lactamases (ESBL): Challenges and Opportunities. *Biomedicines* **2023**, *11*, 2937. [CrossRef] [PubMed]
39. EFSA Panel on Biological Hazards (BIOHAZ). Scientific opinion on the public health risks of bacterial strains producing extended-spectrum β-lactamases and/or AmpC β-lactamases in food and food-producing animals. *EFSA J.* **2011**, *9*, 2322. [CrossRef]
40. Kallau, N.H.G.; Wibawan, I.W.T.; Lukman, D.W.; Sudarwanto, M.B. Detection of multidrug-resistant (MDR) *Escherichia coli* and *tet* gene prevalence at a pig farm in Kupang, Indonesia. *J. Adv. Vet. Anim. Res.* **2018**, *5*, 388–396. [CrossRef]
41. Mbanga, J.; Amoako, D.G.; Abia, A.L.K.; Allam, M.; Ismail, A.; Essack, S.Y. Genomic insights of multidrug-resistant *Escherichia coli* from wastewater sources and their association with clinical pathogens in South Africa. *Front. Vet. Sci.* **2021**, *8*, 636715. [CrossRef]
42. Heng, J.; Zhao, Y.; Liu, M.; Liu, Y.; Fan, J.; Wang, X.; Zhao, Y.; Zhang, X.C. Substrate-bound structure of the *E. coli* multidrug resistance transporter MdfA. *Cell Res.* **2015**, *25*, 1060–1073. [CrossRef]

43. De Jong, A.; Thomas, V.; Simjee, S.; Godinho, K.; Schiessl, B.; Klein, U.; Butty, P.; Valle, M.; Marion, H.; Shryock, T.R. Pan-European monitoring of susceptibility to human-use antimicrobial agents in enteric bacteria isolated from healthy food-producing animals. *J. Antimicrob. Chemother.* **2012**, *67*, 638–651. [CrossRef]
44. Ramos, S.; Silva, N.; Canica, M.; Capelo-Martinez, J.L.; Brito, F.; Igrejas, G.; Poeta, P. High prevalence of antimicrobial-resistant *Escherichia coli* from animals at slaughter: A food safety risk. *J. Sci. Food Agric.* **2013**, *93*, 517–526. [CrossRef]
45. Osterberg, J.; Wingstrand, A.; Nygaard Jensen, A.; Kerouanton, A.; Cibin, V.; Barco, L.; Denis, M.; Aabo, S.; Bengtsson, B. Antibiotic resistance in *Escherichia coli* from pigs in organic and conventional farming in four European countries. *PLoS ONE* **2016**, *11*, e0157049. [CrossRef]
46. Smith, M.G.; Jordan, D.; Gibson, J.S.; Cobbold, R.N.; Chapman, T.A.; Abraham, S.; Trott, D.J. Phenotypic and genotypic profiling of antimicrobial resistance in enteric *Escherichia coli* communities isolated from finisher pigs in Australia. *Aust. Vet. J.* **2016**, *94*, 371–376. [CrossRef] [PubMed]
47. Wang, X.M.; Li, S.G.; Liu, W.J.; Zhang, H.X.; Zhang, W.-J.; Jiang, H.-X.; Zhang, M.-J.; Zhu, H.-Q.; Sun, Y.; Sun, J.; et al. Serotypes, virulence genes, and antimicrobial susceptibility of *Escherichia coli* isolates from pig. *Foodborne Pathog. Dis.* **2011**, *8*, 687–692. [CrossRef] [PubMed]
48. Wu, S.; Dalsgaard, A.; Hammerum, A.M.; Porsbo, L.J.; Jensen, L.B. Prevalence and characterisation of plasmids carrying sulfonamide resistance genes among *Escherichia coli* from pigs, pig carcasses, and humans. *Acta Vet. Scand.* **2010**, *52*, 47. [CrossRef] [PubMed]
49. Mazurek, J.; Bok, E.; Stosik, M.; Baldy-Chudzik, K. Antimicrobial resistance in commensal *Escherichia coli* from pigs during metaphylactic trimethoprim and sulfamethoxazole treatment and in the post-exposure period. *Int. J. Environ. Res. Public Health* **2015**, *12*, 2150–2163. [CrossRef]
50. Hu, J.; Shi, J.; Chang, H.; Li, D.; Yang, M.; Kamagata, Y. Phenotyping and genotyping of antibiotic-resistant *Escherichia coli* isolated from a natural river basin. *Environ. Sci. Technol.* **2008**, *42*, 3415–3420. [CrossRef]
51. Pholwat, S.; Pongpan, T.; Chinli, R.; Rogawski McQuade, E.T.; Thaipisuttikul, I.; Ratanakorn, P.; Liu, J.; Taniuchi, M.; Houpt, E.R.; Foongladda, S. Antimicrobial resistance in swine fecal specimens across different farm management systems. *Front. Microbiol.* **2020**, *11*, 1238. [CrossRef]
52. Frye, J.G.; Lindsey, R.L.; Meinersmann, R.J.; Berrang, M.E.; Jackson, C.R.; Englen, M.D.; Turpin, J.B.; Fedorka-Cray, P.J. Related antimicrobial resistance genes detected in different bacterial species co-isolated from swine fecal samples. *Foodborne Pathog. Dis.* **2011**, *8*, 663–679. [CrossRef]
53. Bhat, B.A.; Mir, R.A.; Qadri, H.; Dhiman, R.; Almilaibary, A.; Alkhanani, M.; Mir, M.A. Integrons in the Development of Antimicrobial Resistance: Critical Review and Perspectives. *Front. Microbiol.* **2023**, *14*, 1231938. [CrossRef]
54. Nguyen, P.M.; Woerther, P.L.; Bouvet, M.; Andremont, A.; Leclercq, R.; Canu, A. *Escherichia coli* as a reservoir for macrolide resistance genes. *Emerg. Infect. Dis.* **2009**, *15*, 1648–1650. [CrossRef]
55. Gomes, C.; Ruiz-Roldan, L.; Mateu, J.; Ochoa, T.J.; Ruiz, J. Azithromycin resistance levels and mechanisms in *Escherichia coli*. *Sci. Rep.* **2019**, *9*, 6089. [CrossRef]
56. Beketskaia, M.S.; Bay, D.C.; Turner, R.J. Outer membrane protein OmpW participates with small multidrug resistance protein member EmrE in quaternary cationic compound efflux. *J. Bacteriol.* **2014**, *196*, 1908–1914. [CrossRef] [PubMed]
57. Guerra, B.; Junker, E.; Schroeter, A.; Malorny, B.; Lehmann, S.; Helmuth, R. Phenotypic and genotypic characterisation of antimicrobial resistance in German *Escherichia coli* isolates from cattle, swine, and poultry. *J. Antimicrob. Chemother.* **2003**, *52*, 489–492. [CrossRef] [PubMed]
58. Abuoun, M.; O'Connor, H.M.; Stubberfield, E.J.; Nunez-Garcia, J.; Sayers, E.; Crook, D.W.; Smith, R.P.; Anjum, M.F. Characterizing antimicrobial-resistant *Escherichia coli* and associated risk factors in a cross-sectional study of pig farms in Great Britain. *Front. Microbiol.* **2020**, *11*, 861. [CrossRef] [PubMed]
59. Kidsley, A.K.; Abraham, S.; Bell, J.M.; O'Dea, M.; Laird, T.J.; Jordan, D.; Mitchell, P.; McDevitt, C.A.; Trott, D.J. Antimicrobial susceptibility of *Escherichia coli* and *Salmonella* spp. isolates from healthy pigs in Australia: Results of a pilot national survey. *Front. Microbiol.* **2018**, *9*, 1207. [CrossRef]
60. Osei Sekyere, J.; Govinden, U.; Bester, L.A.; Essack, S.Y. Colistin and tigecycline resistance in carbapenemase-producing Gram-negative bacteria: Emerging resistance mechanisms and detection methods. *J. Appl. Microbiol.* **2016**, *121*, 601–617. [CrossRef]
61. Yaghoudi, S.; Zekiy, A.O.; Krutov, M.; Gholami, M.; Kouhsari, E.; Sholeh, M.; Ghafouri, Z.; Maleki, F. Tigecycline antibacterial activity, clinical effectiveness, and mechanisms and epidemiology of resistance: Narrative review. *Eur. J. Clin. Microbiol. Infect. Dis.* **2021**, *40*, 1003–1022.
62. Yehouenou, C.L.; Bogaerts, B.; De Keersmaecker, S.C.J.; Roosens, N.H.C.; Marchal, K.; Tchiakpe, E.; Affolabi, D.; Simon, A.; Dossou, F.M.; Vanneste, K.; et al. Whole-Genome Sequencing-Based Antimicrobial Resistance Characterization and Phylogenomic

Investigation of 19 Multidrug-Resistant and Extended-Spectrum Beta-Lactamase-Positive *Escherichia coli* Strains Collected from Hospital Patients in Benin in 2019. *Front. Microbiol.* **2021**, *12*, 752883. [CrossRef]
63. Martinez-Vazquez, A.V.; Rivera-Sanchez, G.; Lira-Mendez, K.; Reyes-Lopez, M.A.; Bocanegra-Garcia, V. Prevalence, antimicrobial resistance, and virulence genes of *Escherichia coli* isolated from retail meat in Tamaulipas, Mexico. *J. Glob. Antimicrob. Resist.* **2018**, *14*, 266–272. [CrossRef]
64. Reid, C.J.; Wyrsch, E.R.; Roy Chowdhury, P.; Zingali, T.; Liu, M.; Darling, A.E.; Chapman, T.A.; Djordjevic, S.P. Porcine commensal *Escherichia coli*: A reservoir for class 1 integrons associated with IS26. *Microb. Genom.* **2017**, *3*, e000143. [CrossRef]
65. Brilhante, M.; Perreten, V.; Dona, V. Multidrug resistance and multivirulence plasmids in enterotoxigenic and hybrid Shiga toxin-producing/enterotoxigenic *Escherichia coli* isolated from diarrheic pigs in Switzerland. *Vet. J.* **2019**, *244*, 60–68. [CrossRef]
66. World Health Organization. *Integrated Surveillance of Antimicrobial Resistance in Foodborne Bacteria: Application of a One Health Approach: Guidance from the WHO Advisory Group on Integrated Surveillance of Antimicrobial Resistance (AGISAR)*; WHO: Geneva, Switzerland, 2017.
67. Sahibzada, W.A.; Sahibzadi, A.G.; Sana, F.; Tayba, K.; Adila, S.; Sahibzadi, S.G.; Umair, A. Detection of *Escherichia coli* and total microbial population in River Siran water of Pakistan using EMB and TPC agar. *Afr. J. Microbiol. Res.* **2018**, *12*, 908–912. [CrossRef]
68. Souvorov, A.; Agarwala, R.; Lipman, D.J. SKESA: Strategic k-mer extension for scrupulous assemblies. *Genome Biol.* **2018**, *19*, 153. [CrossRef] [PubMed]
69. Ahmed, S.; Olsen, J.E.; Herrero-Fresno, A. The genetic diversity of commensal *Escherichia coli* strains isolated from non-antimicrobial treated pigs varies according to age group. *PLoS ONE* **2017**, *12*, e0178623. [CrossRef] [PubMed]
70. Feldgarden, M.; Brover, V.; Haft, D.H.; Prasad, A.B.; Slotta, D.J.; Tolstoy, I.; Tyson, G.H.; Zhao, S.; Hsu, C.-H.; McDermott, P.F.; et al. Validating the AMRFinder tool and resistance gene database by using antimicrobial resistance genotype-phenotype correlations in a collection of NARMS isolates. *Antimicrob. Agents Chemother.* **2019**, *63*, e00483-19. [CrossRef]
71. McArthur, A.G.; Waglechner, N.; Nizam, F.; Yan, A.; Azad, M.A.; Baylay, A.J.; Bhullar, K.; Canova, M.J.; De Pascale, G.; Ejim, L.; et al. The comprehensive antibiotic resistance database. *Antimicrob. Agents Chemother.* **2013**, *57*, 3348–3357. [CrossRef]
72. Kleinheinz, K.A.; Joensen, K.G.; Larsen, M.V. Applying the ResFinder and VirulenceFinder web-services for easy identification of acquired antibiotic resistance and *E. coli* virulence genes in bacteriophage and prophage nucleotide sequences. *Bacteriophage* **2014**, *4*, e27943. [CrossRef]
73. Zhou, Y.; Liang, Y.; Lynch, K.H.; Dennis, J.J.; Wishart, D.S. PHAST: A fast phage search tool. *Nucleic Acids Res.* **2011**, *39*, W347–W352. [CrossRef]
74. Siguier, P.; Perochon, J.; Lestrade, L.; Mahillon, J.; Chandler, M. ISfinder: The reference centre for bacterial insertion sequences. *Nucleic Acids Res.* **2006**, *34*, D32–D36. [CrossRef]
75. Moura, A.; Soares, M.; Pereira, C.; Leitão, N.; Henriques, I.; Correia, A. INTEGRALL: A database and search engine for integrons, integrases, and gene cassettes. *Bioinformatics* **2009**, *25*, 1096–1098. [CrossRef]
76. Ahrenfeldt, J.; Skaarup, C.; Hasman, H.; Pedersen, A.G.; Aarestrup, F.M.; Lund, O. Bacterial whole genome-based phylogeny: Construction of a new benchmarking dataset and assessment of some existing methods. *BMC Genom.* **2017**, *18*, 19. [CrossRef]

antibiotics

MDPI

Article

Resistance of *Pseudomonas aeruginosa* to Antibiotics During Long-Term Persistence in Patients with Cystic Fibrosis

Natalia Belkova [1,*,†], Uliana Nemchenko [1,†], Elizaveta Klimenko [1], Nadezhda Smurova [1], Raisa Zugeeva [1], Marina Sukhoreva [2], Viacheslav Sinkov [1] and Evgenij Savilov [1]

1 Federal State Budgetary Scientific Institution 'Scientific Centre for Family Health and Human Reproduction Problems', Epidemiology and Microbiology Institute, 3, K. Marks Str., 664003 Irkutsk, Russia; umnemch@mail.ru (U.N.); klimenko.elizabet@gmail.com (E.K.); nadinasmurova@mail.ru (N.S.); raya.zugeeva@mail.ru (R.Z.); vsinkov@yandex.ru (V.S.); savilov47@gmail.com (E.S.)

2 Regional State Autonomous Healthcare Institution 'Ivano-Matreninskaya City Children's Clinical Hospital', 57, Sovetskaya Str., 664009 Irkutsk, Russia; imdkb-bak@mail.ru

* Correspondence: belkovanl@sbamsr.irk.ru; Tel.: +7-3952240352

† These authors contributed equally to this work.

Abstract: *Pseudomonas aeruginosa* is one of the leading causes of nosocomial respiratory tract infections, significantly affecting morbidity and mortality. It can persist in the lungs of patients with cystic fibrosis (CF) for extended periods because of its adaptive capacity. The main aim of this study was to determine the phenotypic and genotypic resistance to antibiotics of clinical isolates of *P. aeruginosa* that persist in patients with CF receiving long-term antimicrobial therapy. The study included nine strains of *P. aeruginosa* isolated from the sputum of patients with CF admitted to the hospital. Susceptibility to antibiotics was determined using the European Committee on Antimicrobial Susceptibility Testing (EUCAST) criteria. Whole-genome sequencing was performed for phylogeny, sequence typing, and to identify antibiotic-resistant genes. The study showed that during long-term persistence in the lungs of patients receiving antibacterial therapy, the restoration of susceptibility to antibiotics occurred in some cases. Multilocus sequence typing and phylogeny revealed six sequence types. Functional annotation identified 72 genes responsible for resistance to antibacterial and chemical substances, with either chromosomal or plasmid localisation.

Keywords: *Pseudomonas aeruginosa*; cystic fibrosis; antibiotic resistance; whole-genome sequencing; multilocus sequence typing

1. Introduction

The high morbidity associated with persistent multidrug-resistant bacterial infections is a global public health burden. Persistent infections limit therapeutic options, creating a clinical scenario in which antibiotics are ineffective in treating certain infections. This represents a significant mechanism contributing to widespread antibiotic resistance in infectious agents [1–3]. Considering antibiotics as one of the environmental stress factors to which infectious agents are exposed, resistance and persistence can be defined as two physiological systems that enable them to adapt to changing conditions [1]. A resistant cell exhibits genetically mediated survival in a stressful environment, whereas a persistent cell is genetically susceptible to antibiotics but survives by slowing its metabolism, resuming growth once the stress is relieved [3]. These persistent cells are largely responsible for the challenges in treating chronic and recurrent infections. Biofilms, which harbour the highest number of persistent cells, provide protective and barrier functions, enhancing tolerance to stress factors [2].

Cystic fibrosis (CF) is a hereditary disease affecting the glands that secrete external substances, manifesting in pathological conditions of the gastrointestinal tract and respiratory system. The increased viscosity of glandular secretions leads to a chronic inflammatory process in the lungs, exocrine pancreatic insufficiency, hepatobiliary pathology, and an abnormally high electrolyte content in sweat. The respiratory disease associated with CF is characterised by chronic infection and inflammation driven by neutrophils, resulting in progressive airway damage and early mortality.

Pseudomonas aeruginosa is one of the leading causes of nosocomial respiratory tract infections, significantly impacting morbidity and mortality and necessitating early detection and aggressive antibiotic treatment [4–6]. Experimental data indicate that *P. aeruginosa* is detected in the lower respiratory tract of children with CF under the age of 4 years with a frequency of 31%, while among patients over 18 years of age, the prevalence ranges from 60% to 80% of cases [7]. According to the Cystic Fibrosis Foundation Patient Registry, infection rates with multidrug-resistant *P. aeruginosa* are highest among older adolescents and adults with CF, reflecting the cumulative effects of antibiotic exposure [8].

Pseudomonas aeruginosa can persist in the lungs of patients with CF for extended periods because of its strong adaptive capacity [9]. In this regard, the stability and constancy of the microorganism population, as well as changes in the properties of the pathogen that may necessitate adjustments in treatment strategies, are of particular interest [10]. It has been shown that the variability of infectious agents in CF occurs through three mechanisms: phenotypic heterogeneity; microevolution, resulting from the acquisition of mobile genetic elements by the pathogen genome and variability in chromosomal genes; and changes in the genotype of the pathogen within the same species, or the emergence of a new strain or species [10]. Based on findings from various longitudinal studies, the key genetic adaptations of *P. aeruginosa* to the CF lung environment have been summarised, and the phenotypes of these CF-adapted *P. aeruginosa* that contribute to persistent infection have been described [9].

The primary aim of the present study was to determine the genetic variability of clinical isolates of *P. aeruginosa* that persist in patients with CF undergoing continuous antimicrobial therapy.

2. Results

2.1. Evaluation of the Susceptibility of Pseudomonas aeruginosa Strains to Antibiotics: Phenotypic Aspect

This study included strains of *P. aeruginosa* isolated from paediatric patients aged 8 to 12 years who were hospitalised for treatment of CF. All strains obtained from patients via sputum culture were analysed. All patients received a course of antibiotic therapy according to the recommended regimens both in hospital and after discharge.

Nine strains of *P. aeruginosa* were obtained (Table 1). In one patient (P1), strains were collected at the initiation of treatment and again after 5 months. From another patient (P2), four strains were obtained—at the start of treatment and after 2, 7, and 11 months while receiving antibiotic therapy. Three additional strains were obtained from different patients (P3, P4, and P5) during antibiotic therapy.

Strains isolated at different times from patients with CF exhibited variations in phenotypic properties. In the pair of strains obtained from P1, strain IMB25 was resistant to carbapenems (meropenem) and aminoglycosides (amikacin), but showed susceptibility with increased exposure to cephalosporins (ceftazidime), penicillins (piperacillin/tazobactam), and fluoroquinolones (ciprofloxacin). By contrast, strain IMB100, isolated 5 months later, was susceptible to almost all antibiotics except piperacillin/tazobactam.

Table 1. Summary of clinical strains of *Pseudomonas aeruginosa* and results of antibiotic susceptibility assessment. Susceptibility classifications: susceptible (S), indicated in yellow; susceptible with increased exposure (I), indicated in orange; and resistant (R), marked in red.

Indicators	IMB25	IMB100	IMB54	IMB82	IMB102	IMB105	IMB101	IMB103	IMB104
Strain description									
Date of isolation	20 May 2021	15 October 2021	6 August 2021	15 October 2021	29 March 2022	19 July 2022	29 March 2022	29 March 2022	19 July 2022
Patient	P1	P1	P2	P2	P2	P2	P3	P4	P5
Susceptibility to antibiotics									
Piperacillin/tazobactam	I	R	I	I	I	I	R	I	I
Ceftazidime	I	I	I	I	I	I	I	I	I
Meropenem	R	S	R	S	S	S	S	S	S
Ciprofloxacin	I	I	I	I	I	I	I	I	I
Amikacin	R	S	S	S	S	S	S	S	S

Strains isolated from P2 exhibited a similar spectrum of antibiotic susceptibility and were either susceptible or not susceptible with increased exposure to all antibiotics used, except for meropenem. The first strain, IMB54, was resistant to this antibiotic; however, the subsequent isolates—IMB82, IMB102, and IMB105—later obtained from this patient, were all susceptible to meropenem.

Strains isolated from the three other patients (P3, P4, and P5) while undergoing antibiotic therapy were also either susceptible or not susceptible with increased exposure to all antibiotics, except for piperacillin/tazobactam. Strain IMB101 was resistant to piperacillin/tazobactam, whereas strains IMB103 and IMB104 were susceptible with increased exposure.

Thus, it is worth noting that during long-term persistence in patients undergoing antibacterial therapy, a restoration of antibiotic susceptibility was observed in some cases. However, the analysis of phenotypic manifestations alone does not provide sufficient insight into the duration of *P. aeruginosa* persistence in patients with CF or the variability of its biological properties.

2.2. *Whole-Genome Sequencing of* Pseudomonas aeruginosa *Strains: Analysis of Draft Genomes and Genotyping*

The sequencing of *P. aeruginosa* genomes yielded between 13,477,896 and 50,189,120 reads, with a mean whole-genome coverage of 1328 $\pm$ 896-fold. The draft genomes consisted of 249.5 $\pm$ 13.72 contigs and contained between 5493 and 5935 coding DNA sequences (CDSs), with a mean of 5522.5 $\pm$ 25.48 (Table 2). The GC content in the genomes of all strains was 66%, except for strain IMB104, in which it was 65%. The number of rRNA genes was 12 in most strains, except for IMB101 and IMB104, where 11 rRNA genes were identified. The number of tRNA genes varied significantly, ranging from 64 to 72.

Based on multilocus sequence typing (MLST) (Table 2), six different sequence types were identified. Strains IMB25 and IMB100 (from P1) were assigned to sequence type ST1641, while strains IMB101 (P3), IMB103 (P4), and IMB104 (P5) were classified as ST554, ST379, and ST970, respectively. Strains obtained from P2 were assigned to different sequence types, deviating from the typical pattern of 'one patient—*P. aeruginosa* strains of a single sequence type'. During the first isolation, strain IMB54 was classified as ST532; however, in subsequent isolations, all three strains—IMB82, IMB102, and IMB105—were assigned to ST555.

Table 2. Brief description of the draft genomes of *Pseudomonas aeruginosa* strains: results of assembly, genome annotation, and MLST.

Indicators	**IMB25**	**IMB100**	**IMB54**	**IMB82**	**IMB102**	**IMB105**	**IMB101**	**IMB103**	**IMB104**
Patient	P1		P2				P3	P4	P5
				Genome assembly					
Number of reads	50,189,120	36,044,976	37,282,953	33,293,656	5,815,755	13,477,896	12,369,102	3,737,929	14,328,761
Number of scaffolds	235	258	241	264	322	274	121	278	487
N50	67,136	65,489	63,297	56,717	328,046	298,770	255,738	550,311	238,191
				Genome annotation					
GC, %	66	66	66	66	66	66	66	66	65
Number of CDSs	5539	5548	5493	5510	5874	5874	5895	5910	5935
Number of rRNA genes	12	12	12	12	12	12	11	12	11
Number of tRNA genes	72	72	70	72	69	69	64	66	64
Number of plasmids	3	3	2	2	0	0	2	0	1
				MLST					
ST	1641	1641	532	555	555	555	554	379	970
acsA	11	11	5	16	16	16	16	39,345	6
aroE	10	10	4	5	5	5	5	5	5
guaA	1	1	5	30	30	30	12	11	11
mutL	3	3	5	11	11	11	3	28	3
nuoD	27	27	5	3	3	3	2	4	4
ppsA	4	4	20	20	20	20	15	4	3

Phylogenetic analysis revealed that the studied genomes clustered with the reference genomes of the corresponding sequence types (Figure 1). Notably, at the time of analysis, there were no validated genomes of *P. aeruginosa* ST555 in the PubMLST database; therefore, in the phylogenetic tree, this sequence type is represented solely by the genomes obtained in the present study.

Significant differences in the plasmid profiles of the strains were identified (Table 2). The number of plasmids per strain varied from zero to three. The most striking case was observed in P2, with strains classified as ST555. In the genome of strain IMB82, obtained during the second isolation, two plasmids (AA531 and AE982) were identified; however, in subsequent isolations, plasmids were not detected in the genomes of IMB102 and IMB105. Both strains isolated from P1 carried three plasmids in their genomes, while strains from P3, P4, and P5 contained two, zero, and one plasmid, respectively. It is worth noting that with long-term persistence under antibacterial therapy, a reduction in the genome size of clinical strains of *P. aeruginosa* was observed, likely due to the elimination of plasmids.

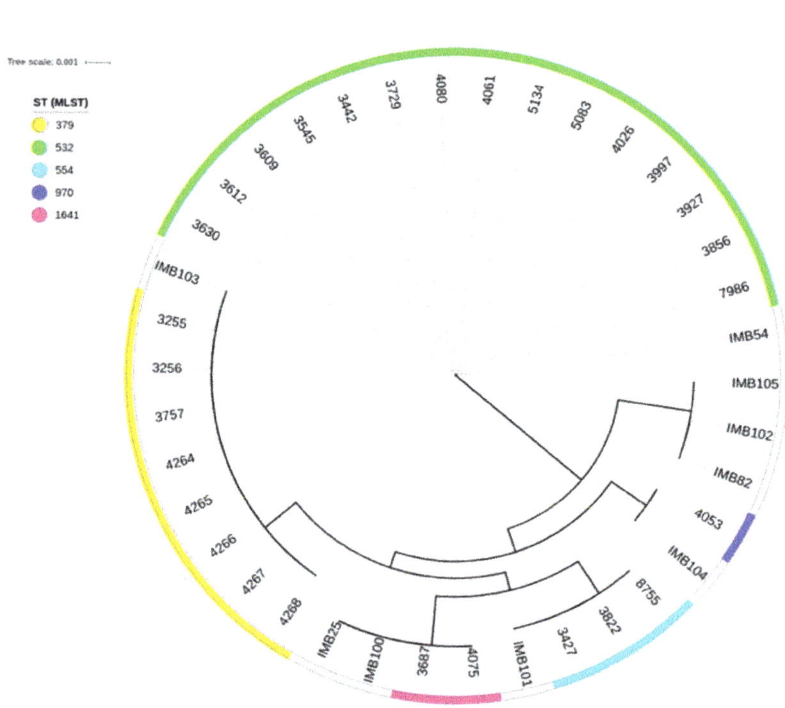

Figure 1. Minimum-spanning trees from allelic profiles obtained for the studied *Pseudomonas aeruginosa* strains and reference genomes of the corresponding sequence types. The labelling of the reference genomes in the tree corresponds to the IDs of these genomes in the PubMLST database.

2.3. Evaluation of Susceptibility of Pseudomonas aeruginosa Strains to Antibiotics: Functional Annotation of Genomes

Based on the results of functional annotation, 72 genes responsible for resistance to antimicrobial and chemical substances were identified. The following AMR gene families were detected in the genomes of all analysed strains (see Supplementary Table S1 for details):

- Aminoglycoside-modifying enzymes: APH(3′) (*aph(3′)-IIb*);
- Chloramphenicol resistance: chloramphenicol acetyltransferase (*catB7*);
- Fosfomycin resistance: fosfomycin thiol transferase (*fosA*);
- Beta-lactam resistance: *blaPAO*;
- Major facilitator superfamily antibiotic efflux pump: *bcr-1*;
- Small multidrug resistance antibiotic efflux pump: *emrE*;
- Pmr phosphoethanolamine transferase: *arnA*, *arnT*, *basS*, *cprR*, and *cprS*;
- Resistance–nodulation–cell division (RND) antibiotic efflux pump: *cpxR*, *mexABCDEFGHIJKLNPQRSTVWY*, *muxABC*, *nalCD*, *opmBDEH*, *oprJM*, *rsmA*, *triABC*, *Type A nfxB*, and *yajC*;
- Outer membrane porin: *parRS*;
- PDC beta-lactamase: *PDC-3*, *PDC-1*, and *PDC-59*;
- Multidrug and toxic compound extrusion transporter: *pmpM*;
- ATP-binding cassette antibiotic efflux pump;
- Major facilitator superfamily antibiotic efflux pump;
- RND antibiotic efflux pump: *soxR*;
- Glycopeptide resistance gene cluster: *vanW* gene in the *vanG* cluster.

Antibiotic resistance is achieved through various mechanisms targeting different classes of antibiotics and chemical compounds. For example, the inactivation of aminoglycoside antibiotics occurs through the involvement of the *aph(3′)-IIb* gene, the inactivation of phenicol antibiotics with the *catB7* gene, inactivation of phosphonic acid antibiotics with the *fosA* gene, and inactivation of monobactam, carbapenem, and cephalosporin antibiotics with the *PDC-3*, *PDC-1*, or *PDC-59* genes. In the genomes of clinical strains, the *arnAT* and *vanW* genes (in the *vanG* cluster) were identified, both of which contribute to target alteration—*arnAT* for peptide antibiotics and *vanW* for glycopeptide antibiotics.

Genes such as *basS*, *cprRS*, *mexR*, and *soxR* contribute not only to antibiotic target alteration, but also to antibiotic efflux. *parR* and *parS* are involved in both reduced antibiotic permeability and antibiotic efflux. Notably, the largest number of genes play a role in the efflux mechanism, which affects not only antibiotics, but also other chemicals, such as disinfectants and antiseptics. These genes include *mexGHIJKLMPQVWYZ*, *opmDEH*, *oprMN*, *parRS*, *pmpM*, *soxR*, *triABC*, and *yajC*.

The differences in genotypic antibiotic resistance among strains are summarised in Table 3. The main distinction was the identification of the *blaOXA* gene loci, with strains of different sequence types exhibiting distinct allelic variants of this locus. Sequence type (ST)-dependent allelic variants of the *PDC* and *armR* genes were also identified. All genes were chromosomally located. Additionally, the *adeF* gene was found in the chromosome of the IMB103 strain carrying ST379, whereas *crpP* was localised in the chromosomes of IMB54, IMB101, and IMB104, which carry ST532, ST554, and ST970, respectively.

Table 3. Differences in genotypic antibiotic resistance among clinical strains of *Pseudomonas aeruginosa*. The presence of chromosomal DNA (^{ch}) is indicated in yellow, plasmid DNA (^{pl}) in orange, and the absence of the gene (nd) in green.

Indicators	IMB25	IMB100	IMB54	IMB82	IMB102	IMB105	IMB101	IMB103	IMB104
Patient	P1		P2				P3	P4	P5
BlaOXA	396 *	396	906	486	486	486	494	904	50
PDC-3	$+^{ch}$	$+^{ch}$	nd	$+^{ch}$	$+^{ch}$	$+^{ch}$	$+^{ch}$	$+^{ch}$	nd
PDC-1	nd	nd	nd	nd	nd	nd	nd	nd	$+^{ch}$
PDC-59	nd	nd	$+^{ch}$	nd	nd	nd	nd	nd	nd
adeF	nd	nd	nd	nd	nd	nd	nd	$+^{ch}$	nd
crpP	nd	nd	$+^{ch}$	nd	nd	nd	$+^{ch}$	nd	$+^{ch}$
armR	nd	nd	nd	$+^{ch}$	$+^{ch}$	$+^{ch}$	nd	$+^{ch}$	$+^{ch}$
kpnF	nd	$+^{ch}$	nd	nd	nd	nd	nd	nd	nd
aph(6)-Id	$+^{pl}$	nd	nd	nd	nd	nd	nd	nd	nd
aph(3″)-Ib	$+^{pl}$	nd	nd	nd	nd	nd	nd	nd	nd
aac(6′)-Ib9	$+^{pl}$	nd	nd	nd	nd	nd	nd	nd	nd
armA	$+^{pl}$	nd	nd	nd	nd	nd	nd	nd	nd
dfrB4	$+^{pl}$	nd	nd	nd	nd	nd	nd	nd	nd
fosA8	$+^{pl}$	nd	nd	nd	nd	nd	nd	nd	nd
sul1	$+^{pl}$	nd	nd	nd	nd	nd	nd	nd	nd
qacEdelta1	$+^{pl}$	nd	nd	nd	nd	nd	nd	nd	nd
qacJ	nd	nd	nd	$+^{pl}$	nd	nd	nd	nd	nd

Note: *: beta-lactamase gene name. nd: genes were marked as 'not detected' when a search using three databases (ResFinder 4.6.0, RGI 6.0.3, and CARD 3.3.0) showed no results.

We note that the *armA* gene was identified in the plasmid of the IMB25 strain, which was obtained during the first isolation from P1. However, in the genome of the IMB100 strain, isolated from the same patient 5 months later, this gene was not detected—neither through search services using online databases nor via BLAST v2.16.0 analysis. A similarly thorough search was conducted to identify the *kpnF* gene, which encodes the *KpnF* subunit of the *KpnEF* protein in the *Klebsiella pneumoniae* genome. In the genome of the *P. aeruginosa* IMB100 strain, this gene was found to have chromosomal localisation. It appears that during persistence within the polymicrobial community formed in patients with CF undergoing long-term antibiotic therapy, horizontal gene transfer and genetic exchange occur, facilitating the development of stable resistance to AMP.

The genes *aph(6)-Id*, *aph(3″)-Ib*, *aac(6′)-Ib9*, *dfrB4*, *fosA8*, and *sul1* were identified in the IMB25 genome and had plasmid localisation. Notably, the genes *qacEdelta1* and *qacJ*, which confer resistance to antiseptic and disinfectant agents such as quaternary ammonium compounds (e.g., benzalkonium chloride and chlorhexidine) were also identified. These genes had plasmid localisation and were detected only in the strains isolated during the first and second sampling from P1 and P2—IMB25 and IMB82, respectively. The *qacEdelta1* and *qacJ* genes were localised in the AE788 and AE982 plasmids within the genomes of strains IMB25 and IMB82, respectively.

3. Discussion

In this study, we compared strains isolated at different times from five patients with CF who were receiving continuous antibacterial treatment. Bacterial cells employ various mechanisms to survive in adverse environmental conditions. With prolonged exposure to antibiotics, cells can not only develop resistance, but also utilise another mechanism: persistence. Unlike a resistant cell, a persistent cell is genetically susceptible to antibiotics but is phenotypically tolerant, exhibiting reduced metabolism and becoming metabolically inactive [1–3]. This phenotypic heterogeneity in chronically colonising *P. aeruginosa* populations is well documented and complicates direct comparisons of phenotypic properties between specific strains [4,11]. The phenomenon of persistence has been observed in both Gram-positive and Gram-negative bacteria under the influence of various classes of antibiotics [1]. Studies have shown that the frequency of persistent cell formation depends on the bacterial growth phase and is higher in the stationary phase, during which cells experience increased stress due to high population density, the reduced availability of nutrients, and the presence of additional stress factors such as active secondary metabolites [1]. The passive aspect of persistence should also be noted, as a bacterial cell in response to an antibiotic may exhibit no growth, remain unresponsive, and enter a dormant state without modifying or resisting the antibiotic. However, in the absence of an antibiotic, the persistent cell resumes growth and displays a phenotype susceptible to antibiotics. Experimental evidence has demonstrated that a persistent cell can reversibly shift between persistence and active metabolism depending on the environmental conditions, a phenomenon known as 'resuscitation' [12].

We demonstrated that strains isolated from a single patient exhibit different phenotypic antibiotic susceptibility properties. For example, strains IMB25 and IMB100, isolated from the same patient 5 months apart, differed in their resistance to penicillins (piperacillin/tazobactam), carbapenems (meropenem), and aminoglycosides (amikacin). Moreover, upon repeated isolation from the same patient, both the emergence of resistance to antibiotics such as piperacillin/tazobactam and the restoration of susceptibility to antibiotics such as meropenem and amikacin were observed. Additionally, the recovery of susceptibility to meropenem was noted for strains IMB54 and IMB82, though this may have been due to the different sequence types assigned to these strains. Clinical studies have shown that, in most cases, a single strain of *P. aeruginosa* is isolated from patients with chronic infection and persists over time [13–15]. However, even in individuals with CF who are chronically infected with only one strain of *P. aeruginosa*, significant genetic diversity among related colonising strains has been identified [9]. At the genetic level, these microorganisms undergo point mutations, insertions, and even large-scale deletions, leading to the development of distinct clonal lineages that compete within the complex environment of the lower respiratory tract in patients with CF [16]. As a result, clonal isolates of the same strain from individual patients exhibit phenotypic heterogeneity [17].

We found that strains isolated from a single patient may belong to either the same or different sequence types. The isolation of different sequence types of *P. aeruginosa* from one patient supports a previously proposed adaptation mechanism, in which complete eradication of the pathogen during antibiotic therapy may lead to its replacement by either another sequence type of the same pathogen or a different species altogether [9,10]. Shaginyan et al. analysed 1800 throat swabs and sputum samples from 300 children with CF over a 10-year period (2008–2017) and identified three main patterns of pathogen variability: population heterogeneity, pathogen microevolution, and replacement with another genotype of the same species [10]. The authors highlighted the epidemiological significance of molecular mechanisms driving pathogen variation, primarily due to the ability of strains to form epidemiologically significant clones [10]. Camus et al. reviewed various longitudinal studies that attempted to compare strains from early infections with those adapted to chronic infections [9]. Their findings provide an updated description of the key genetic adaptations of *P. aeruginosa* to the CF lung environment.

Our studies have shown that strains of *P. aeruginosa* isolated from patients with CF undergoing long-term antibiotic therapy, despite exhibiting different phenotypic properties, share similarities in genome structure and the presence of key genes responsible for resistance to antibacterial drugs and chemicals.

Of greatest interest, in our opinion, are the differences observed among the analysed strains in genes with chromosomal localisation, such as *adeF*, *crpP*, *armR*, and *kpnF*. It has been shown that the *adeF* gene is involved in resistance to fluoroquinolone and tetracycline antibiotics through the antibiotic efflux mechanism [18]. AdeF is a membrane fusion protein within the multidrug efflux complex AdeFGH. Overexpression of intrinsic efflux systems, such as RND efflux pumps, is a known mechanism contributing to multidrug resistance in opportunistic bacteria [19–23]. The RND efflux pumps AdeABC and AdeIJK were initially studied in *Acinetobacter baumannii* strains, which are nosocomial pathogens with multidrug resistance [19]. However, the application of next-generation sequencing methods has enabled the identification of homologous regions with both chromosomal and plasmid localisation in the complete genomes of various opportunistic microorganisms, including *P. aeruginosa*, as well as in whole-genome shotgun assemblies [18].

The *crpP* gene was found at a chromosomal localisation in the genomes of three *P. aeruginosa* types: strains with different sequence IMB54 (ST532), IMB101 (ST554), and IMB104 (ST970). This gene is involved in peptide antibiotic resistance through the mechanisms of antibiotic target alteration and antibiotic efflux [24]. There are at least three main mechanisms of quinolone resistance in bacteria: mutations in genes encoding quinolone-targeted proteins, changes in the expression of efflux pumps or porin channels, and the acquisition of quinolone resistance genes such as *crpP* [25]. The CrpP enzyme, capable of phosphorylating ciprofloxacin (CIP), was first described in 2018 [26]. It has been experimentally demonstrated that the pUM505 plasmid, isolated from a clinical strain of *P. aeruginosa* carrying the *crpP* gene, confers resistance to CIP when transferred to the *P. aeruginosa* PU21 strain. When the *P. aeruginosa* PAO1 strain was transformed with this plasmid, the transformant exhibited a fourfold increase in MIC values for CIP, norfloxacin, and moxifloxacin compared with the receptor strain *P. aeruginosa* PAO1 [26]. Whole-genome sequencing has revealed a high diversity of *crpP* homologues in the genomes of *P. aeruginosa* [27–33]. Notably, clinical strains of *P. aeruginosa* carrying the *crpP* gene have been isolated from various biotopes, including different parts of the respiratory tract [32,34,35]. Currently, homologous variants are being actively described, and data are being systematised for the typing and nomenclature of *crpP* variants in genomes, as well as in integrative and conjugative elements carrying *crpP* [31,33]. The typing of *crpP* variants, their geographic distribution, presence in mobile genetic elements, and potential ancestral origin are of great interest in epidemiological studies of antibiotic resistance.

The *armR* gene is part of the system regulating the formation of the MexR-ArmR complex and plays a role in controlling the expression of the *mexAB-oprM* operon—one of the four efflux systems of the RND family—contributing to multidrug resistance [23,36]. Experimental studies highlight the key role of genes in the NalC regulatory pathway (*nalC*, *armR*, and *mexR*), as well as *AmgRS*, in providing protection against aldehyde biocides, underscoring their importance in *P. aeruginosa* resistance [37,38]. Notably, the *armR* gene has been proposed as part of a panel of molecular markers for analysing *P. aeruginosa* clonal lineages to assess their virulence and antibiotic resistance [39,40]. The detection of markers associated with identifying high-risk clones during molecular surveillance necessitates additional precautions for infection control.

Equally noteworthy is the chromosomal localisation of the *kpnF* gene in the genome of the IMB100 strain, which was repeatedly isolated from P1, but was not detected via any of the tested methods in the genome of the IMB25 strain obtained during the first isolation. This suggests that horizontal gene transfer occurred within a heterogeneous microbial community during persistent infection in a patient with CF. The *kpnF* gene encodes the KpnF subunit of the KpnEF protein, which is homologous to the EbrAB protein of *Escherichia coli* and performs similar functions [41]. KpnEF belongs to the small multidrug resistance family of efflux pumps [41]. The direct involvement of *kpnEF* in capsule synthesis in mucoid strains of *Klebsiella pneumoniae* has been demonstrated, with mutations in KpnEF leading to increased susceptibility to antibiotics such as cefepime, ceftriaxone, colistin, erythromycin, rifampin, tetracycline, and streptomycin. Additionally, mutations resulted in heightened sensitivity to structurally related chemicals, such as sodium dodecyl sulphate, deoxycholate, dyes, benzalkonium chloride, chlorhexidine, and triclosan, as well as increased vulnerability to hyperosmotic conditions and high concentrations of bile [41].

The main limitation of our study was the small number of strains that we were able to analyse. In the year of the study, we received only nine isolates from patients with CF in a paediatric hospital. We will continue to select strains from patients with CF to identify genetic determinants that distinguish them from other clinically significant isolates of *P. aeruginosa*. Another important aspect of future studies could be the comparison of *P. aeruginosa* genomes that circulate in different hospitals with the same profile, or in the hospital from patients with different diseases.

4. Materials and Methods

4.1. Strain Collection and Identification

The study examined nine clinical strains of *P. aeruginosa* from the culture collection of the Laboratory of Microbiome and Microecology at the Institute of Epidemiology and Microbiology of the Scientific Centre for Family Health and Human Reproduction Problems [42]. The strains were obtained from patients with CF aged 8 to 12 years who were treated at the Ivano-Matreninskaya City Children's Clinical Hospital (Irkutsk, Russia).

The strains were isolated from the sputum of hospitalised patients with CF. Identification was performed using a bacteriological method that assessed the morphological, cultural, tinctorial, and biochemical properties of the isolates, with confirmation provided via MALDI-TOF direct protein profiling [43].

4.2. *Antibiotic Susceptibility Testing*

Antibiotic susceptibility was determined using standard methods based on the criteria of the European Committee on Antimicrobial Susceptibility Testing (EUCAST), version 11.0 (valid from 1 January 2021) and version 12.0 (valid from 1 January 2022) [44]. The *P. aeruginosa* type strain ATCC 27853 was used for quality control. The study included discs containing piperacillin–tazobactam (30–6 μg), ceftazidime (10 μg), meropenem (10 μg), amikacin (30 μg), and ciprofloxacin (5 μg) (HiMedia Laboratories Pvt. Limited, Maharashtra, India; Bio-Rad Laboratories, Hercules, CA, USA).

To assess antibiotic susceptibility, a bacterial suspension was prepared according to the standard method, with an optical density of 0.5 on the McFarland scale. Susceptibility was determined using the disc diffusion method on Mueller–Hinton medium (HiMedia Laboratories Pvt. Limited, Maharashtra, India). Strains were classified as resistant (R), susceptible with increased exposure (I), or susceptible (S) based on the defined criteria.

4.3. *Whole-Genome Sequencing and Bioinformatical Analyses*

The biomass of daily cultures was used for DNA extraction, as described previously [45]. Genomic DNA was isolated using the Quick-DNA Fungal/Bacterial Miniprep Kit (Zymo Research, Tustin, CA, USA). The quality and quantity of DNA were assessed using a Nano-500 spectrophotometer and a Qubit 4 fluorometer with a Qubit ds-DNA HS Assay Kit (Invitrogen, Carlsbad, CA, USA). A DNA concentration of >4 ng/μL was considered acceptable for the whole-genome sequencing of the sample.

Whole-genome sequencing of the strains was performed on Illumina NextSeq 550 equipment using the Illumina® DNA Prep Tagmentation, IDT® for Illumina® DNA/RNA UD Indexes Set Tagmentation, and NextSeq 500/550 High Output Kit v2.5 (300 Cycles) library preparation reagent kits, following the manufacturer's recommendations.

Genome assembly was performed using SPAdes v3.11.1 software [46]. Contig alignment and orientation correction were carried out using MAUVE v2.4.0 [47] and the *P. aeruginosa* PAO1 reference genome (GenBank AE 004091.2).

Strain typing based on genome sequences was performed using the PubMLST database [48], and antibiotic-resistant genes were identified using the ResFinder v4.6.0 database [49], RGI v6.0.3, and CARD v3.3.0 [50–52]. The localisation (plasmid or chromosomal) of antibiotic-resistant genes was determined using BLAST+ v2.16.0 [53]. Viral sequences were identified using the PHASTER tool [54].

Functional annotation was performed using Prokka v1.14.6 [55].

To construct the phylogeny, reference strains of *P. aeruginosa* sequence types ST379, ST532, ST554, ST970, and ST1641 from the PubMLST database, along with those obtained during this study, were used. The GrapeTree tool [56] and MLST gene sequences were used to build the tree. The nearest neighbour algorithm was applied to calculate the phylogeny.

Supplementary Materials: The following supporting information can be downloaded at: https://www.mdpi.com/article/10.3390/antibiotics14030302/s1. Table S1: Genes, responsible for the resistance to antibacterial and chemical substances detected in the genomes of clinic strains of Pseudomonas aeruginosa.

Author Contributions: Conceptualization, N.B.; methodology, U.N. and M.S.; software, E.K.; validation, N.B., U.N. and E.K.; formal analysis, N.S., R.Z. and V.S.; data curation, E.S.; writing—original draft preparation, N.B.; writing—review and editing, N.B.; visualisation, E.K. and N.S.; supervision, N.B.; project administration, E.S. All authors have read and agreed to the published version of the manuscript.

Funding: This research was conducted in accordance with medical technology №123051600027-1.

Institutional Review Board Statement: Not applicable.

Informed Consent Statement: Not applicable.

Data Availability Statement: The raw data have been deposited in the NCBI database as Bioproject PRJNA1026796.

Acknowledgments: The work was carried out using the equipment of the Centre for Collective Use 'Centre for the Development of Progressive Personalised Health Technologies' and the 'Human Microbiota Collection of the Irkutsk Region' of the Federal State Budgetary Scientific Institution Scientific Centre for Family Health and Human Reproduction Problems (Irkutsk).

Conflicts of Interest: The authors declare no conflicts of interest.

References

1. Jung, S.H.; Ryu, C.M.; Kim, J.S. Bacterial persistence: Fundamentals and clinical importance. *J. Microbiol.* **2019**, *57*, 829–835. [CrossRef] [PubMed]
2. Kunnath, A.P.; Suodha Suoodh, M.; Chellappan, D.K.; Chellian, J.; Palaniveloo, K. Bacterial Persister Cells and Development of Antibiotic Resistance in Chronic Infections: An Update. *Br. J. Biomed. Sci.* **2024**, *81*, 12958. [CrossRef] [PubMed]
3. Niu, H.; Gu, J.; Zhang, Y. Bacterial persisters: Molecular mechanisms and therapeutic development. *Signal Transduct. Target. Ther.* **2024**, *9*, 174. [CrossRef] [PubMed]
4. Parkins, M.D.; Somayaji, R.; Waters, V.J. Epidemiology, Biology, and Impact of Clonal *Pseudomonas aeruginosa* Infections in Cystic Fibrosis. *Clin. Microbiol. Rev.* **2018**, *31*, e00019-18. [CrossRef]
5. Chandler, C.E.; Hofstaedter, C.E.; Hazen, T.H.; Rasko, D.A.; Ernst, R.K. Genomic and Functional Characterization of Longitudinal *Pseudomonas aeruginosa* Isolates from Young Patients with Cystic Fibrosis. *Microbiol. Spectr.* **2023**, *11*, e0155623. [CrossRef]
6. Cohen-Cymberknoh, M. Infection and Inflammation in the Cystic fibrosis (CF) airway. *Pediatr. Pulmonol* 2024. *Early View*. [CrossRef]
7. Shaginyan, I.A.; Kapranov, N.I.; Chernukha, M.Y.; Alekseeva, G.V.; Semykin, S.Y.; Avetisyan, L.R.; Kashirskaya, N.Y.; Pivkina, N.V.; Danilina, G.A.; Batov, A.B.; et al. Mikrobnyy peyzazh nizhnikh dykhatel'nykh putey u razlichnykh vozrastnykh grupp detey, bol'nykh mukovistsidozom. *J. Microbiol. Epidemiol. Immunobiol.* **2010**, *87*, 15–20. (In Russian)
8. Cystic Fibrosis Foundation Patient Registry. *Annual Data Report*; Cystic Fibrosis Foundation Patient Registry: Bethesda, ML, USA, 2021.
9. Camus, L.; Vandenesch, F.; Moreau, K. From genotype to phenotype: Adaptations of *Pseudomonas aeruginosa* to the cystic fibrosis environment. *Microb. Genom.* **2021**, *7*, mgen000513. [CrossRef]
10. Shaginyan, I.A.; Avetisyan, L.R.; Chernukha, M.Y.; Siyanova, E.A.; Burmistrov, E.M.; Voronkova, A.Y.; Kondratieva, E.I.; Chuchalin, A.G.; Gintzburg, A.L. Epidemiological significance of genome variations in *Pseudomonas aeruginosa* causing chronic lung infection in patients with cystic fibrosis. *Clin. Microbiol. Antimicrob. Chemother.* **2019**, *21*, 340–351. (In Russian) [CrossRef]
11. Folkesson, A.; Jelsbak, L.; Yang, L.; Johansen, H.K.; Ciofu, O.; Hoiby, N.; Molin, S. Adaptation of *Pseudomonas aeruginosa* to the cystic fibrosis airway: An evolutionary perspective. *Nat. Rev. Microbiol.* **2012**, *10*, 841–851. [CrossRef]
12. Kim, J.S.; Yamasaki, R.; Song, S.; Zhang, W.; Wood, T.K. Single cell observations show persister cells wake based on ribosome content. *Environ. Microbiol.* **2018**, *20*, 2085–2098. [CrossRef] [PubMed]
13. Fothergill, J.L.; Mowat, E.; Ledson, M.J.; Walshaw, M.J.; Winstanley, C. Fluctuations in phenotypes and genotypes within populations of *Pseudomonas aeruginosa* in the cystic fibrosis lung during pulmonary exacerbations. *J. Med. Microbiol.* **2010**, *59*, 472–481. [CrossRef] [PubMed]
14. Fernández-Olmos, A.; García-Castillo, M.; Alba, J.M.; Morosini, M.I.; Lamas, A.; Romero, B.; Galán, J.C.; del Campo, R.; Cantón, R. Population structure and antimicrobial susceptibility of both non persistent and persistent *Pseudomonas aeruginosa* isolates recovered from cystic fibrosis patients. *J. Clin. Microbiol.* **2013**, *51*, 2761–2765. [CrossRef] [PubMed]

15. Kalferstova, L.; Vilimovska Dedeckova, K.; Antuskova, M.; Melter, O.; Drevinek, P. How and why to monitor *Pseudomonas aeruginosa* infections in the long term at a cystic fibrosis centre. *J. Hosp. Infect.* **2016**, *92*, 54–60. [CrossRef]
16. Cramer, N.; Klockgether, J.; Wrasman, K.; Schmidt, M.; Davenport, C.F.; Tümmler, B. Microevolution of the major common *Pseudomonas aeruginosa* clones C and PA14 in cystic fibrosis lungs. *Environ. Microbiol.* **2011**, *13*, 1690–1704. [CrossRef]
17. Iwańska, A.; Trafny, E.A.; Czopowicz, M.; Augustynowicz-Kopeć, E. Phenotypic and genotypic characteristics of *Pseudomonas aeruginosa* isolated from cystic fibrosis patients with chronic infections. *Sci. Rep.* **2023**, *13*, 11741. [CrossRef]
18. The Comprehensive Antibiotic Resistance Database. Available online: https://card.mcmaster.ca/aro/3000777#resistomes-table (accessed on 4 February 2025).
19. Coyne, S.; Rosenfeld, N.; Lambert, T.; Courvalin, P.; Périchon, B. Overexpression of resistance-nodulation-cell division pump AdeFGH confers multidrug resistance in *Acinetobacter baumannii*. *Antimicrob. Agents Chemother.* **2010**, *54*, 4389–4393. [CrossRef]
20. Puzari, M.; Chetia, P. RND efflux pump mediated antibiotic resistance in Gram-negative bacteria *Escherichia coli* and *Pseudomonas aeruginosa*: A major issue worldwide. *World J. Microbiol. Biotechnol.* **2017**, *33*, 24. [CrossRef]
21. Zwama, M.; Nishino, K. Ever-Adapting RND Efflux Pumps in Gram-Negative Multidrug-Resistant Pathogens: A Race against Time. *Antibiotics* **2021**, *10*, 774. [CrossRef]
22. Athar, M.; Gervasoni, S.; Catte, A.; Basciu, A.; Malloci, G.; Ruggerone, P.; Vargiu, A.V. Tripartite efflux pumps of the RND superfamily: What did we learn from computational studies? *Microbiology* **2023**, *169*, 001307. [CrossRef]
23. Kavanaugh, L.G.; Dey, D.; Shafer, W.M.; Conn, G.L. Structural and functional diversity of Resistance-Nodulation-Division (RND) efflux pump transporters with implications for antimicrobial resistance. *Microbiol. Mol. Biol. Rev.* **2024**, *88*, e0008923. [CrossRef] [PubMed]
24. The Comprehensive Antibiotic Resistance Database. Available online: https://card.mcmaster.ca/aro/3005063 (accessed on 4 February 2025).
25. Ruiz, J. Transferable Mechanisms of Quinolone Resistance from 1998 Onward. *Clin. Microbiol. Rev.* **2019**, *32*, e00007-19. [CrossRef] [PubMed]
26. Chávez-Jacobo, V.M.; Hernández-Ramírez, K.C.; Romo-Rodríguez, P.; Pérez-Gallardo, R.V.; Campos-García, J.; Gutiérrez-Corona, J.F.; García-Merinos, J.P.; Meza-Carmen, V.; Silva-Sánchez, J.; Ramírez-Díaz, M.I. CrpP Is a Novel Ciprofloxacin-Modifying Enzyme Encoded by the *Pseudomonas aeruginosa* pUM505 Plasmid. *Antimicrob. Agents Chemother.* **2018**, *62*, e02629-17. [CrossRef]
27. Ruiz, J. CrpP, a passenger or a hidden stowaway in the *Pseudomonas aeruginosa* genome? *J. Antimicrob. Chemother.* **2019**, *74*, 3397–3399. [CrossRef] [PubMed]
28. Khan, M.; Summers, S.; Rice, S.A.; Stapleton, F.; Willcox, M.D.P.; Subedi, D. Acquired fluoroquinolone resistance genes in corneal isolates of *Pseudomonas aeruginosa*. *Infect. Genet. Evol.* **2020**, *85*, 104574. [CrossRef]
29. Ortiz de la Rosa, J.M.; Nordmann, P.; Poirel, L. Pathogenicity Genomic Island-Associated CrpP-Like Fluoroquinolone-Modifying Enzymes among *Pseudomonas aeruginosa* Clinical Isolates in Europe. *Antimicrob. Agents Chemother.* **2020**, *64*, e00489-20. [CrossRef] [PubMed]
30. Xu, Y.; Zhang, Y.; Zheng, X.; Yu, K.; Sun, Y.; Liao, W.; Jia, H.; Xu, C.; Zhou, T.; Shen, M. The prevalence and functional characteristics of CrpP-like in *Pseudomonas aeruginosa* isolates from China. *Eur. J. Clin. Microbiol. Infect. Dis.* **2021**, *40*, 2651–2656. [CrossRef]
31. Zhu, Z.; Yang, H.; Yin, Z.; Jing, Y.; Zhao, Y.; Fu, H.; Du, H.; Zhou, D. Diversification and prevalence of the quinolone resistance *crpP* genes and the *crpP*-carrying Tn*6786*-related integrative and conjugative elements in *Pseudomonas aeruginosa*. *Virulence* **2021**, *12*, 2162–2170. [CrossRef]
32. Ruiz, J.; Ocampo, K.; Salvador-Luján, G.; Reyes, Y.V.; Gómez, A.C.; Valera-Krumdieck, C.; Baca-Cumpa, A.D.; Soza, G.; Pinto, J.A.; Ramos-Chirinos, M.; et al. First report of CrpP prevalence in a South American country. *New Microbes New Infect.* **2023**, *51*, 101082. [CrossRef]
33. Hu, Z.; Zhou, L.; Tao, X.; Li, P.; Zheng, X.; Zhang, W.; Tan, Z. Antimicrobial resistance survey and whole-genome analysis of nosocomial *P. aeruginosa* isolated from eastern Province of China in 2016-2021. *Ann. Clin. Microbiol. Antimicrob.* **2024**, *23*, 12. [CrossRef]
34. Hernández-García, M.; García-Castillo, M.; García-Fernández, S.; López-Mendoza, D.; Díaz-Regañón, J.; Romano, J.; Pássaro, L.; Paixão, L.; Cantón, R. Presence of Chromosomal *crpP-like* Genes Is Not Always Associated with Ciprofloxacin Resistance in *Pseudomonas aeruginosa* Clinical Isolates Recovered in ICU Patients from Portugal and Spain. *Microorganisms.* **2021**, *9*, 388. [CrossRef]
35. Bharadwaj, V.G.; Suvvari, T.K.; Kandi, V.; Chitra Rajalakshmi, P.; Dharsandia, M.V. Molecular Characterization of *Pseudomonas aeruginosa* Clinical Isolates Through Whole-Genome Sequencing: A Comprehensive Analysis of Biotypes, Sequence Types, and Antimicrobial and Virulence Genes. *Cureus* **2024**, *16*, e71118. [CrossRef] [PubMed]

36. Wu, W.; Huang, J.; Xu, Z. Antibiotic influx and efflux in *Pseudomonas aeruginosa*: Regulation and therapeutic implications. *Microb. Biotechnol.* **2024**, *17*, e14487. [CrossRef] [PubMed]
37. Tetard, A.; Zedet, A.; Girard, C.; Plésiat, P.; Llanes, C. Cinnamaldehyde induces expression of efflux pumps and multidrug resistance in *Pseudomonas aeruginosa*. *Antimicrob. Agents Chemother.* **2019**, *63*, e01081-19. [CrossRef]
38. Dubois, E.; Spasovski, V.; Plésiat, P.; Llanes, C. Role of the two-component system AmgRS in early resistance of *Pseudomonas aeruginosa* to cinnamaldehyde. *Microbiol. Spectr.* **2025**, *13*, e0169924. [CrossRef]
39. Nageeb, W.; Amin, D.H.; Mohammedsaleh, Z.M.; Makharita, R.R. Novel Molecular Markers Linked to *Pseudomonas aeruginosa* Epidemic High-Risk Clones. *Antibiotics* **2021**, *10*, 35. [CrossRef] [PubMed]
40. Nageeb, W.M.; Hetta, H.F. The predictive potential of different molecular markers linked to amikacin susceptibility phenotypes in *Pseudomonas aeruginosa*. *PLoS ONE* **2022**, *17*, e0267396. [CrossRef]
41. Srinivasan, V.B.; Rajamohan, G. KpnEF, a new member of the *Klebsiella pneumoniae* cell envelope stress response regulon, is an SMR-type efflux pump involved in broad-spectrum antimicrobial resistance. *Antimicrob. Agents Chemother.* **2013**, *57*, 4449–4462. [CrossRef]
42. Nemchenko, U.M.; Belkova, N.L.; Klimenko, E.S.; Smurova, N.E.; Savilov, E.D. Phenotypic and genetic variability of clinical isolates of Pseudomonas aeruginosa persisting in patients with cystic fibrosis. *Klin. Mikrobiol. Antimikrobn. Himioter.* **2024**, *26*, 42. (In Russian)
43. Voropaeva, N.M.; Belkova, N.L.; Nemchenko, U.M.; Grigorova, E.V.; Kungurtseva, E.A.; Noskova, O.A.; Chemezova, N.N.; Savilov, E.D. Identification of Infectious Diseases Patterns in the Combined Use of Bacteriological Diagnostics and MALDI Biotyper. *Acta Biomed. Sci.* **2020**, *5*, 88–94. (In Russian) [CrossRef]
44. The European Committee on Antimicrobial Susceptibility Testing—EUCAST. Available online: http://www.eucast.org (accessed on 24 January 2025).
45. Belkova, N.L.; Klimenko, E.S.; Nemchenko, U.M.; Grigorova, E.V.; Sitnikova, K.O.; Zugeeva, R.E.; Smurova, N.E.; Chemezova, N.N.; Savilov, E.D. Biological properties and genetic structure of clinic isolates of *Klebsiella pneumoniae* species. *Acta Biomed. Sci.* **2024**, *9*, 53–63. (In Russian) [CrossRef]
46. Bankevich, A.; Nurk, S.; Antipov, D.; Gurevich, A.A.; Dvorkin, M.; Kulikov, A.S.; Lesin, V.M.; Nikolenko, S.I.; Pham, S.; Prjibelski, A.D.; et al. SPAdes: A new genome assembly algorithm and its applications to single-cell sequencing. *J. Comput. Biol.* **2012**, *19*, 455–477. [CrossRef]
47. Rissman, A.I.; Mau, B.; Biehl, B.S.; Darling, A.E.; Glasner, J.D.; Perna, N.T. Reordering contigs of draft genomes using the Mauve aligner. *Bioinformatics* **2009**, *25*, 2071–2073. [CrossRef] [PubMed]
48. Jolley, K.A.; Bray, J.E.; Maiden, M.C.J. Open-access bacterial population genomics: BIGSdb software, the PubMLST.org website and their applications. *Wellcome Open Res.* **2018**, *3*, 124. [CrossRef]
49. Bortolaia, V.; Kaas, R.S.; Ruppe, E.; Roberts, M.C.; Schwarz, S.; Cattoir, V.; Philippon, A.; Allesoe, R.L.; Rebelo, A.R.; Florensa, A.F.; et al. ResFinder 4.0 for predictions of phenotypes from genotypes. *J. Antimicrob. Chemother.* **2020**, *75*, 3491–3500. [CrossRef]
50. The Comprehensive Antibiotic Resistance Database. Available online: https://card.mcmaster.ca/analyze/rgi (accessed on 24 January 2025).
51. McArthur, A.G.; Waglechner, N.; Nizam, F.; Yan, A.; Azad, M.A.; Baylay, A.J.; Bhullar, K.; Canova, M.J.; De Pascale, G.; Ejim, L.; et al. The comprehensive antibiotic resistance database. *Antimicrob. Agents Chemother.* **2013**, *57*, 3348–3357. [CrossRef]
52. Alcock, B.P.; Raphenya, A.R.; Lau, T.T.Y.; Tsang, K.K.; Bouchard, M.; Edalatmand, A.; Huynh, W.; Nguyen, A.V.; Cheng, A.A.; Liu, S.; et al. CARD 2020: Antibiotic resistome surveillance with the comprehensive antibiotic resistance database. *Nucleic Acids Res.* **2020**, *48*, D517–D525. [CrossRef] [PubMed]
53. Camacho, C.; Coulouris, G.; Avagyan, V.; Ma, N.; Papadopoulos, J.; Bealer, K.; Madden, T.L. BLAST+: Architecture and applications. *BMC Bioinform.* **2009**, *10*, 421. [CrossRef]
54. Arndt, D.; Grant, J.R.; Marcu, A.; Sajed, T.; Pon, A.; Liang, Y.; Wishart, D.S. PHASTER: A better, faster version of the PHAST phage search tool. *Nucleic Acids Res.* **2016**, *44*, W16–W21. [CrossRef]
55. Seemann, T. Prokka: Rapid prokaryotic genome annotation. *Bioinformatics.* **2014**, *30*, 2068–2069. [CrossRef]
56. Zhou, Z.; Alikhan, N.F.; Sergeant, M.J.; Luhmann, N.; Vaz, C.; Francisco, A.P.; Carriço, J.A.; Achtman, M. GrapeTree: Visualization of core genomic relationships among 100,000 bacterial pathogens. *Genome Res.* **2018**, *28*, 1395–1404. [CrossRef] [PubMed]

antibiotics

MDPI

Article

Determination of Antimicrobial Resistance Megaplasmid-Like pESI Structures Contributing to the Spread of *Salmonella* Schwarzengrund in Japan

Kanako Ishihara [1,*], Suzuka Someno [1], Kaoru Matsui [1], Chisato Nakazawa [1], Takahiro Abe [1], Hayato Harima [1], Tsutomu Omatsu [2], Manao Ozawa [3], Eriko Iwabuchi [4] and Tetsuo Asai [5]

1 Laboratory of Veterinary Public Health, Faculty of Agriculture, Tokyo University of Agriculture and Technology, 3-5-8 Saiwai-cho, Fuchu 183-8509, Tokyo, Japan; matsuika1341@gmail.com (K.M.); chisato.nakazawa.0218@gmail.com (C.N.); taka.vet.2024@gmail.com (T.A.)
2 Center for Infectious Disease Epidemiology and Prevention Research, Tokyo University of Agriculture and Technology, 3-5-8 Saiwai-cho, Fuchu 183-8509, Tokyo, Japan; tomatsu@cc.tuat.ac.jp
3 Assay Division I, National Veterinary Assay Laboratory, Ministry of Agriculture, Forestry, and Fisheries, 1-15-1 Tokura, Kokubunji 185-8511, Tokyo, Japan; manao_ozawa500@maff.go.jp
4 Department of Nutrition, School of Nursing and Nutrition, Tenshi College, Kita 13 Higashi 3, Higashi-ku, Sapporo 065-0013, Hokkaido, Japan; iwabuti@tenshi.ac.jp
5 United Graduate School of Veterinary Science, Gifu University, 1-1 Yanagido, Gifu-shi 501-1193, Gifu, Japan; asai.tetsuo.x3@f.gifu-u.ac.jp
* Correspondence: kanako-i@cc.tuat.ac.jp

Abstract: Background/Objectives: The acquisition of antimicrobial resistance by food-borne pathogens is a serious human health concern. In Japan, combinations of antimicrobial resistance genes in *Salmonella* from chicken meat were common among several serovars. Therefore, we hypothesized that different *S. enterica* serovars share a common antimicrobial resistance plasmid. Methods: Antimicrobial resistance transfer was tested in *S.* Infantis and *S.* Schwarzengrund, the major serovars used as donors. The plasmid structure was determined by subjecting *S.* Infantis Sal_238 and *S.* Schwarzengrund Sal_249 to short- and long-read sequencing. Results: The high homology between pSal_249Sch and pSal_238Inf suggests they have a common ancestor. Because the sequences of pSal_238Inf and pSal_249Sch were highly homologous to pESI (a plasmid for emerging *S.* Infantis), pSal_238Inf and pSal_249Sch were identified as pESI-like plasmids. *S.* Schwarzengrund is the third *Salmonella* serovar to expand its distribution related to pESI-like plasmid acquisition. Core-genome multilocus sequence-type analysis revealed that *S.* Schwarzengrund isolates with pESI-like plasmids from Japan (core-genome sequence-type [cgST] 167363 and cgST287831), the UK (cgST167363), and the USA (cgST167363, cgST196045, and cgST287831) were closely related; they are also suggested to share a common ancestor. The transfer of antimicrobial resistance was observed in combinations of both serovars. Specifically, the tentative plasmid sequence obtained via short-read sequencing, PCR, and conjugation experiments identified deletions of antimicrobial resistance genes (*aadA*, *sul1*, and *tetA*), class 1 integron, mercury resistance operon, and/or plasmid transfer region in the pESI-like plasmid. Conclusion: These data on the structural diversity of pESI-like plasmids suggest that some time has passed since *S.* Schwarzengrund acquired them.

Keywords: *Salmonella* Schwarzengrund; antimicrobial resistance; pESI; cgMLST

1. Introduction

Salmonella spp. is a common cause of diarrheal diseases globally. Most diarrhea cases occur due to the consumption of contaminated food of animal origin [1]. Over 2500 *Salmonella* serovars have been identified to date [2]. Most serovars are found in various hosts. However, almost all serovars cause diseases in humans. *Salmonella* typically causes gastroenteritis, which often does not cause complications. Antimicrobial therapy is needed for some high-risk groups, such as infants, the elderly, and immunocompromised patients, but not for mild or moderate cases in healthy individuals [3].

In recent years, the prevalence of *Salmonella enterica* serovar Infantis (*S.* Infantis) infections has increased worldwide, owing to contamination of chicken and other foods. With multilocus sequence typing, *S.* Infantis sequence type (ST) 32 has significantly increased broiler production, leading to its spread through the poultry food chain [4]. *S.* Infantis isolates obtained from patients and chickens were resistant to multiple antimicrobials [4]. Plasmids are important vectors for the dissemination of antimicrobial resistance genes in bacteria. A conjugative megaplasmid named pESI (plasmid for emerging *S.* Infantis) was previously identified and characterized [5]. pESI had a significant impact on the spread of *S.* Infantis [5].

The area of *S.* Schwarzengrund-contaminated broiler chickens expanded from Western Japan to Eastern Japan in 2015 [6]. Since *S.* Schwarzengrund isolates obtained from chicken meat in Japan between 2008 and 2019 were assigned as ST241, *S.* Schwarzengrund ST241 may have spread from Western to Eastern Japan [7]. In a previous study, 157 *Salmonella* isolates were obtained from chicken meat samples collected between 2015–2017 [6]. Of these, *S.* Schwarzengrund isolates have been reported to show antimicrobial resistance and have been genetically characterized [7]. The antimicrobial susceptibility of the remaining 101 *Salmonella* isolates is reported in this study. The same antimicrobial resistance gene combinations in *Salmonella* isolates were often observed in several serovars. However, it remains unknown whether they carry a common antimicrobial resistance plasmid. To confirm that different *S. enterica* serovars carry a common antimicrobial resistance plasmid, antimicrobial resistance transfer was examined in selected *S.* Infantis and *S.* Schwarzengrund isolates, the major serovars obtained from chicken meat. Whole-genome sequencing (WGS) was performed on representative isolates.

2. Results

2.1. Antimicrobial Susceptibilities of the Isolates and Their Antimicrobial Resistance Genes

Table 1 shows the 16 antimicrobial resistance gene patterns detected via PCR among the 101 *Salmonella* isolates investigated in this study. The most predominant pattern was *aac(6′)-Iaa*, *aadA1*, *tetA*, *sul1*, and *dfrA14*, as confirmed in 32 *S.* Infantis isolates, followed by *aac(6′)-Iaa*, *aadA1*, *tetA*, *sul1*, *dfrA14*, and *aphA1*, as confirmed in 18 *S.* Infantis isolates. The third most common pattern, *aac(6′)-Iaa*, *aadA1*, *tetA*, and *sul1*, was found in six *S.* Manhattan, five *S.* Infantis, three *S.* Agona, and two *S.* Yovokome isolates.

Antimicrobial resistance was observed depending on the antimicrobial resistance genes detected via PCR. However, *sul1* and *dfrA14* did not completely confer antimicrobial resistance. We also found that 27 *sul1*-positive and 46 *dfrA14*-positive isolates were susceptible to trimethoprim–sulfamethoxazole and trimethoprim, respectively. Eleven isolates (four *S.* Typhimurium, three *S.* Infantis, two *S.* Yovokome, one *S.* Manhattan, and one serovar Untypable [OUT: r: 1,5]) were resistant to nalidixic acid. All 101 isolates were susceptible to ciprofloxacin (minimum inhibitory concentration [MIC], 0.25 to 0.5 μg/mL).

Table 1. Antimicrobial resistance genes present in *Salmonella* isolates obtained from chicken meats in Japan.

Antimicrobial Resistance Genes	No. of Isolates	S. Infantis			S. Typhimurium		S. Manhattan		Other	
		1 kb [a]		No [b]	1 kb	No	1 kb	No	1 kb	No
		aadA1 [c]	aadA2		aadA1		aadA1		aadA1	
aac(6′)-Iaa, aadA1, tetA, sul1, dfrA14, aphA1, bla$_{TEM}$	1	1								
aac(6′)-Iaa, aadA1, tetA, sul1, aphA2, bla$_{TEM}$	1				1					
aac(6′)-Iaa, aadA1, tetA, sul1, dfrA14, aphA1	18	18								
aac(6′)-Iaa, aadA1, tetA, sul1, dfrA14, bla$_{CMY-2}$	3	3								
aac(6′)-Iaa, aadA1, tetA, sul1, dfrA14, bla$_{TEM}$	1				1					
aac(6′)-Iaa, aadA1, tetA, sul1, aphA1	5	5								
aac(6′)-Iaa, aadA1, tetA, sul1, dfrA14	32	32								
aac(6′)-Iaa, aadA1, tetA, sul1, bla$_{TEM}$	1				1					
aac(6′)-Iaa, aadA2, tetA, sul1, bla$_{CTX-M}$	2		2							
aac(6′)-Iaa, aadA1, tetA, sul1	16	5					6		5 [d]	
aac(6′)-Iaa, dfrA14, aphA1, bla$_{TEM}$	1					1				
aac(6′)-Iaa, aphA2, bla$_{TEM}$	6					6				
aac(6′)-Iaa, dfrA14, aphA1	5			4						1 [e]
aac(6′)-Iaa, aphA1	1			1						
aac(6′)-Iaa, dfrA14	5			5						
aac(6′)-Iaa	3			1				1		1 [f]
total	101	64	2	11	3	7	6	1	5	2

[a] The length of the class 1 integron; [b] the absence of class 1 integrons; [c] antimicrobial resistance genes located on class 1 integron. [d] three *S.* Agona isolates and two *S.* Yovokome; [e] O untypable: r: 1,5; [f] *S.* Kedougou.

Antimicrobial resistance was observed depending on the antimicrobial resistance genes detected via PCR. However, *sul1* and *dfrA14* did not completely confer antimicrobial resistance. We also found that 27 *sul1*-positive and 46 *dfrA14*-positive isolates were susceptible to trimethoprim–sulfamethoxazole and trimethoprim, respectively. Eleven isolates (four *S.* Typhimurium, three *S.* Infantis, two *S.* Yovokome, one *S.* Manhattan, and one serovar Untypable [OUT: r: 1,5]) were resistant to nalidixic acid. All 101 isolates were susceptible to ciprofloxacin (minimum inhibitory concentration [MIC], 0.25 to 0.5 μg/mL).

All 80 *aadA*-positive isolates detected via PCR harbored 1.0-kb class 1 integrons (Table 1). Agarose gel electrophoresis of the FastDigest TaqI-digested PCR products for gene cassettes of class 1 integrons showed that the restriction fragment length polymorphism (RFLP) patterns of 78 isolates matched those of the *aadA1*-positive isolate (Sal_G1) sequenced in a previous study [7]. The RFLP patterns of the remaining two *S.* Infantis isolates (Sal_45 and Sal_157) were the same but different from those of the *aadA1*-positive isolate. The amplicons of these two isolates were sequenced, and DNA alignment showed 100% (918/918) identity with the class 1 integron containing the *aadA2* gene cassette (accession no. CP040321; position, 4,110,362 to 4,111,279). The results of the discrimination of the harboring *aadA* detected by PCR into *aadA1* and *aadA2* according to the RFLP patterns are shown in Table 1.

2.2. Horizontal Transfer of Antimicrobial Resistance

Table 2 shows the results of the transfer of antimicrobial resistance tests using 29 *Salmonella* isolates as donors. Antimicrobial resistances were transferred from 20 of the 29 donor isolates to the recipient *Escherichia coli* DH5α-R3. The antimicrobial resistance genes detected via PCR in donors and transconjugants are shown in Table 2. Although the antimicrobial resistance gene that the donor harbored was tested via PCR in transconjugants, the *aac(6′)-Iaa* on a chromosome was excluded from testing. All harbored antimicrobial resistance genes were transferred from most donors. Sal_235 yielded transconjugants with four different antimicrobial resistance patterns. Transconjugants were obtained using deoxycholate–hydrogen sulfide–lactose (DHL) agar medium supplemented with streptomycin, tetracycline, or kanamycin. From three donor isolates (Sal_287, Sal_289, and Sal_291) harboring *strA*, *strB*, *sul2*, *dfrA14*, and *aphA1*, only *aphA1* was transferred. Transconjugants that acquired only *aphA1* were then obtained using a DHL agar medium supplemented with kanamycin. Supplementary Table S1 shows the antimicrobial susceptibility of the transconjugants. The transconjugants were resistant to the corresponding antimicrobial agents according to the acquired antimicrobial resistance genes. However, seven transconjugants that acquired *dfrA14* but were susceptible to trimethoprim were also observed. In addition, one transconjugant (TC73) did not acquire *aphA1*, but the MIC of kanamycin for TC73 (32 μg/mL) was higher than those of DH5α-R3 (1 μg/mL). The MICs of kanamycin for the other transconjugants that acquired *aphA1* were higher, with values above 256 μg/mL. No transconjugants were obtained from the six donor isolates (*S.* Infantis, Sal_102; *S.* Schwarzengrund, Sal_83, Sal_145, Sal_266, Sal_290, and Sal_294) that harbored only *aphA1*. Moreover, the transfer of antimicrobial resistance genes could not be confirmed from *S.* Infantis Sal_286, which harbored *aadA1*, *tetA*, *sul1*, *dfrA14* and $bla_{CMY\text{-}2}$; *S.* Infantis Sal_256, which harbored *aadA1*, *tetA*, *sul1*, *dfrA14* and *aphA1*; and *S.* Infantis Sal_80, which harbored *aadA1*, *tetA*, *sul1*, and *dfrA14*. For *S.* Infantis Sal_80, Sal_102, Sal_256, and Sal_286, and *S.* Schwarzengrund Sal_83, Sal_145, Sal_266, Sal_290, and Sal_294 as donors, conjugation experiments were carried out using the broth mating method and/or the filter mating method, with one to six strains of rifampicin-resistant mutants selected from six antimicrobial-susceptible *E. coli* as recipients. The transfer of antimicrobial resistance was not observed in the donor–recipient combinations.

Table 2. Antimicrobial resistance genes transferred from *Salmonella* Schwarzengrund and *Salmonella* Infantis obtained from chicken in Japan.

Serovar	Donar		Transconjugant
	Isolate ID	Antimicrobial Resistance Genes	Antimicrobial Resistance Genes
S. Infantis	Sal_235	*aadA1, tetA, sul1, dfrA14, aphA1, bla_{TEM}*	*aadA* **, *tetA, sul1, dfrA14, aphA1, bla_{TEM}* *aadA, tetA, sul1, dfrA14, aphA1* *aadA, sul1* *aadA, bla_{TEM}*
S. Infantis	Sal_286	*aadA1, tetA, sul1, dfrA14, bla_{CMY-2}*	No transfer
S. Infantis	Sal_31	*aadA1, tetA, sul1, dfrA14, aphA1*	*aadA, tetA, sul1, dfrA14, aphA1*
S. Infantis	Sal_238 *	*aadA1, tetA, sul1, dfrA14, aphA1*	*aadA, tetA, sul1, dfrA14, aphA1*
S. Infantis	Sal_180	*aadA1, tetA, sul1, dfrA14, aphA1*	*aadA, tetA, sul1, dfrA14, aphA1*
S. Infantis	Sal_181	*aadA1, tetA, sul1, dfrA14, aphA1*	*aadA, tetA, sul1, dfrA14, aphA1*
S. Infantis	Sal_256 *	*aadA1, tetA, sul1, dfrA14, aphA1*	No transfer
S. Schwarzengrund	Sal_63	*aadA1, tetA, sul1, dfrA14, aphA1*	*aadA, tetA, sul1, dfrA14, aphA1*
S. Schwarzengrund	Sal_15	*aadA1, tetA, sul1, dfrA14, aphA1*	*aadA, tetA, sul1, dfrA14, aphA1*
S. Schwarzengrund	Sal_249 *	*aadA1, tetA, sul1, dfrA14, aphA1*	*aadA, tetA, sul1, dfrA14, aphA1*
S. Schwarzengrund	Sal_278 *	*aadA1, tetA, sul1, dfrA14, aphA1*	*aadA, tetA, sul1, dfrA14, aphA1*
S. Infantis	Sal_51	*aadA1, tetA, sul1, aphA1*	*aadA, tetA, sul1, aphA1*
S. Schwarzengrund	Sal_272	*aadA1, tetA, sul1, aphA1*	*aadA, tetA, sul1, aphA1*
S. Infantis	Sal_80	*aadA1, tetA, sul1, dfrA14*	No transfer
S. Infantis	Sal_25	*aadA1, tetA, sul1, dfrA14*	*aadA, tetA, sul1, dfrA14*
S. Schwarzengrund	Sal_159	*aadA1, tetA, sul1, dfrA14*	*aadA, tetA, sul1, dfrA14*
S. Schwarzengrund	Sal_167 *	*aadA1, tetA, sul1, dfrA14*	*aadA, tetA, sul1, dfrA14*
S. Infantis	Sal_157	*aadA2, tetA, sul1, bla_{CTX-M}*	*aadA, tetA, sul1, bla_{CTX-M}*
S. Infantis	Sal_36	*aadA1, tetA, sul1*	*aadA, tetA, sul1*
S. Schwarzengrund	Sal_82	*aadA1, tetA, sul1*	*aadA, tetA, sul1*
S. Schwarzengrund	Sal_287	*strA, strB, sul2, dfrA14, aphA1*	*aphA1*
	Sal_289	*strA, strB, sul2, dfrA14, aphA1*	*aphA1*
	Sal_291 *	*strA, strB, sul2, dfrA14, aphA1*	*aphA1*
S. Infantis	Sal_102	*aphA1*	No transfer
S. Schwarzengrund	Five isolates †	*aphA1*	No transfer

* Whole-genome sequencing was performed. ** The distinction between *aadA1* and *aadA2* acquired by the transconjugant was not determined. But, it is assumed that the transferred gene corresponds to *aadA1* or *aadA2* present in the donor. † Sal_83, Sal_145, Sal_266, Sal_290, and Sal_294.

2.3. WGS

The number of reads obtained from the short- and long-read sequencing performed in this study is shown in Supplementary Table S2. Short-read sequences for the five *S.* Schwarzengrund isolates, as previously reported [7], were also analyzed in this study, as shown in Supplementary Table S2.

Short-read sequencing was performed on two *S.* Infantis isolates (Sal_238 and Sal_256). The two *S.* Infantis isolates were assigned to ST32 via multilocus sequence-typing (MLST) analysis and core-genome sequence-type (cgST) 40031 or cgST18978 via core-genome multilocus sequence-type (cgMLST) analysis [7–9] (Table 3). Six antimicrobial resistance genes and chromosomal point mutation in *parC* (T57S) were detected using the ResFinder (Table 3). IncFIB for Sal_238 and Sal_256 and IncX4 for Sal_256 were determined using PlasmidFinder (Table 3).

Although our previous study reported the presence of antimicrobial resistance genes [7] in five *S.* Schwarzengrund isolates, a chromosomal point mutation in *parC* (T57S) was also confirmed in this study. PlasmidFinder identified an IncFIB plasmid in five *S.* Schwarzengrund isolates (Table 3).

Long-read sequencing was performed on *S.* Infantis Sal_238 and *S.* Schwarzengrund Sal_249. The hybrid assembly yielded contigs of the chromosomes (4,680,940 and 4,659,199 bp) and a circular sequence of one plasmid (288,159 and 288,207 bp) for Sal_238 and Sal_249, respectively. The plasmid sequences in Sal_238 and Sal_249 were registered in the DDBJ database as pSal_238Inf (accession no. LC 785392) and pSal_249Sch (accession no. LC785393), respectively. A comparison of the plasmid sequences of the two isolates showed 99.58% (287,568/288,780 bases) sequence identity. The pSal_238Inf and pSal_249Sch sequences were compared with sequences from the international nucleotide sequence database using BLAST (GENETYX version 15). Although no plasmids matched the full-length sequences of these plasmids, their sequences were highly homologous to those of pESI (CP047882) and pN17S0349 (CP052814). The sequences of pSal_238Inf and pSal_249Sch were 91.94% (274,576/298,632 bases) and 91.93% (274,578/298,679 bases) identical to the pESI sequence, respectively. Thus, pSal_238Inf and pSal_249Sch were identified as pESI-like plasmids. Similarly, the sequences of pSal_238Inf and pSal_249Sch were 82.59% (264,241/319,935 bases) and 82.58% (264,246/319,979 bases) identical to the pN17S0349 sequence, respectively.

Table 3. Genetic characteristics of *Salmonella* harboring pESI-like plasmid in Japan, the UK, and the USA.

Serovar	Country	Isolation of Year	Isolate ID	ST	cgST	Antimicrobial Resistance Genes *	Chromosomal Point Mutation	Plasmid Replicons	Reference
S. Schwarzengrund									
	Japan								
		2016	Sal_167	241 **	287831	*aadA1, sul1, tetA, dfrA14* **	*parC* (T57S)	IncFIB	[7]
		2016	Sal_249	241 **	167363	*aadA1, sul1, tetA, dfrA14, aphA1* **	*parC* (T57S)	IncFIB	[7]
		2017	Sal_266	241 **	167363	*aphA1* **	*parC* (T57S)	IncFIB	[7]
		2017	Sal_278	241 **	167363	*aadA1, sul1, tetA, dfrA14, aphA1* **	*parC* (T57S)	IncFIB	[7]
		2019	Sal_291	241 **	167363	*aphA1, strA/strB, dfrA14, sul2* **	*parC* (T57S)	IncFIB	[7]
	USA								
		No data	GCA_004237325	241 **	287831	*aadA1, sul1, tetA, dfrA14, aphA1* **	*parC* (T57S) **	IncFIB **	[8]
		No data	GCA_005730955	241 **	167363	Not detected**	*parC* (T57S) **	IncFIB **	[8]
		No data	GCA_007854735	241 **	196045	*aadA1, sul1, tetA, dfrA14, aphA1* **	*parC* (T57S) **	IncFIB **	[8]
	UK								
		2018	GCA_007418385	241 **	167363	*aadA1, sul1, tetA, dfrA14, aphA1* **	*parC* (T57S) **	IncFIB **	[8]
		2018	GCA_007589355	241 **	167363	*aadA1, sul1, tetA, aphA1***	*parC* (T57S) **	IncFIB, IncI1-I **	[8]
S. Infantis									
	Japan								
		2016	Sal_238	32	40031	*aadA1, sul1, tetA, dfrA14, aphA1*	*parC* (T57S)	IncFIB	[6]
		2016	Sal_256	32	18978	*aadA1, sul1, tetA, dfrA14, aphA1*	*parC* (T57S)	IncFIB, IncX4	[6]

ST, sequence type by multilocus sequence type analysis; cgST, core-genome sequence type by core-genome multilocus sequence-type analysis. * Antimicrobial resistance genes were detected through resFinder. The gene of *aadA1*, *strB* and *aphA1* are aliases for *ant(3′)-Ia*, *aph(6)-Id* and *aph(3′)-Ia*, respectively. The names of the former genes, the same as previously reported [7], are listed in this table. ** These data were already reported in the reference [7].

The sequences of four plasmids (pSal_238Inf, pSal_249Sch, pESI, and pN17S0349) were annotated using the DDBJ Fast Annotation and Submission Tool (DFAST), and the plasmid structures were drawn and compared using Easyfig (Figure 1A). In the region between *dfrA14* and the MFS transporter, the *lspA* and *aphA1* genes were not found in the pESI but were inserted at the same position in pSal_238Inf, pSal_249Sch, and pN17S0349 (Figure 1B). Moreover, two *dfrA14* genes were identified in pSal_238Inf (Figure 1B).

Next, the short-read sequences of *S.* Schwarzengrund Sal_167, Sal_266, Sal_278, and Sal_291 were aligned to the reference pSal_249Sch sequence to obtain each consensus sequence. Short-read sequences of *S.* Infantis Sal_256 were also aligned with the sequence of pSal_238Inf as a reference. Contigs were obtained via de novo assembly using short-read sequences from these five isolates. Contigs with high homology to the consensus sequences of the pSal_249Sch or pSal_238Inf isolates were selected using local BLAST with GENETYX. The number of contigs, coverage, and base match rates for the consensus sequence are shown in Supplementary Table S3. Because the coverage and concordance rates were at least 94.29% and 94.31%, respectively, the consensus sequence was used as the tentative plasmid sequence. The tentative plasmid sequences were then annotated using DFAST, and the plasmid structure was constructed using Easyfig (Supplementary Figure S1). The sequences of pSal_278 and pSal_167 showed high homology to the pSal_249Sch sequence [99.07% (285,562/288,216 bases) and 98.71% (284,521/288,215 bases), respectively], whereas the sequences of pSal_291 and pSal_266 were less identical [88.78% (255,898/288,214 bases) and 77.97% (224,729/288,218 bases), respectively] to the pSal_249Sch sequence. Similarly, the sequences of pSal_256 showed high homology with those of pSal_238Inf [99.21% (285,921/288,170 bases)].

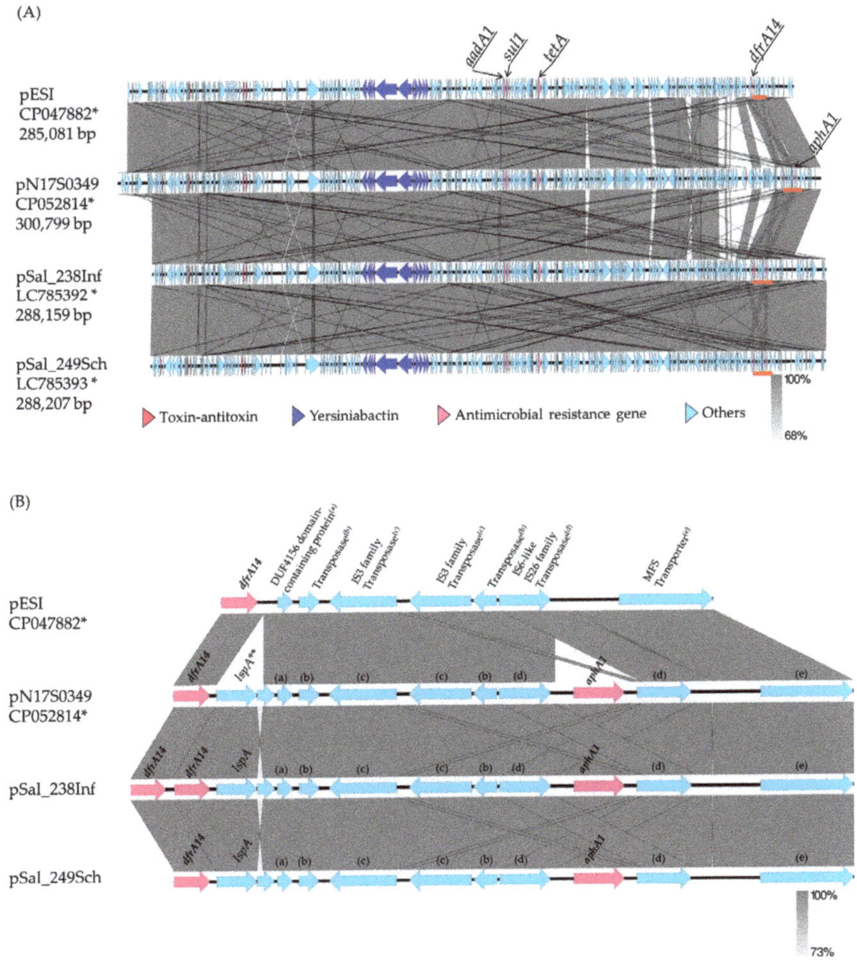

Figure 1. Comparison of plasmid structures among *Salmonella enterica* serovar Infantis and *Salmonella enterica* serovar Schwarzengrund. (**A**) Comparison of the entire plasmid structures of pESI, pN17S0349, and pSal_238Inf in *Salmonella* Infantis isolates obtained in Israel, USA, and Japan, respectively, and pSal_249Sch in *Salmonella* Schwarzengrund obtained in Japan. (**B**) Comparison of gene regions between *dfrA14* and MFS transporter. This figure (**B**) only shows the positions of the following plasmids: pESI, 266456–272779; pN17S0349, 283940–292690; pSal_238Inf, 256081–265387; pSal_249Sch, 256685–265435. The regions marked by the red bars are shown on a larger scale in Figure 1B. * Accession no., ** *lspA*, signal peptidase II; (a) DUF4156 domain-containing protein; (b) transposase; (c) IS3 family transposase; (d) IS6-like IS26 family transposase; (e) MFS transporter.

Two plasmid sequences (pSal_238Inf and pSal_249Sch) and five tentative plasmid sequences (pSal_278, pSal_167, pSal_291, pSal_266, and pSal_256) contained the yersiniabactin biosynthetic, regulatory, and transport operons and the toxin–antitoxin system (Figure 1A and Figure S1). The presence of yersiniabactin siderophore genes, which are characteristic of pESI, was tested for all isolates using PCR targeting the *irp2* gene [9]. All *S.* Schwarzengrund, *S.* Infantis, *S.* Manhattan, *S.* Agona, and *S.* Yovokome isolates and one serovar UT (OUT: r: 1,5) isolate were positive for *irp2* (Table 4). Four of the ten *S.* Typhimurium isolates were also positive for *irp2*; one *S.* Kedougou isolate was negative for *irp2* (Table 4).

Table 4. Percentage of pESI-positive *Salmonella* isolates in each serovar estimated by *irp2* detected by PCR.

Serovar	Year of Isolation	No. of Isolates		(%)
		Tested	Positive	
S. Schwarzengrund	2008	37	37	(100%)
	2015–2019	87	87	(100%)
S. Infantis	2015–2017	77	77	(100%)
S. Typhimurium	2015–2016	10	4	(40%)
S. Manhattan	2016	7	7	(100%)
S. Agona	2015–2017	3	3	(100%)
S. Yovokome	2015	2	2	(100%)
S. Kedougou	2016	1	0	
OUT: r: 1,5	2016	1	1	(100%)
Total		225	218	(96.9%)

UT, untypable.

Although the presence of *acc(6′)-Iaa* was confirmed on the chromosome, the remaining antimicrobial resistance genes confirmed on two plasmid sequences and five tentative plasmid sequences matched the genes confirmed to be harbored by those isolates, as determined using PCR (Figure 1A and Table 2). As an exception, *strA*, *strB*, and *sul2*, which were detected via PCR and ResFinder for Sal_291, were not on the tentative pSal_291 sequence, as pSal_249Sch, which was used as a reference, did not contain them.

All seven isolates for which the plasmid structures were determined or tentatively determined were tested for antimicrobial resistance transfer (Table 2). The determined or tentatively determined plasmid structures were then compared to evaluate antimicrobial resistance transfer. No transfer of antimicrobial resistance was observed for Sal_266 and Sal_256. The region from position 170,499 to 221,059 in pSal_249Sch included genes involved in plasmid transfer. Although the region from position 149,229 to 203,510 of pSal_249Sch, including part of the transfer region, was not confirmed in pSal_266, the entire transfer region was confirmed in pSal_256, similar to pSal_238Inf (Supplementary Figure S1). For nine isolates, including Sal_266 and Sal_256, which did not transfer antimicrobial resistance (Table 2), the presence of *pilV*, *traU*, and *traW* in the transfer region was tested using PCR. Moreover, 32 *S.* Schwarzengrund isolates and two *S.* Infantis isolates that were not tested for antimicrobial resistance transfer were tested for *pilV*, *traU*, and *traW*. Three isolates that exhibited the transfer of antimicrobial resistance (Sal_287, Sal_289, and Sal_291) were used as positive controls. Sal_266, which contained *traW* but not *pilV* or *traU* on the tentative plasmid, was also positive for *traW* and negative for *pilV* and *traU* via PCR. Sal_256 and Sal_291, which contain three genes on each tentative plasmid, were also positive for all of the *pilV*, *traU*, and *traW* genes. Eight isolates with no confirmed antimicrobial resistance transfer, other than Sal_266, tested positive for these three genes using PCR. Of the 34 isolates not tested for antimicrobial resistance transfer, one isolate (Sal_57) was negative for *pilV* only, and one isolate (Sal_G68) was negative for both *pilV* and *traU*.

Although one *dfrA14* was detected on pSal_249Sch, pESI, and pN17S0349, two *dfrA14* positioned side by side were detected on pSal_238Inf (Figure 1A,B). Therefore, a PCR was carried out on the other *dfrA14*-positive isolates to test whether they contained two *dfrA14* genes adjacent to each other. The primers used here were the same as those used for *dfrA14* detection. The extension time for amplification was increased to 90 s, and the number of cycles was increased to 40. Three *S.* Infantis isolates, from which a 1,021-bp product was amplified, were presumed to be two sequential *dfrA14* genes as Sal_238. The three *S.* Infantis isolates and Sal_238 were obtained from chicken meat in the same prefecture. The remaining 14 *S.* Infantis isolates from chicken meat from the same prefecture, as well as *S.* Infantis and *S.* Schwarzengrund isolates from chicken meat produced in other prefectures, harbored one *dfrA14* gene that yielded an amplification product that is 455 bp long.

2.4. cgMLST Analysis for *S.* Schwaruzengrund Isolates Carrying pESI-Like Plasmids in Japan, the UK, and the USA

A cgMLST analysis was performed for five *S.* Schwarzengrund isolates from Japan and the genomes of *S.* Schwarzengrund strains carrying a pESI-like plasmid [8]; strains 478612 (Assembly No. GCA_007418385) and 490272 (GCA_007589355) obtained from clinical samples from the UK; and PNUSAS062103 (GCA_004237325), PNUSAS066835 (GCA_005730955), and PNUSAS084711 (GCA_007854735) obtained from clinical samples from the USA. The assigned cgSTs are listed in Table 3. The major cgST gene is cgST167363. A phylogenetic tree based on cgMLST allele data of these *S.* Schwarzengrund isolates was then visualized using MSTree V2 algorithms with GrapeTree software version 1.1. A minimum spanning tree based on the cgMLST data is shown in Figure 2. The minimum spanning tree branched out from a node that included only two UK strains, whereas the Japanese and USA strains were classified into six nodes. Each node included an isolate from Japan and a strain from the USA. All five isolates from Japan were included in different nodes.

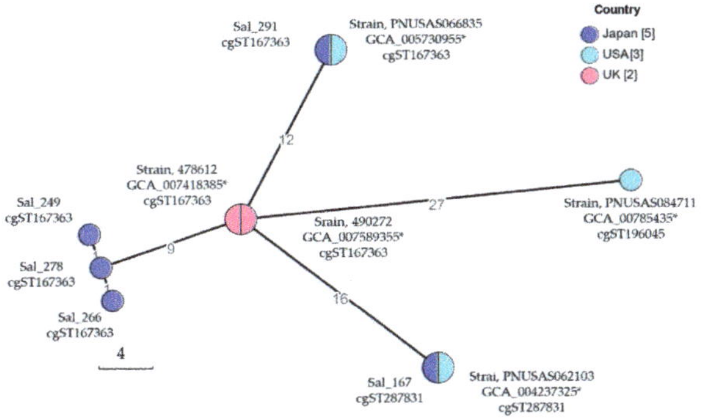

Figure 2. Minimum spanning tree of core-genome multi-locus sequence typing (cgMLST) of *Salmonella enterica* serovar Schwaruzengrund carrying pESI-like plasmids. cgMLST analysis was performed through cgMLSTFinder [8]. The core genomic relationships of *Salmonella* Schwaruzengrund isolates obtained in Japan, the UK, and the USA were visualized using the algorithms of MSTree V2 with GrapeTree software. Nodes are color-coded according to the country where each isolate was obtained. The isolate ID or assembly no. and the assigned core-genome sequence type (cgST) of the isolate included in that node are listed near the node. The number of isolates in each country analyzed is shown in square parentheses. * Assembly No.

3. Discussion

S. Infantis is one of the most commonly isolated strains from human patients and chickens and has recently emerged worldwide [4,10–13]. In Israel, *S.* Infantis emerged rapidly in 2006–2007; the emerging *S.* Infantis carried a mosaic megaplasmid (~280 kb). The resulting plasmid was named pESI [5]. Comparative analyses between pre-emergent and emergent *S.* Infantis suggested that pESI played an important role in the spread of *S.* Infantis in Israel for 2–3 years [5]. The increase of *S.* Infantis among broiler chickens and human patients worldwide has also been associated with a pESI-like plasmid [4].

In 2018, *S.* Muenchen showed an increase in Israel. Then, in 2019–2020, *S.* Muenchen was the dominant serovar isolated from clinical, poultry, and food sources in Israel. The emerging *S.* Muenchen carried a pESI-like plasmid with a 99.96% nucleotide sequence identity to the pESI carried by *S.* Infantis. This is the second reported case of pathogen emergence and spread associated with pESI-like plasmid acquisition [14]. We previously reported that *S.* Schwarzengrund ST241 expanded its distribution from Western to Eastern Japan [7]. The present study revealed that antimicrobial-resistant *S.* Schwarzengrund, which has expanded its distribution, carries a pESI-like plasmid. PCR targeting the *irp2* gene suggested that, in 2008, when the distribution of *S.* Schwarzengrund in Japan was restricted to Western Japan [15], it already carried a pESI-like plasmid. Although previous bioinformatics analyses of *Salmonella* genomes revealed a pESI-like plasmid structure in *S.* Schwarzengrund in the UK and USA [8], to our knowledge, there are no reports on the distribution of *S.* Schwarzengrund strains with pESI-like plasmids. The present study reported that *S.* Schwarzengrund was the third serovar of *Salmonella* to expand, whose distribution expanded in association with pESI-like plasmid acquisition in Japan.

S. Infantis carrying pESI emerged rapidly in Israel in 2006–2007 [5]. The prevalence of *S.* Muenchen carrying pESI also increased in 2018 and rapidly became the most predominant serovar isolated in Israel in 2019–2020 [14]. The present study revealed that *S.* Schwarzengrund already carried a pESI-like plasmid in 2008. However, seven years later, in 2015, its distribution was reported to have expanded to the entire country for the first time, with an isolation rate of 23.3% (56/240), lower than that of *S.* Infantis (77/240, 32.1%) [6]. From 2018 to 2021, *S.* Schwarzengrund was the most commonly isolated serovar from chicken meat products collected in Japan (146/235, 62.1%), followed by *S.* Infantis (30/235, 12.8%) [16]. The expansion of the distribution of *S.* Schwarzengrund containing pESI-like plasmids and the increase in its contamination rate did not occur as rapidly as those of *S.* Infantis [5] and *S.* Muenchen [14] in Israel. Although *S.* Typhimurium (4/10), *S.* Manhattan (7/7), *S.* Agona (3/3), and *S.* Yovokome (2/2) obtained from chicken meat collected between 2015 and 2017 were also suggested to carry a pESI-like plasmid following the results of *irp2*-targeted PCR, the isolation percentage of these serovars did not increase in Japan from 2018 to 2021 [16]. There may be differences in the changes of the characteristics of pathogens by pESI depending on the serovar, such as increasing the ability of *S.* Infantis to spread infection. The presence of pESI-like plasmids in *S.* Agona isolated in Japan in 2014 has already been reported [8]. However, to the best of our knowledge, there are no reports of *S.* Manhattan and *S.* Yovokome carrying a pESI-like plasmid. In a previous study using a mouse model, following *S.* Infantis infection, pESI was horizontally transferred to the gut microbiota and then to *S.* Typhimurium [17]. However, there have been no reports of *S.* Typhimurium pESI-carrying isolates from clinical, animal, or food samples.

By performing a genomic analysis of *Salmonella* genomes from the National Center for Biotechnology Information (NCBI), the structure of the pESI-like plasmid was confirmed in five *S.* Schwarzengrund isolates obtained from clinical samples in the UK in 2018 and in the USA (isolation date unknown) [8]. All five *S.* Schwarzengrund isolates with pESI-like plasmids from the UK and USA were classified as ST241 via MLST [8]. The five *S.* Schwarzengrund isolates from Japan were confirmed to carry pESI-like plasmids via genome analysis in this study and were assigned to the same ST, ST241 [7]. The cgMLST performed in this study revealed that 10 *S.* Schwarzengrund isolates with a pESI-like plasmid were classified into three cgSTs that were closely related. However, differences were identified in their core genomes. From the minimum spanning tree obtained using cgMLST, all five isolates from Japan were classified into different nodes, indicating that they acquired more genetic variation than the two UK strains classified in the same node. In Japan, *S.* Schwarzengrund has been isolated from broiler [18,19] and chicken meat [6,15], and the number of imported and exported chickens was confirmed through animal quarantine statistics [20]. Japan exported an average of 6,954 tons of poultry meat annually between 2019 and 2022, with yearly exports ranging from 3289 to 10,016 tons. During this period, no live chickens were exported. The export destinations for poultry meat from Japan have not been published. However, the amount of exported poultry meat is not large compared to the amount of chicken meat and number of live chickens that Japan imported. Japan is an established importer of one-day-old chicks. In 2014, 37.1% (170,245/459,443) of the one-day-old chicks imported to Japan came from the UK and 7.6% (34,825/459,443) from the USA. In 2019, the percentage of one-day-old chicks imported from the UK and the USA increased to 60.5% (186,563/308,592) and 11.4% (35,182/308,592), respectively. The predominant serovars of *Salmonella* isolated from retail chickens in the USA during 2013–2020 were *S.* Enteritidis, *S.* Infantis, *S.* Kentucky, and *S.* Typhimurium; *S.* Schwarzengrund was neither isolated nor considered a minor serovar [21]. Among *Salmonella* isolates from foodborne outbreaks in the USA from 1998 to 2021, the most predominant serovar was *S.* Enteritidis (28.42%); *S.* Schwarzengrund (0.52%) was a minor serovar [22]. In the UK, only *S.* Infantis (3/310) and *S.* Java (2/310) were detected in samples of frozen, raw, or partly cooked coated chicken products collected from retailers in 2021 [23]. These data suggest that *S.* Schwarzengrund was a minor serovar in the UK and USA, and no evidence of pESI-carrying *S.* Schwarzengrund entering Japan from both countries was found.

Over the last 10 years, multidrug-resistant *S.* Infantis has spread globally. Specifically, *S.* Infantis isolates in recent years have been reported to carry a pESI-like plasmid [4]. In Japan, *S.* Infantis was the predominant serovar in *Salmonella*-contaminated chicken meat from 1995 to 1998. The antimicrobial resistance and plasmids of these *S.* Infantis isolates have not been examined [24]. *S.* Infantis isolates from chickens in Japan in 2011 were subjected to PCR targeting the *ipr2* gene. Consequently, *S.* Infantis isolates positive for the *ipr2* gene were considered to carry a pESI-like plasmid [25]. In the present study, complete plasmid sequences were identified for *S.* Infantis Sal_238 and *S.* Schwarzengrund Sal_249, which were isolated from chicken meat in Japan in 2016. The high homology between pSal_249Sch and pSal_238Inf suggests that their ancestors were common. pSal_249Sch and pSal_238Inf showed high homology to pESI in *S.* Infantis [5] but differed from pSal_249Sch and pSal_238Inf carrying *aphA1*. The pN17S0349 (accession no. CP052814) in *S.* Infantis CVM N17S349 isolated from a ground turkey was identical to pSal_249Sch and pSal_238Inf carrying *aphA1*, but the homology of pN17S0349 to pSal_249Sch and pSal_238Inf was lower than that of pESI. The structure of pESI-like plasmids varies. Nonetheless, antimicrobial resistance genes were inserted into pESI as follows: *aph(3′)-Ia* (*aphA1*), $bla_{\text{CTX-M-65}}$, *floR*, and *fosA3* in the USA [26]; $bla_{\text{CTX-M-1}}$, $bla_{\text{CTX-M-65}}$, *fosA*, and *floR* in Italy [10]; and $bla_{\text{CTX-M-14}}$ in Russia [27]. In contrast, high homology was observed between pSal_249Sch and pSal_238Inf. Their genetic characteristics suggest the horizontal transfer of a pESI-like plasmid from *S.* Infantis, which was previously widely distributed in Japan, to *S.* Schwarzengrund, which was once a rare serovar in Japan [24,28].

The tentative plasmid sequences of four *S.* Schwarzengrund isolates with different antimicrobial resistance gene patterns and one *S.* Infantis isolate were also determined in this study. Although their plasmid structures were diverse, the tentative plasmid sequences contained K88-like and *Ipf* fimbria; yersiniabactin biosynthetic, regulatory, and transport operons; and type II toxin–antitoxin systems identified on pESI [5]. The type II toxin–antitoxin system found in pESI was suggested to increase the stability of pESI in the host [5]. This suggests that these pESI-like plasmids, which contained the toxin–antitoxin system, were also stably maintained in *S.* Schwarzengrund and *S.* Infantis in Japan. The structure of the pESI-like plasmid carried by *S.* Schwarzengrund in Japan changed with the deletion of the plasmid transfer region (pSal_266 and pSal_291) or class 1 integrons, antimicrobial resistance genes (*aadA1*, *sul1*, and *tetA*), and the mercury resistance operon (pSal_266 and pSal_291). Although *Salmonella* carrying pESI-like plasmids lacking these genes would not be able to increase under antimicrobial selection pressure, the percentage of isolates harbored *aadA1*, *sul1*, and *tetA* significantly decreased from 2008 to 2015–2019 [7]. The prudent use of antimicrobials is widely practiced, and *Salmonella* that have lost these genes may be able to continue to survive in chicken flocks. The diversity in the structure of pESI-like plasmids suggests that some time had passed after *S.* Schwarzengrund acquired the pESI-like plasmid, not immediately.

Salmonella carrying pESI-like plasmids in Japan typically contained *aac(6′)-Iaa*, *aadA1*, *sul1*, *tetA*, and *dfrA14*. In addition, many isolates harbored *aphA1*. Based on the characteristics of the antimicrobial resistance gene pattern, we hypothesized that different serovars among *Salmonella* isolates shared a common plasmid. However, pESI-like plasmids were detected via PCR even if *S.* Schwarzengrund, *S.* Infantis, and other isolates harbored only *aac(6′)-Iaa* or only one or two additional antimicrobial resistance genes. Therefore, the presence of a pESI-like plasmid should not be assumed based on the combination of antimicrobial resistance genes alone. Furthermore, pESI was reported to be a self-transmissible plasmid [5]. However, deletion of the plasmid transfer region in a pESI-like plasmid was confirmed. The plasmid lost its ability to be transferred. This suggested that the transfer region was deleted after the pESI-like plasmid was transferred between the *Salmonella* isolates from different serovars.

4. Materials and Methods

4.1. Salmonella Isolates

From the *Salmonella* isolates obtained from chicken meat in a previous report [6], 101 isolates, excluding *S.* Schwarzengrund, were tested for antimicrobial susceptibility in this study, including *Salmonella* Infantis (77 isolates), *S.* Typhimurium (10 isolates), *S.* Manhattan (seven isolates), *S.* Agona (three isolates), *S.* Yovokome (two isolates), *S.* Kedougou (one isolate), and an Untypable isolate (O Untypable [UT]: r: 1,5) (one isolate). Although the antimicrobial resistances of 124 *S.* Schwarzengrund isolates obtained from chicken meat collected in 2008 and between 2015 and 2019 were reported in our previous study [7], these *S.* Schwarzengrund isolates were used in this study for analyses other than antimicrobial resistance.

4.2. Antimicrobial Susceptibility Tests

The minimum inhibitory concentrations (MICs) of nine antimicrobial agents [ampicillin (breakpoint, 32 μg/mL), cefazolin (8 μg/mL), gentamicin (16 μg/mL), kanamycin (64 μg/mL), tetracycline (16 μg/mL), chloramphenicol (32 μg/mL), nalidixic acid

(32 μg/mL), ciprofloxacin (1 μg/mL), and trimethoprim–sulfamethoxazole (4/76 μg/mL)] on the 101 investigated isolates were determined with the broth microdilution method using Frozen Plate (Eiken Chemical Co., Ltd., Tokyo, Japan). In addition, the MICs for streptomycin (16 μg/mL) (FUJIFILM Wako Pure Chemical Co., Osaka, Japan) and trimethoprim (16 μg/mL) (FUJIFILM Wako Pure Chemical Co.) were determined using the agar dilution method in Mueller–Hinton agar (Thermo Fisher Scientific K. K., Osaka, Japan), according to the Clinical Laboratory Standards Institute guidelines [29].

4.3. Antimicrobial Resistance Gene and Integron Detection

The major genes for resistance to streptomycin [*aac(6′)-Iaa*, *aadA* (another name, *ant(3″)-Ia*) and *strA* (*aph(3″-Ib*)), *strB* (*aph(6)-Id*)], kanamycin [*aphA1* (*aph(3′)-Ia*), and *aphA2* (*aph(3′)-IIa*)], tetracycline (*tetA* and *tetB*), sulfamethoxazole (*sul1* and *sul2*), and trimethoprim (*dfrA14*) were tested using simplex or multiplex PCR, as reported previously [7,30]. Moreover, the genes conferring resistance to ampicillin (bla_{TEM}) [30] and cefazolin ($bla_{CTX\text{-}M}$ [31] and $bla_{CMY\text{-}2}$ [30] were tested using multiplex PCR.

To determine the inserted gene cassettes in the class 1 integron, the corresponding regions in each isolate were classified using RFLP analysis, as previously described [7,32]. The primer sequences used for PCR in this step are listed in Supplementary Table S4.

4.4. Conjugation Assay

Next, conjugation experiments were performed using the broth mating method to determine the transfer of antimicrobial resistance genes. A rifampicin-resistant *E. coli* strain DH5α-3R was used as the recipient strain [33]. Briefly, the donor and recipient strains were each inoculated into 3 mL of Mueller–Hinton broth (Thermo Fisher Scientific K. K.) and pre-cultured for 6 h at 35 °C. One milliliter of the recipient strain culture was centrifuged at 12,000 rpm for 3 min, and the recipient cells were resuspended in 3 mL of fresh Mueller–Hinton broth. Next, 0.1 mL of the donor culture medium was inoculated into this recipient culture and co-cultured overnight at 35 °C. Subsequently, 1 mL of the resulting co-culture centrifuged at 12,000 rpm for 3 min was suspended in 100 μL of sterile saline, inoculated onto Mueller–Hinton or DHL agar (Shimadzu Diagnostics Corporation, Tokyo, Japan) supplemented with rifampicin (100 μg/mL) and either streptomycin (5 μg/mL), tetracycline (12.5 μg/mL) (FUJIFILM Wako Pure Chemical Co.), or kanamycin (100 μg/mL) (Tokyo Chemical Industry Co., Ltd., Tokyo, Japan), and incubated at 35 °C for 24 h. Up to four colonies of candidate transconjugant strains were selected per donor and subcultured on additive-free Mueller–Hinton agar.

The agar dilution method was used to determine the MICs of ampicillin, cefazolin, streptomycin, kanamycin, tetracycline, and trimethoprim for the transconjugants. The antimicrobial resistance genes in the transconjugants were also detected via PCR, as described above.

4.5. WGS Analysis

Short-read sequencing for five *S.* Schwarzengrund isolates (Sal_167, Sal_249, Sal_266, Sal_278, and Sal_291) was performed as in our previous study [7]. In the present study, for representative *S.* Infantis isolates, cDNA libraries were prepared using a Nextera XT Library Prep Kit (Illumina K. K., San Diego, CA, USA) according to the manufacturer's instructions and sequenced on an iSeq System (Illumina K. K.), as previously reported [7,34]. To accurately characterize the structure of the plasmids, DNA was purified using a FavorPrep tissue genomic DNA extraction mini kit (Chiyoda Science Co., Ltd., Tokyo, Japan) following the manufacturer's instructions. cDNA libraries were prepared using a Rapid Barcoding Sequencing kit (SQK-RBK001; Oxford Nanopore Technologies, Oxford, UK), and long-read sequencing was performed on a MinION flow cell R9.4 system (Oxford Nanopore Technologies).

Trimming of raw short reads and long-read data was performed using FastP (v 0.23.3) [35] and filtlong v 0.2.1 (https://github.com/rrwick/Filtlong, accessed on 25 May 2023). The long-read sequences obtained using MinION were corrected with short reads obtained via iSeq using LoRDEC [36]. Complete plasmid sequences were constructed via de novo assembly of corrected long-read data using Flye v 2.9.2 [37]. DFAST was used to annotate the genome [38]. The Nucleotide BLAST program was used to compare DNA alignments with data from the NCBI using GENETYX version 15 (Genetyx Corporation, Tokyo, Japan). Multiple alignment using the Fast Fourier Transform (MAFFT) version 7 (https://mafft.cbrc.jp/alignment/software/, accessed on 5 April 2024) was used for nucleotide sequence alignment. The tentative plasmid sequence was also determined from the short-read sequences of the isolates that were not long-sequenced. Short-read sequences were aligned to the determined plasmid sequence as a reference to obtain each consensus sequence with the mapping function using GENETYX-NGS version 5 (Genetyx Corporation). Contigs were also obtained via the de novo assembly of short-read sequences using GENETYX-NGS. Contigs with high homology to the consensus sequences of each isolate were selected using local BLAST with GENETYX. The percentage of tentative plasmid sequences covered by the contig obtained in the de novo assembly and the percentage of nucleotide sequence matches were further confirmed. To compare the plasmid structures, Easyfig was used for map generation [39]. Bacterial genome contigs were scanned against the ResFinder, PointFinder, PlasmidFinder, and PubMLST databases using staramr [40]. Some genes detected by the genome analysis were tested using PCR for other isolates.

Core-genome multilocus sequence-type (cgMLST) analysis was performed using cgMLSTFinder [41,42] for *S.* Schwaruzengrund isolates obtained in Japan (Sal_167, Sal_249, Sal_266, Sal_278, and Sal_291 from chicken meat) and genome data obtained from the NCBI database (https://www.ncbi.nlm.nih.gov/datasets/genome/, accessed on 1 November 2024). Genetic relationships were visualized using MSTree V2 algorithms in GrapeTree software [43].

5. Conclusions

In this study, complete plasmid sequences were identified for *S. enterica* serovar Schwarzengrund Sal_249 and *S. enterica* serovar Infantis Sal_238 isolated from chicken meat in Japan in 2016. High homology was observed among pSal_249Sch, pSal_238Inf, and pESI. The distribution, expansion, and contamination rate of *S.* Schwarzengrund with pESI-like plasmids increased more slowly compared with those of *S.* Infantis and *S.* Muenchen in Israel. cgMLST analysis showed that *S.* Schwarzengrund isolates with pESI-like plasmids from Japan, the UK, and the USA were closely related. Moreover, the ancestors of these *S.* Schwarzengrund isolates harboring the pESI-like plasmid are suggested to be common. Genomic analysis of more isolates is required to identify the genetic characteristics of pESI-carrying *S.* Schwarzengrund and the factors contributing to the increase in its prevalence.

Supplementary Materials: The following supporting information can be downloaded at https://www.mdpi.com/article/10.3390/antibiotics14030288/s1: Figure S1: Tentative plasmid structure of *Salmonella enterica* serovar Schwarzengrund and serovar Infantis obtained from chicken meat in Japan; Table S1: Antimicrobial susceptibilities and antimicrobial resistance genes for transconjugants; Table S2: Number of reads obtained in short- and long-read sequencing for *Salmonella* isolates from chicken in Japan; Table S3: Summary of tentative plasmid sequences for *Salmonella* Schwarzengrund and *Salmonella* Infantis obtained from chicken in Japan; Table S4: PCR conditions for detecting antimicrobial resistance genes, class 1 integrons, and virulence and transfer-related genes in pESI.

Author Contributions: Conceptualization, K.I.; methodology, K.I., M.O., T.O. and T.A. (Tetuso Asai); data curation, K.I., S.S., K.M., C.N. and T.A. (Takahiro Abe); validation, K.I., S.S. and K.M.; formal analysis, K.I. and S.S.; investigation, K.I., S.S., K.M, C.N., T.A. (Takahiro Abe), H.H., T.O., M.O., E.I. and T.A. (Tetsuo Asai); resources, K.I., S.S., K.M., C.N., E.I. and T.A. (Tetsuo Asai); writing—original draft preparation, K.I. and S.S.; writing—review and editing, K.I., S.S., K.M., H.H., T.O., M.O. and T.A. (Tetsuo Asai). All authors have read and agreed to the published version of the manuscript.

Funding: This study was partially supported by a grant in aid (R5-25) from the Ito Foundation of Japan.

Institutional Review Board Statement: Not applicable.

Informed Consent Statement: Not applicable.

Data Availability Statement: The data presented in this study are available upon request from the corresponding author.

Acknowledgments: We are grateful to Sakiko Ueda of the Women's Future Development Organization, Institute of Global Innovation Research, Tokyo University of Agriculture and Technology, for her assistance.

Conflicts of Interest: The authors declare no conflicts of interest.

Abbreviations

The following abbreviations are used in this manuscript:

DFAST	DDBJ Fast Annotation and Submission Tool
MAFFT	Multiple Alignment using Fast Fourier Transform
MIC	Minimum inhibitory concentration
NCBI	National Center for Biotechnology Information
RFLP	Restriction fragment length polymorphism
WGS	Whole-genome sequencing

References

1. Han, J.; Aljahdali, N.; Zhao, S.; Tang, H.; Harbottle, H.; Hoffmann, M.; Frye, G.J.; Foley, L.S. Infection biology of *Salmonella enterica*. *EcoSal Plus* **2024**, *12*, eesp-0001. [CrossRef]
2. Grimont, A.D.P.; Weill, F.-X. *Antigenic Formulae of the Salmonella Serovars*, 9th ed.; WHO Collaborating Center for Reference and Research on Salmonella: Paris, France, 2007.
3. World Health Organization. Salmonella (Non-Typhoidal). Available online: https://www.who.int/en/news-room/fact-sheets/detail/salmonella-(non-typhoidal) (accessed on 18 February 2025).
4. Alvarez, D.; Barrón-Montenegro, R.; Conejeros, J.; Rivera, D.; Undurraga, E.; Moreno-Switt, A. A review of the global emergence of multidrug-resistant *Salmonella enterica* subsp. *enterica* serovar Infantis. *Int. J. Food Microbiol.* **2023**, *403*, 110297. [CrossRef] [PubMed]
5. Aviv, G.; Tsyba, K.; Steck, N.; Salmon-Divon, M.; Cornelius, A.; Rahav, G.; Grassl, G.A.; Gal-Mor, O. A unique megaplasmid contributes to stress tolerance and pathogenicity of an emergent *Salmonella enterica* serovar Infantis strain. *Environ. Microbiol.* **2014**, *16*, 977–994. [CrossRef] [PubMed]
6. Ishihara, K.; Nakazawa, C.; Nomura, S.; Elahi, S.; Yamashita, M.; Fujikawa, H. Effects of climatic elements on Salmonella contamination in broiler chicken meat in Japan. *J. Vet. Med. Sci.* **2020**, *82*, 646–652. [CrossRef]
7. Matsui, K.; Nakazawa, C.; Khin, S.; Iwabuchi, E.; Asai, T.; Ishihara, K. Molecular Characteristics and antimicrobial resistance of *Salmonella enterica* serovar Schwarzengrund from chicken meat in Japan. *Antibiotics* **2021**, *10*, 1336. [CrossRef] [PubMed]

8. Dos Santos, A.M.P.; Panzenhagen, P.; Ferrari, R.G.; Conte-Junior, C.A. Large-scale genomic analysis reveals the pESI-like megaplasmid presence in *Salmonella* Agona, Muenchen, Schwarzengrund, and Senftenberg. *Food Microbiol.* **2022**, *108*, 104112. [CrossRef]
9. Karch, H.; Schubert, S.; Zhang, D.; Zhang, W.; Schmidt, H.; Ölschlager, T.; Hacker, J. A genomic island, termed high-pathogenicity island, is present in certain non-O157 shiga toxin-producing *Escherichia coli* clonal lineages. *Infect. Immun.* **1999**, *67*, 5994–6001. [CrossRef]
10. Franco, A.; Leekitcharoenphon, P.; Feltrin, F.; Alba, P.; Cordaro, G.; Iurescia, M.; Tolli, R.; D'Incau, M.; Staffolani, M.; Di Giannatale, E.; et al. Emergence of a clonal lineage of multidrug-resistant ESBL-producing *Salmonella* Infantis transmitted from broilers and broiler meat to humans in Italy between 2011 and 2014. *PLoS ONE* **2015**, *10*, e0144802. [CrossRef]
11. García-Soto, S.; Abdel-Glil, M.; Tomaso, H.; Linde, J.; Methner, U. Emergence of multidrug-resistant Salmonella enterica subspecies enterica serovar Infantis of multilocus sequence Type 2283 in German broiler farms. *Front. Microbiol.* **2020**, *11*, 1741. [CrossRef]
12. Alba, P.; Carfora, V.; Feltrin, F.; Diaconu, E.; Sorbara, L.; Dell'Aira, E.; Cerci, T.; Ianzano, A.; Donati, V.; Franco, A.; et al. Evidence of structural rearrangements in ESBL-positive pESI(like) megaplasmids of *S. Infantis*. *FEMS Microbiol. Lett.* **2023**, *370*, fnad014. [CrossRef]
13. Bogomazova, A.N.; Gordeeva, V.D.; Krylova, E.V.; Soltynskaya, I.V.; Davydova, E.E.; Ivanova, O.E.; Komarov, A.A. Megaplasmid found worldwide confers multiple antimicrobial resistance in *Salmonella* Infantis of broiler origin in Russia. *Int. J. Food Microbiol.* **2020**, *319*, 108497. [CrossRef] [PubMed]
14. Cohen, E.; Kriger, O.; Amit, S.; Davidovich, M.; Rahav, G.; Gal-Mor, O. The emergence of a multidrug resistant *Salmonella* Muenchen in Israel is associated with horizontal acquisition of the epidemic pESI plasmid. *Clin. Microbiol. Infect.* **2022**, *28*, 1499.e7–1499.e14. [CrossRef]
15. Iwabuchi, E.; Yamamoto, S.; Endo, Y.; Ochiai, T.; Hirai, K. Prevalence of *Salmonella* isolates and antimicrobial resistance patterns in chicken meat throughout Japan. *J. Food Prot.* **2011**, *74*, 270–273. [CrossRef] [PubMed]
16. Sasaki, Y.; Kakizawa, H.; Baba, Y.; Ito, T.; Haremaki, Y.; Yonemichi, M.; Ikeda, T.; Kuroda, M.; Ohya, K.; Hara-Kudo, Y.; et al. Antimicrobial resistance in *Salmonella* isolated from food workers and chicken products in Japan. *Antibiotics* **2021**, *10*, 1541. [CrossRef]
17. Aviv, G.; Rahav, G.; Gal-Mor, O. Horizontal transfer of the *Salmonella enterica* serovar Infantis resistance and virulence plasmid pESI to the gut microbiota of warm-blooded hosts. *mBio* **2016**, *7*, e01395-16. [CrossRef]
18. Sasaki, Y.; Ikeda, A.; Ishikawa, K.; Murakami, M.; Kusukawa, M.; Asai, T.; Yamada, Y. Prevalence and antimicrobial susceptibility of *Salmonella* in Japanese broiler flocks. *Epidemiol. Infect.* **2012**, *140*, 2074–2081. [CrossRef] [PubMed]
19. Duc, V.M.; Shin, J.; Nagamatsu, Y.; Fuhiwara, A.; Toyofuku, H.; Obi, T.; Chuma, T. Increased Salmonella Schwarzengrund prevalence and antimicrobial susceptibility of Salmonella enterica isolated from broiler chickens in Kagoshima Prefecture in Japan between 2013 and 2016. *J. Vet. Med. Sci.* **2020**, *82*, 585–589. [CrossRef]
20. Animal Quarantine Service, Ministry of Agriculture, Forestry and Fisheries. Annual Report on Animal Quarantine. Available online: https://www.maff.go.jp/aqs/tokei/toukeinen.html (accessed on 28 November 2024).
21. Sodagari, H.; Sohail, M.; Varga, C. Differences in the prevalence and antimicrobial resistance among non-typhoidal *Salmonella* serovars isolated from retail chicken meat across the United States of America, 2013–2020. *Food Control* **2024**, *165*, 110701. [CrossRef]
22. Buyrukoglu, G.; Moreira, J.; Topalcengiz, Z. Causal mediation analysis of foodborne *Salmonella* outbreaks in the United States: Serotypes and food vehicles. *Pathogens* **2024**, *13*, 1134. [CrossRef]
23. Willis, C.; Jorgensen, F.; Cawthraw, S.; Aird, H.; Lai, S.; Kesby, M.; Chattaway, M.; Lock, I.; Quill, E.; Raykova, G. A survey of *Salmonella*, *Escherichia coli*, and antimicrobial resistance in frozen, part-cooked, breaded, or battered chicken products on retail sale in the UK. *J. Appl. Microbiol.* **2023**, *134*, lxad093. [CrossRef]
24. Murakami, K.; Horikawa, K.; Ito, T.; Otsuki, K. Environmental survey of salmonella and comparison of genotypic character with human isolates in Western Japan. *Epidemiol. Infect.* **2001**, *126*, 159–171. [CrossRef] [PubMed]
25. Yokoyama, E.; Ando, N.; Ohta, T.; Kanada, A.; Shiwa, Y.; Ishige, T.; Murakami, K.; Kikuchi, T.; Murakami, S. A novel subpopulation of *Salmonella enterica* serovar Infantis strains isolated from broiler chicken organs other than the gastrointestinal tract. *Vet. Microbiol.* **2015**, *175*, 312–318. [CrossRef]
26. Tate, H.; Folster, J.P.; Hsu, C.H.; Chen, J.; Hoffmann, M.; Li, C.; Morales, C.; Tyson, G.H.; Mukherjee, S.; Brown, A.C.; et al. Comparative analysis of extended-spectrum-β-lactamase CTX-M-65-producing *Salmonella enterica* serovar Infantis isolates from humans, food animals, and retail chickens in the United States. *Antimicrob. Agents Chemother.* **2017**, *61*, e00488-17. [CrossRef]
27. Egorova, A.; Mikhaylova, Y.; Saenko, S.; Tyumentseva, M.; Tyumentsev, A.; Karbyshev, K.; Chernyshkov, A.; Manzeniuk, I.; Akimkin, V.; Shelenkov, A. Comparative whole-genome analysis of Russian foodborne multidrug-resistant *Salmonella* Infantis isolates. *Microorganisms* **2022**, *10*, 89. [CrossRef]

28. Ishihara, K.; Takahashi, T.; Morioka, A.; Kojima, A.; Kijima, M.; Asai, T.; Tamura, Y. National surveillance of *Salmonella enterica* in food-producing animals in Japan. *Acta Vet. Scand.* **2009**, *51*, 35. [CrossRef]
29. *M100-30*; Performance Standards for Antimicrobial Susceptibility Testing. Clinilac and Laboratry Standards Institute: Wayne, PA, USA, 2019.
30. Kozak, G.K.; Boerlin, P.; Janecko, N.; Reid-Smith, R.J.; Jardine, C. Antimicrobial resistance in *Escherichia coli* isolates from swine and wild small mammals in the proximity of swine farms and in natural environments in Ontario, Canada. *Appl. Environ. Microbiol.* **2009**, *75*, 559–566. [CrossRef] [PubMed]
31. Edelstein, M.; Pimkin, M.; Palagin, I.; Edelstein, I.; Stratchounski, L. Prevalence and molecular epidemiology of CTX-M extended-spectrum beta-lactamase-producing *Escherichia coli* and *Klebsiella pneumoniae* in Russian hospitals. *Antimicrob. Agents Chemother.* **2003**, *47*, 3724–3732. [CrossRef] [PubMed]
32. Ng, L.K.; Mulvey, M.R.; Martin, I.; Peters, G.A.; Johnson, W. Genetic characterization of antimicrobial resistance in Canadian isolates of Salmonella serovar Typhimurium DT104. *Antimicrob. Agents Chemother.* **1999**, *43*, 3018–3021. [CrossRef]
33. Lin, T.; Nomura, S.; Someno, S.; Abe, T.; Nishiyama, M.; Shiki, S.; Harima, H.; Ishihara, K. Role of multidrug resistance and co-resistance on a high percentage of streptomycin resistance in *Escherichia coli* isolated from chicken meats in Japan. *J. Vet. Med. Sci.* **2023**, *85*, 832–836. [CrossRef]
34. Sekizuka, T.; Yatsu, K.; Inamine, Y.; Segawa, T.; Nishio, M.; Kishi, N.; Kuroda, M. Complete genome sequence of a bla_{KPC-2}-Positive *Klebsiella pneumoniae* strain isolated from the effluent of an urban sewage treatment plant in Japan. *Msphere* **2018**, *3*, e00314-18. [CrossRef]
35. Chen, S.; Zhou, Y.; Chen, Y.; Gu, J. fastp: An ultra-fast all-in-one FASTQ preprocessor. *Bioinformatics* **2018**, *34*, i884–i890. [CrossRef] [PubMed]
36. Salmela, L.; Rivals, E. LoRDEC: Accurate and efficient long read error correction. *Bioinformatics* **2014**, *30*, 3506–3514. [CrossRef] [PubMed]
37. Kolmogorov, M.; Yuan, J.; Lin, Y.; Pevzner, P.A. Assembly of long, error-prone reads using repeat graphs. *Nat. Biotechnol.* **2019**, *37*, 540–546. [CrossRef]
38. Tanizawa, Y.; Fujisawa, T.; Kaminuma, E.; Nakamura, Y.; Arita, M. DFAST and DAGA: Web-based integrated genome annotation tools and resources. *Biosci. Microbiota. Food Health* **2016**, *35*, 173–184. [CrossRef]
39. Sullivan, M.J.; Petty, N.K.; Beatson, S.A. Easyfig: A genome comparison visualizer. *Bioinformatics* **2011**, *27*, 1009–1010. [CrossRef]
40. Bharat, A.; Petkau, A.; Avery, B.P.; Chen, J.C.; Folster, J.P.; Carson, C.A.; Kearney, A.; Nadon, C.; Mabon, P.; Thiessen, J.; et al. Correlation between phenotypic and in silico detection of antimicrobial resistance in *Salmonella enterica* in Canada using staramr. *Microorganisms* **2022**, *10*, 292. [CrossRef]
41. Clausen, P.; Aarestrup, F.; Lund, O. Rapid and precise alignment of raw reads against redundant databases with KMA. *BMC Bioinformatics* **2018**, *19*, 1–8. [CrossRef] [PubMed]
42. Alikhan, N.; Zhou, Z.; Sergeant, M.; Achtman, M. A genomic overview of the population structure of *Salmonella*. *PLoS Genet.* **2018**, *14*, e1007261. [CrossRef]
43. Zhou, Z.; Alikhan, N.; Sergeant, M.; Luhmann, N.; Vaz, C.; Francisco, A.; Carriço, J.; Achtman, M. GrapeTree: Visualization of core genomic relationships among 100,000 bacterial pathogens. *Genome Res.* **2018**, *28*, 1395–1404. [CrossRef]

antibiotics

MDPI

Article

Genomic Characterization of Extended-Spectrum β-Lactamase-Producing and Third-Generation Cephalosporin-Resistant *Escherichia coli* Isolated from Stools of Primary Healthcare Patients in Ethiopia

Deneke Wolde [1,2,3], Tadesse Eguale [2,4], Girmay Medhin [2], Aklilu Feleke Haile [2], Haile Alemayehu [2], Adane Mihret [5,6], Mateja Pirs [7], Katja Strašek Smrdel [7], Jana Avberšek [8], Darja Kušar [8], Tjaša Cerar Kišek [9], Tea Janko [9], Andrej Steyer [9] and Marjanca Starčič Erjavec [3,*]

1 Department of Medical Laboratory Science, College of Medicine and Health Sciences, Wachemo University, Hossana P.O. Box 667, Ethiopia; deneke2024@wcu.edu.et
2 Aklilu Lemma Institute of Pathobiology, Addis Ababa University, Addis Ababa P.O. Box 1176, Ethiopia; tadesse.eguale@aau.edu.et (T.E.); girmay.medhin@aau.edu.et (G.M.); aklilu.feleke@aau.edu.et (A.F.H.); haile.alemayehu@aau.edu.et (H.A.)
3 Department of Microbiology, Biotechnical Faculty, University of Ljubljana, 1000 Ljubljana, Slovenia
4 Ohio State Global One Health, Addis Ababa, Ethiopia
5 College of Health Sciences, Addis Ababa University, Addis Ababa P.O. Box 1176, Ethiopia; adane.mihret@aau.edu.et
6 Armauer Hansen Research Institute, Addis Ababa P.O. Box 1005, Ethiopia
7 Institute of Microbiology and Immunology, Faculty of Medicine, University of Ljubljana, 1000 Ljubljana, Slovenia; mateja.pirs@mf.uni-lj.si (M.P.); katja.strasek@mf.uni-lj.si (K.S.S.)
8 Institute of Microbiology and Parasitology, Veterinary Faculty, University of Ljubljana, 1000 Ljubljana, Slovenia; jana.avbersek@vf.uni-lj.si (J.A.); darja.kusar@vf.uni-lj.si (D.K.)
9 National Laboratory of Health, Environment and Food, 2000 Maribor, Slovenia; tjasa.cerar.kisek@nlzoh.si (T.C.K.); tea.janko@nlzoh.si (T.J.); andrej.steyer@nlzoh.si (A.S.)
* Correspondence: marjanca.starcic.erjavec@bf.uni-lj.si; Tel.: +386-1-3203-402

Abstract: The global spread of antimicrobial resistance genes (ARGs) in *Escherichia coli* is a major public health concern. The aim of this study was to investigate the genomic characteristics of extended-spectrum β-lactamase (ESBL)-producing and third-generation cephalosporin-resistant *E. coli* from a previously obtained collection of 260 *E. coli* isolates from fecal samples of patients attending primary healthcare facilities in Addis Ababa and Hossana, Ethiopia. A total of 29 *E. coli* isolates (19 phenotypically confirmed ESBL-producing and 10 third-generation cephalosporin-resistant isolates) were used. Whole-genome sequencing (NextSeq 2000 system, Illumina) and bioinformatic analysis (using online available tools) were performed to identify ARGs, virulence-associated genes (VAGs), mobile genetic elements (MGEs), serotypes, sequence types (STs), phylogeny and conjugative elements harbored by these isolates. A total of 7 phylogenetic groups, 22 STs, including ST131, and 23 serotypes with different VAGs were identified. A total of 31 different acquired ARGs and 10 chromosomal mutations in quinolone resistance-determining regions (QRDRs) were detected. The isolates harbored diverse types of MGEs, with *IncF* plasmids being the most prevalent (66.7%). Genetic determinants associated with conjugative transfer were identified in 75.9% of the *E. coli* isolates studied. In conclusion, the isolates exhibited considerable genetic diversity and showed a high potential for transferability of ARGs and VAGs. Bioinformatic analyses also revealed that the isolates exhibited substantial genetic diversity in phylogenetic groups, sequence types (ST) and serogroups and were harboring a variety of virulence-associated genes (VAGs). Thus, the studied isolates have a high potential for transferability of ARGs and VAGs.

Keywords: *Escherichia coli*; antimicrobial resistance genes; virulence-associated genes; mobile genetic elements; phylogenetic groups; Ethiopia

1. Introduction

Antimicrobial resistance (AMR) is a major global health problem that poses a significant threat to public health [1]. It is rapidly increasing and is expected to result in high healthcare costs and poor patient outcomes, potentially making AMR the leading cause of global mortality [2]. Africa and Asia are the two regions highly affected by AMR, with the potential for over 4.1 million annual deaths from AMR by 2050 [3]. *Escherichia coli* is one of the most important human pathogens and represents a major public health challenge. It is also one of the widely abundant commensal organisms in the gastrointestinal tracts of humans and animals, known for its ability to easily develop resistance to antimicrobial agents and to serve as a vector for the spread of AMR, which is a major problem for humans and animals [4]. In Ethiopia, there is an emergence and spread of resistance to common antimicrobials among *E. coli* isolates, leading to increased morbidity, mortality, and healthcare costs [5].

Infections with *E. coli* resistant to critically important antimicrobials, such as third-generation cephalosporins, have increased worldwide [6,7] and pose a challenge to clinical antibiotic therapy. *E. coli* strains resistant to third-generation cephalosporin antibiotics are frequently associated with the expression of extended-spectrum β-lactamases (ESBLs) [8]. ESBL production is the most problematic mechanism of antibiotic resistance, as many β-lactamase-encoding genes are typically carried on mobile genetic elements, such as plasmids, which can be easily transferred between different bacterial species [9]. Resistance to third-generation cephalosporins in *E. coli* can complicate the treatment of infections and lead to the use of a last-resort antimicrobial class such as carbapenems [10], which are not readily available in developing countries.

Bacterial populations acquire resistance to antibiotics either through chromosomal mutation or through horizontal gene transfer (HGT) from other bacteria that are either distant or closely related. This transfer is facilitated by mobile genetic elements (MGEs) [11]. MGEs are DNA elements that can move within or between bacterial cells, either from a chromosome to a plasmid or between plasmids through non-conjugative transposons, gene cassettes, and insertion sequence elements, or between DNA molecules through MGE elements capable of self-replication and conjugative transfer, such as plasmids and integrative conjugative elements [12].

MGEs carry genes that can confer resistance to antibiotics and play a critical role in facilitating the acquisition and spread of antimicrobial resistance genes (ARGs) through HGT [13]. The main mechanisms of HGT are conjugation, transduction, and transformation [14]. Conjugation drives the rapid evolution and adaptation of bacterial strains by facilitating the spread of diverse metabolic capabilities, including virulence, biofilm formation, and heavy metal and antibiotic resistance [15]. The conjugative transfer regions of the self-transmissible MGEs typically involve an origin of transfer (*oriT*) region, a conjugative type IV secretion system (T4SS), type IV coupling protein (ATPase), and a relaxase to deliver single-stranded DNA (ssDNA) into the recipient cell [16]. Understanding the profile of virulence-associated genes (VAGs), ARGs, and MGEs among *E. coli* strains circulating in a certain region is important for preventing the dissemination of antibiotic resistance, virulence factors, and other clinically relevant traits through *E. coli* bacterial populations. As there have been only a few studies conducted on this topic in Ethiopia, the aim of this study was to assess the distribution and diversity of ARGs, VAGs, MGEs, sequence types, serotypes and conjugative elements in 29 phenotypically confirmed ESBL-producing and/or third-generation cephalosporin-resistant *E. coli* isolates, obtained from patients attending primary healthcare units in Addis Ababa (n = 11) and Hossana (n = 18) using whole-genome sequencing.

2. Results

2.1. *Phylogenetic Groups of* E. coli *Isolates*

The *E. coli* isolates analyzed in this study belonged to seven phylogenetic groups. The predominant phylogenetic group was phylogroup B1 (n = 13; 44.8%), followed by phylogroup A (n = 9; 31.03%). The other phylogenetic groups identified in this study were phylogroup C and E (n = 2 each), phylogroup B2 (n = 1), phylogroup D (n = 1), and phylogroup G (n = 1) (Figure 1). Phylogroups B2, C, D, and G were obtained from both non-diarrheic and diarrheic patients from Hossana. These phylogroups were found in a slightly higher proportion (60%) among females, and except for one of the phylogroup C isolate, all of them were detected in the age groups of 5–9 and 20–45 years (Table 1).

Table 1. Distribution of different *E. coli* phylogenetic groups among patients from study sites and their relations with other patients' characteristics.

Characteristics	Phylogroup (N = 29)						
	A N (%)	B1 N (%)	B2 N (%)	C N (%)	D N (%)	E N (%)	G N (%)
Study site							
Addis Ababa (n = 11)	7 (63.6)	3 (27.3)	0	0	0	1 (9.1)	0
Hossana (n = 18)	2 (11.1)	10 (55.6)	1 (5.6)	2 (11.1)	1 (5.6)	1 (5.6)	1 (5.6)
Sex							
Female (n = 13)	3 (23.1)	6 (46.2)	1 (7.7)	1 (7.7)	1 (7.7)	1 (7.7)	0
Male (n = 16)	6 (37.5)	7 (43.7)	0	1 (6.3)	0	1 (6.3)	1 (6.3)
Participants						1 (3.7)	
Diarrheic (n = 13)	3 (23.1)	7 (53.8)	0	1 (7.7)	1 (7.7)	1 (7.7)	0
Non-diarrheic (n = 16)	6 (35.7)	6 (35.7)	1 (6.3)	1 (6.3)	0	1 (6.3)	1 (6.3)
Age groups							
0–4 (n = 1)	0	0	0	1 (100)	0	0	0
5–9 (n = 6)	2 (33.3)	2 (33.3)	0	0	1 (16.7)	0	1 (16.7)
10–14 (n = 3)	1 (66.7)	2 (66.7)	0	0	0	0	0
15–19 (n = 2)	1 (50)	1 (50)	0	0	0	0	0
20–45 (n = 14)	5 (35.7)	5 (35.7)	1 (7.1)	1 (7.1)	0	2 (14.3)	0
46–65 (n = 3)	0	3 (100)	0	0	0	0	0

2.2. Sequence Types of E. coli Isolates

Twenty-two different STs were identified, of which one was a novel sequence type designated as ST15980. The most frequently identified ST was ST10 (n = 4; 13.8%), followed by ST48, ST224, ST345, and ST410 (n = 2; 6.9% each). ST131 was also detected in this study in one *E. coli* isolate obtained from a non-diarrheic patient in Hossana. The novel ST was also among isolates from Hossana. Of the total STs identified, only ST10 and ST48 were detected in *E. coli* isolates from both Addis Ababa and Hossana (Figure 1 and Supplementary Table S1).

2.3. Serotypes of E. coli Isolates

A total of 23 different serotypes were recorded among the 29 *E. coli* isolates investigated in this study. Of the total number of tested isolates, 22 (75.8%) were typable for their O antigen and 28 (96.5%) for their H antigen, with 7 (24.1%) being 'rough' (lacking expression of the O antigen) and 1 (3.4%) lacking expression of the H antigen. The most frequently identified serotype was H30 (n = 3; 10.3%), followed by four serotypes, namely H34, O134:H53, O8:H21, and O9a:H30, each represented by two isolates (Figure 1). The serotype O153:H4 detected in this study was identified as ST131 (Supplementary Table S1).

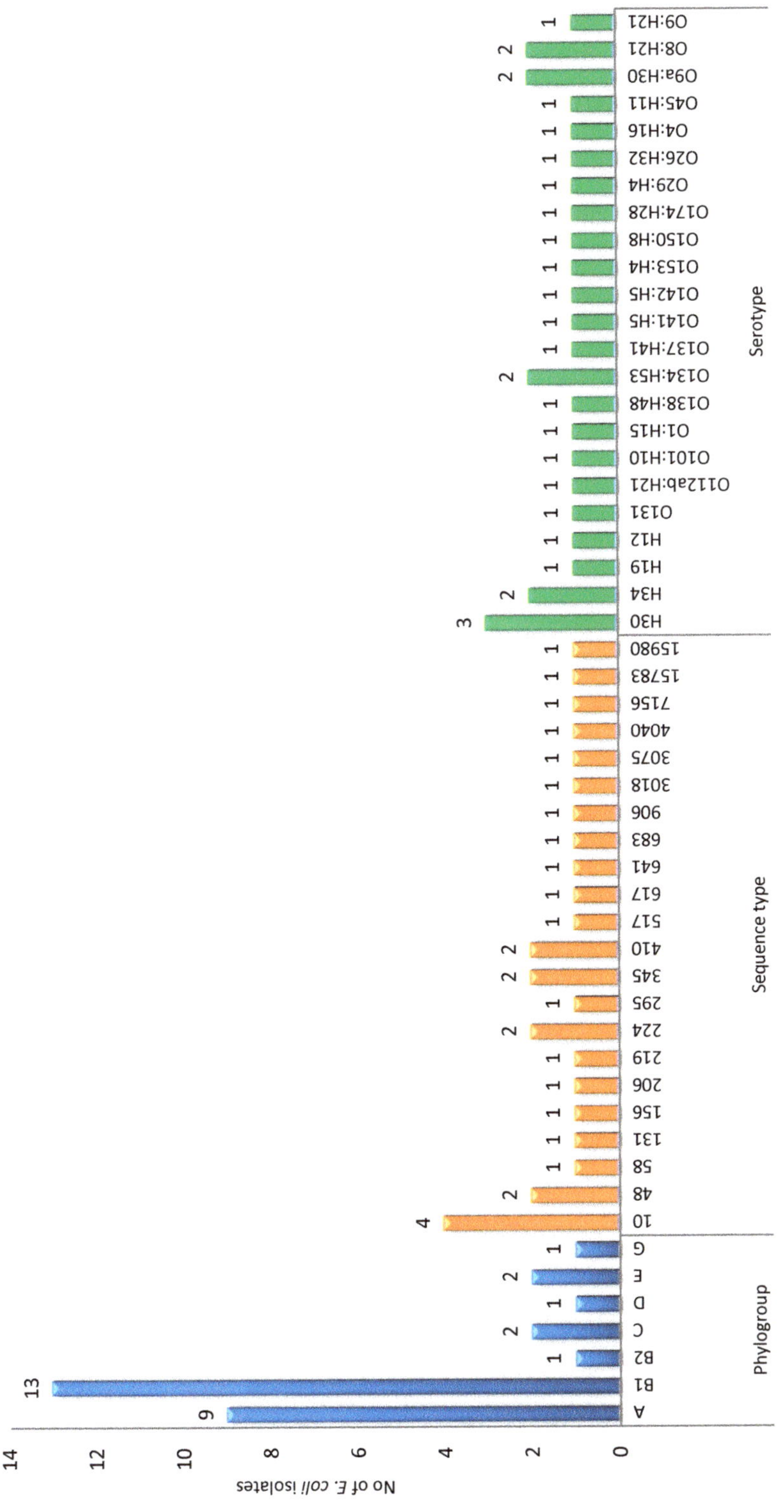

Figure 1. Phylogroups, sequence types, and serotypes of studied *E. coli* isolates.

2.4. *Virulence-Associated Gene Profile of* E. coli *Isolates*

A total of 72 different VAGs involved in adhesion (n = 23), protection against environmental stress and host immune response (n = 32), toxin production (n = 9), and in iron acquisition (n = 8) were identified in the *E. coli* isolates tested in this study. On average, all *E. coli* isolates carried 16 different VAGs, with a minimum of 7 and a maximum of 28 VAGs. The *terC* gene, which encodes for the tellurite resistance, and the *nlpI* gene, which encodes an outer membrane lipoprotein, were identified in all isolates tested. A total of 28 (96.6%) of the isolates contained *yehC* (chaperone, YhcD fimbrial cluster), 27 (93.1%) *csgA* (curlin major subunit CsgA), 26 (89.7%) *hlyE* (avian *E. coli* haemolysin), and 25 (86.2%) *yehB* (usher, YhcD fimbrial cluster). Forty-one of the total virulence-associated genes were detected in isolates from Hossana and Addis Ababa (Supplementary Table S1). Only seven genes, including the *cdt-IIIB* gene, which encodes the CdtB subunit of the cytolethal distending toxin complex, *cib* gene, which encodes a protein that protects *E. coli* cells from the cytotoxic effects of the cloacin DF13 bacteriocin, *colE5* (colicin E5 lysis protein Lys), *etsC* (putative type I secretory outer membrane protein) and *mcbA* (bacteriocin microcin B17) were detected exclusively in isolates from Addis Ababa. However, 27 different genes, including capsular polysaccharide-related genes (*kpsE*, *kpsMII*, and *kpsMII_K5*), the *tia* gene that encodes the Tia (toxigenic invasion locus) protein, which is involved in the adherence and invasion of host epithelial cells, and the *usp* (uropathogenic specific protein), were exclusively detected in isolates from Hossana (Supplementary Table S1). The most frequently detected VAGs (greater than 10%) in *E. coli* isolates from Addis Ababa and Hossana are presented in Figure 2. The study found that 28 VAGs were detected in 18.2% to 100% of the *E. coli* isolates from Addis Ababa, while 36 VAGs were detected in 11.1% to 100% of the isolates from Hossana.

The *afaB*, *afaC*, *afaE*, and *kpsMII_K5* genes were only found in the O153:H4 serotype. This serotype also possessed genes involved in the biosynthesis and uptake of yersiniabactin, aerobactin, and salmochelin siderophores. Seven VAGs (*fimH*, *nlpI*, *terC*, *yehA*, *yehB*, *yehC* and *yehD*) were detected in isolates from all phylogenetic groups. The *afaB*, *afaC*, *afaE*, and *kpsMII_K5* genes were only detected in phylogroup B2 isolates. Fourteen genes, including *usp*, *aslA*, *sigA*, *traT*, *hlyE*, *gad*, *chuA*, *cia*, *csgA*, *iha*, *ireA*, *iucC*, *iutA*, and *ompT*, were detected in phylogroup G. The *neuC* gene, which encodes a polysialic acid capsule biosynthesis protein, the *kpsMII* gene, which encodes a polysialic acid transport protein group 2 capsule and the *usp* gene, which encodes a uropathogenic specific protein, were only detected in phylogroups C, D, and G, respectively. Twenty-eight different VAGs were detected in a single ST131 *E. coli* isolate, among them the *sat* gene, which encodes the serine protease autotransporters of *Enterobacteriaceae* (SPATE), and the *senB* gene, which encodes the plasmid-encoded enterotoxin (Supplementary Table S1).

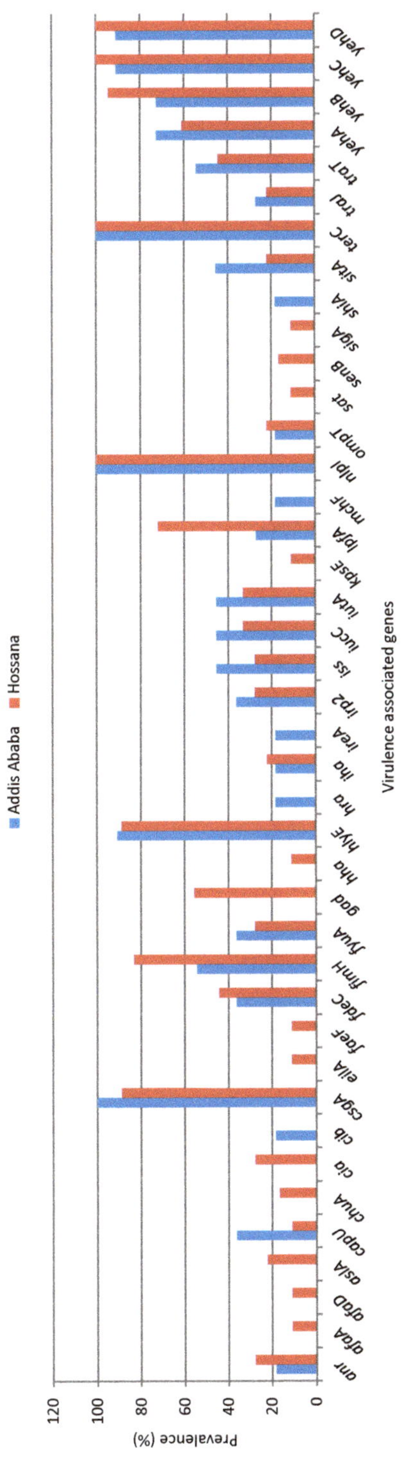

Figure 2. Virulence-associated genes detected with prevalence greater than 10% in studied *E. coli* isolates from Addis Ababa and Hossana.

2.5. Antimicrobial Resistance Genes

The distribution of ARGs among *E. coli* isolates is presented in Supplementary Table S2. ARGs were detected in 28 (96.6%) of *E. coli* isolates tested in this study. Thirty-one different horizontally transmitted acquired ARGs and ten chromosomal mutations in quinolone resistance-determining regions (QRDRs) were detected in *E. coli* isolates. Of the total *E. coli* isolates that carried ARGs, 19 (67.9%) had 5 or more different ARGs. The most prevalent ARGs detected were $bla_{CTX\text{-}M\text{-}15}$ (n = 22 isolates), followed by $bla_{TEM\text{-}1B}$ (n = 15 isolates), and *tet*(A) (n = 14 isolates). The $bla_{CTX\text{-}M\text{-}15}$ gene was identified in 78.9% of phenotypically confirmed ESBL-producing and 70% of third-generation cephalosporin-resistant *E. coli*. In total, 2 *E. coli* isolates from Hossana carried 14 different resistance genes encoding for resistance to antibiotics of different classes. Among third-generation cephalosporin-resistant *E. coli*, 90% carried β-lactamase and quinolone resistance genes. In this study, the plasmid-mediated quinolone resistance (PMQR) determinants, such as *qnrS1*, *aac(6′)-Ib-cr*, *qepA1*, *qepA2*, and *qepA4*, were detected. Among these, *qnrS1* was the most predominant, being detected in 13 (44.8%) of the isolates. Mutations in the QRDR and PMQR genes co-existed in 5 (20%) of the *E. coli* isolates carrying genes that determine resistance to quinolones. Of the *E. coli* isolates tested, 48.3% had chromosomal mutations in *gyrA*, *parC* and/or *parE*, which are associated with resistance to quinolones—ciprofloxacin and nalidixic acid. Double substitutions in *gyrA* (S83L + D87N) were observed in 8 (27.6%) of the *E. coli* isolates and these isolates also had an additional substitution in the *parC* (S80I) and *parE* (S458A) genes. Single amino acid substitutions, including S83L (n = 3), S83A (n = 1) and S83V (n = 1), were also observed in the *gyrA* gene. The substitutions at S80I (n = 6), E84K (n = 1), and A56T (n = 1) in *parC*, as well as S458A (n = 7), L416F (n = 1), and I529L (n = 1) in *parE*, were also observed (Table 2 and Supplementary Table S2).

This study found 7 *E. coli* serotypes (H30, O138:H48, O101:H10, O174:H28, O134:H53, O9a:H30, O8:H21) harboring 10 to 14 different ARGs. The predominant serotypes carrying the highest number (14) of ARGs were O9a:H30 and O8:H21. Furthermore, the *E. coli* isolates across all phylogenetic groups, with the exception of 1 from group B1, carried a range of ARGs, with groups A and B1 sharing 24 different ARGs. The analysis of sequence types revealed that ST224 and ST410 *E. coli* carried a high proportion of ARGs (Supplementary Table S2).

Table 2. Distribution of antimicrobial resistance genes and chromosomal mutations among studied *E. coli* isolates.

Antimicrobial Class	Antimicrobial Resistance Genes/Mutations Detected	No (%) of ESBL-Producing Isolates Possessing the Gene/Mutation (n = 19)	No (%) of 3rd-Generation-Cephalosporin-Resistant Isolates Possessing the Gene/Mutation (n = 10)	Total No (%) of Isolates Possessing the Gene/Mutation (n = 29)
β-lactams	$bla_{CTX\text{-}M\text{-}15}$	15 (78.9)	7 (70)	22 (75.9)
	$bla_{CTX\text{-}M\text{-}3}$	4 (21.1)	0	4 (13.8)
	$bla_{TEM\text{-}1B}$	10 (52.6)	5 (50)	15 (51.7)
	$bla_{TEM\text{-}169}$	2 (10.5)	0	2 (6.9)
	$bla_{TEM\text{-}33}$	2 (10.5)	0	2 (6.9)
	$bla_{SHV\text{-}12}$	0	1 (10)	1 (3.4)
	$bla_{OXA\text{-}1}$	1 (5.3)	1 (10)	2 (6.9)
Aminoglycosides	*aph(6)-Id*	8 (42.1)	3 (30)	11 (37.9)
	aph(3″)-Ib	8 (42.1)	3 (30)	11 (37.9)
	aac(6′)-Ib-cr	1 (5.3)	1 (10)	2 (6.9)
	aac(3)-IId	2 (10.5)	0	2 (6.9)
	aadA1	2 (10.5)	0	2 (6.9)
	aadA2	5 (26.3)	0	5 (17.2)
	aadA5	2 (10.5)	2 (20)	4 (13.8)

Table 2. *Cont.*

Antimicrobial Class	Antimicrobial Resistance Genes/Mutations Detected	No (%) of ESBL-Producing Isolates Possessing the Gene/Mutation (*n* = 19)	No (%) of 3rd-Generation-Cephalosporin-Resistant Isolates Possessing the Gene/Mutation (*n* = 10)	Total No (%) of Isolates Possessing the Gene/Mutation (*n* = 29)
Quinolones	*qnrS1*	8 (42.1)	5 (50)	13 (44.8)
	qepA1	0	1 (10)	1 (3.4)
	qepA2	0	1 (10)	1 (3.4)
	qepA4	1 (5.3)	1 (10)	2 (6.9)
	gyrA:p.S83L	7 (36.8)	4 (40)	11 (37.9)
	gyrA:p.D87N	5 (26.3)	3 (30)	8 (27.6)
	gyrA:p.S83A	1 (5.3)	0	1 (3.4)
	gyrA:p.S83V	1 (5.3)	0	1 (3.4)
	parC:p.S80I	5 (26.3)	2 (20)	7 (24.1)
	parC:p.E84K	0	1 (10)	1 (3.4)
	parC:p.A56T	1 (5.3)	0	1 (3.4)
	parE:p.S458A	5 (26.3)	2 (20)	7 (24.1)
	parE:p.L416F	0	1 (10)	1 (3.4)
	parE:p.I529L	0	1 (10)	1 (3.4)
Sulfonamides	*sul1*	7 (36.8)	3 (30)	10 (34.4)
	sul2	8 (42.1)	3 (30)	11 (37.9)
Trimethoprim	*dfrA1*	2 (10.5)	0	2 (6.9)
	dfrA12	5 (26.3)	0	5 (17.2)
	dfrA14	4 (21.1)	2 (20)	6 (20.7)
	dfrA17	2 (10.5)	2 (20)	4 (13.8)
	dfrB4	0	1 (10)	1 (3.4)
Tetracyclines	*tet*(A)	9 (47.4)	5 (50)	14 (48.3)
	tet(B)	4 (21.1)	1 (10)	5 (17.2)
Macrolides	*mph*(A)	5 (26.3)	2 (20)	7 (24.1)
	erm(B)	1 (5.3)	2 (20)	3 (10.3)
Amphenicols	*catA1*	2 (10.5)	1 (10)	3 (10.3)

2.6. Co-Occurrence of Antimicrobial Resistance Genes

Several ARGs were detected together in most of the isolates in the current study. All *E. coli* isolates carrying at least one aminoglycoside antibiotic resistance gene were shown to carry β-lactam, quinolone, sulfonamide, and trimethoprim resistance genes simultaneously. *E. coli* isolates that carried at least one β-lactam antibiotic resistance gene also carried quinolone resistance genes and sulfonamide resistance genes in 89.3% and 75% of the cases, respectively. The β-lactam antibiotic resistance genes were also detected in 39.3% of *E. coli* isolates that carried at least one trimethoprim resistance gene, 32.1% of *E. coli* isolates that carried at least one macrolide resistance gene and 10.7% of *E. coli* isolates that carried at least one amphenicol resistance gene (Supplementary Table S2).

2.7. Plasmids and Other Mobile Genetic Elements

In this study, *E. coli* isolates were found to harbor diverse MGEs, including plasmids, insertion sequences (IS), integrative conjugative elements (ICEs), and transposons. A total of 86 MGEs were detected in the *E. coli* isolates. Among these, 26 were identified as plasmids, 50 as IS, and 6 as composite transposons and 2 as unit transposons. The remaining two were ICEs and miniature inverted repeats. Except for one isolate, all tested *E. coli* isolates were confirmed to carry one to eleven plasmids. The *IncF* plasmid was the most abundant plasmid in *E. coli* isolates, with different replicon types and replicon variants, including *IncFII*, *IncFII(29)*, *IncFIB(AP001918)*, *IncFII(pRSB107)*, *IncFII(pAMA1167-NDM-5)*, *IncFII(pHN7A8)*, *IncFII(pCoo)*, *IncFIB(S)*, *IncFIC(FII)*, *IncFIA(HI1)*, *IncFII(pSE11)*, *IncFIA*, and *IncFIB(K)*. The isolates also carried other incompatibility group plasmids, including *IncI1*

(n = 11), *IncI2* (n = 1), *IncY* (n = 9), *IncX1* (n = 3), *IncX3* (n = 1), and *IncX4* (n = 2). The *Col* plasmids detected in the *E. coli* isolates in this study were *Col156* (n = 7), *Col(MG828)* (n = 5), *ColRNAI* (n = 2), and *Col(BS512)* (n = 4). A total of 26 plasmids' combination patterns were detected in *E. coli* isolates. Nine isolates carried two plasmids, while four isolates carried three plasmids. Nine isolates carried four or more plasmids, while a single isolate obtained from patient at Hossana carried eleven plasmids (Figure 3).

A total of 24 different plasmids were detected in *E. coli* isolates carrying genes resistant to quinolones, while 22 plasmids were detected in *E. coli* isolates carrying genes resistant to antibiotics of the aminoglycoside class. The *IncI2* plasmid was detected only in *E. coli* isolates harboring genes with resistance to the quinolones, whereas the *IncHI1A* and *IncHI1B(CIT)* plasmids were detected only in *E. coli* isolates harboring genes with resistance to the β-lactams. The dominant plasmid detected in all classes of antibiotic resistance gene-carrying *E. coli* was the *IncFIB(AP001918)* plasmid type.

Similar to plasmids, the *E. coli* isolates also carried several IS elements, miniature inverted repeats, integrative conjugative elements, and transposons. At least four MGEs, other than plasmids, were detected. The MITEEc1 was found in all tested *E. coli* isolates. A total of fifty different IS elements belonging to twenty-one IS families were identified. The most abundant IS family was IS3, followed by ISAs1 and IS630. The most prevalent IS elements were ISEc1 and IS609, detected in 72.4% and 58.6% of all *E. coli* isolates, respectively. Additionally, ISEc38, IS26, and ISEc9 were found in 37.9%, 34.5%, and 34.5% of the isolates, respectively. Isolates carrying twelve different MGEs were the most predominant (14.8%), followed by isolates carrying seven, eight, eleven, and thirteen different MGEs (11.1%) each. Notably, 1 *E. coli* isolate carried 17 different MGEs, while 15 isolates carried 10 to 17 MGEs (Figure 4 and Supplementary Table S3). There was a significant association between genes determining resistance to aminoglycosides and the presence of IS*3* (Fisher exact $p = 0.01$), IS*5075* (Fisher exact $p = 0.01$) and IS*629* ($X2 = 4.821$, $p = 0.028$). Similarly, a significant association was observed between genes determining β-lactam resistance and ISEc*31* (Fisher exact $p = 0.016$), ISEc*9* (Fisher exact $p = 0.048$), and ISSfl*8* (Fisher exact $p = 0.015$). In addition, resistance to quinolones was significantly associated with the presence of ISEc*9* (Fisher exact $p = 0.001$) and ISKpn*19* (Fisher exact $p = 0.003$).

2.8. Determination of Conjugative Transferable Elements

Of the 29 *E. coli* isolates tested, 22 (75.9%) had all the essential components necessary for conjugation, such as *oriT*s, relaxases, T4CPs, and T4SSs, indicating their potential for self-transferability. Fourteen (63.6%) of the *E. coli* isolates with conjugative transferable elements were obtained from eight diarrheic patients and six non-diarrheic patients in Hossana, while the remaining eight (three from diarrheic and five from non-diarrheic) were obtained from Addis Ababa. Four (13.8%) of the *E. coli* isolates were found not to carry *oriT*, whereas relaxases were not detected in three (10.3%) isolates. Only T4CP was detected in one *E. coli* isolate; *oriT*, relaxase, and T4SS were absent. Two (6.9%) isolates had both *oriT* and T4CP, but no relaxase and/or T4SS. Twenty-one (95.4%) *E. coli* with conjugative transferable elements had acquired ARGs. Virulence-associated genes were detected in all *E. coli* isolates with self-transferable and non-self-transferable MGEs.

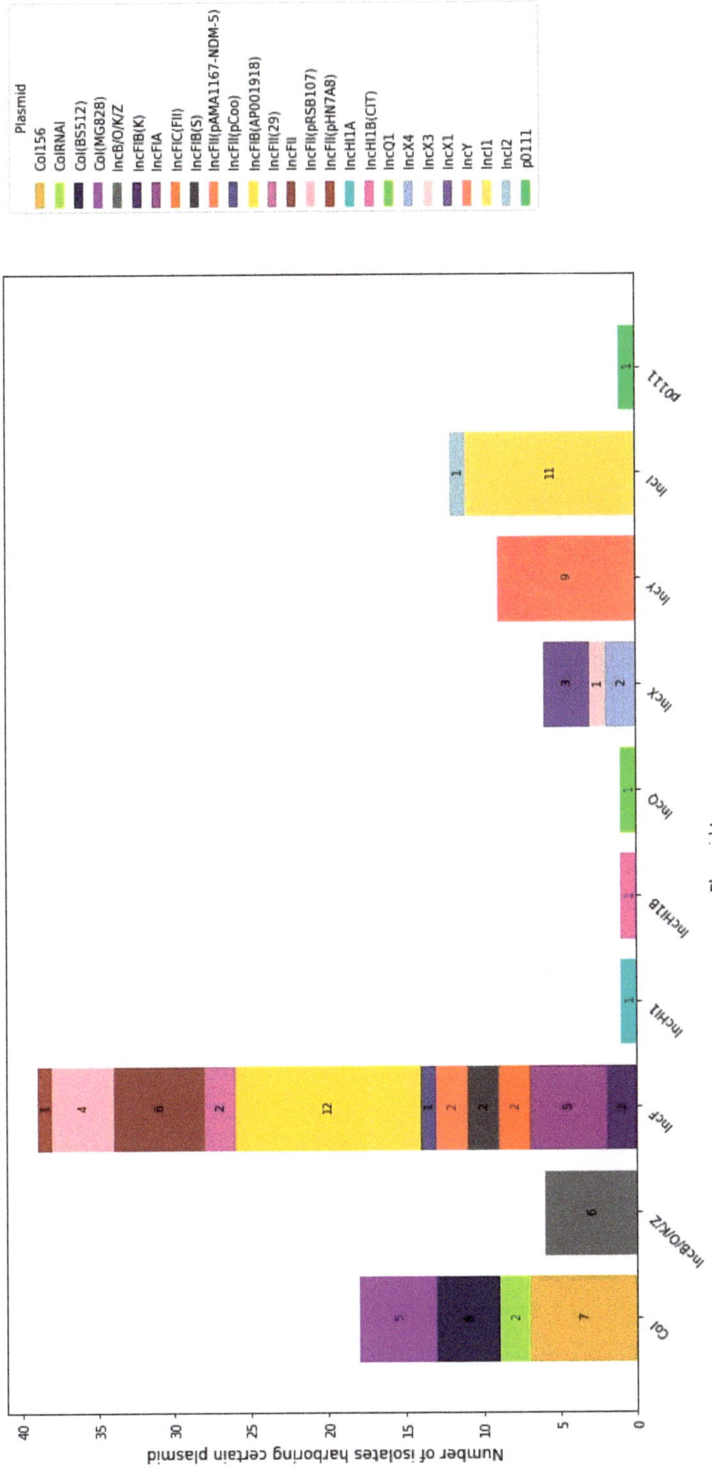

Figure 3. Number of studied *E. coli* isolates harboring certain plasmid types.

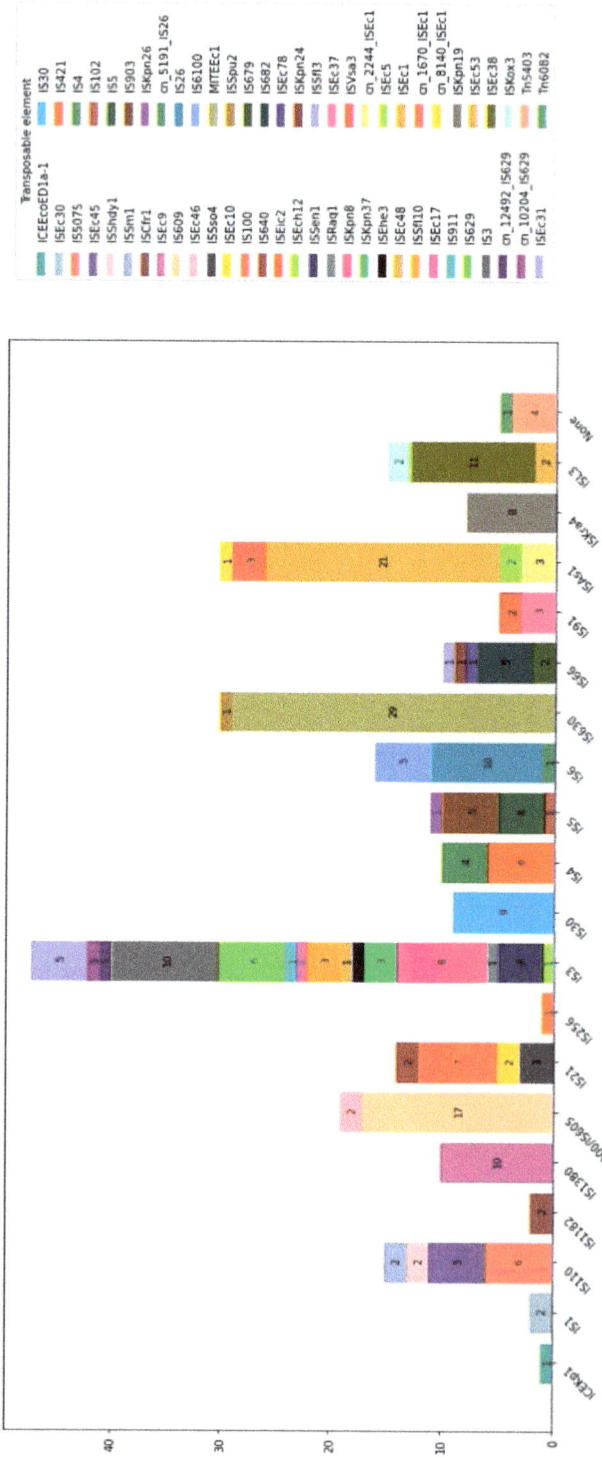

Figure 4. Number of studied *E. coli* isolates harboring certain types of transposable elements.

3. Discussion

E. coli is an extremely diverse bacterial species in which only about 6% of the genes are shared by all strains. The remaining genes, accounting for more than 90%, are variable "accessory genes" that are differentially present in the various *E. coli* strains [17]. The results of the study revealed a high level of genetic diversity among the *E. coli* isolates, which were classified into seven distinct phylogenetic groups. The findings showed that 82.8% of the *E. coli* isolates belonged to phylogroups A, B1, and C, which are generally regarded as phylogenetic groups in which most intestinal pathogenic *E. coli* strains are found in humans [18]. Of all the phylogenetic groups, B1 was the predominant one. This is consistent with previous findings from southwestern Nigeria [19]. In contrast, a study conducted in South Africa among diarrheic children confirmed that a predominant strain characterized by increased virulence and the ability to cause a wide range of infections belonged to phylogroup B2 [20]. The differences can be explained by health, diet and environmental, social, and geographical conditions [21]. The phylogroup B2 *E. coli* strain identified in our study belonged to the ST131. In addition, phylogroup G, which is known to contain highly virulent and AMR strains, was also detected in the current study [22], which implies a higher likelihood of severe and invasive *E. coli* infections.

In the current study, the *E. coli* isolates showed 22 different STs. One of them was a novel ST designated as ST15980, which was isolated from a non-diarrheic patient in Hossana. This isolate showed a broader spectrum of resistance to antimicrobials. The isolate carried resistance genes (*aph(6)-Id, aph(3'')-Ib,* $bla_{\text{CTX-M-15}}$, $bla_{\text{TEM-1B}}$, *qnrS1, sul2, dfrA14*) conferring resistance to aminoglycosides, β-lactams, quinolones, sulfonamides, and trimethoprim, which complicates treatment with a wide range of common antibiotics and poses a serious health problem. This isolate also typically carried plasmids from incompatibility groups (*IncI1, IncY*) and other MGEs such as (*IS26, IS30, ISEc38*), which often harbor genes conferring resistance to various antibiotics and have the ability to be efficiently transferred between bacterial cells [23]. Many VAGs have also been detected in ST15980, enabling it to adhere to host tissues and cells, evade the host immune system, and cause tissue damage. These include *hlyF*, the gene encoding hemolysin, which may be associated with increased production of outer membrane vesicles and contributes to the release of cytolethal distending toxin and other chemicals [24]; and the *terC* gene, which encodes a subunit of the tellurite resistance protein complex, which may contribute to its fitness and allow it to evade the host's primary immune response [25], making the strain more virulent and resistant to immune defense.

The prominent ST131, which is a highly virulent and extensively antimicrobial-resistant strain that has spread explosively throughout the world [26], was also found in our study. It is known to cause extraintestinal infections, including urinary tract and bloodstream infections [17]. The ST131 isolated in this study carried 28 VAGs, including *afa* (Dr binding adhesins), *iutA* (aerobactin receptor), and *kpsMT II* (group 2 capsule synthesis), which are characteristic of extraintestinal pathogenic *E. coli* (ExPEC) [18], demonstrating its ability to colonize and persist in the intestine, as confirmed by other studies [19]. Additionally, this isolate contained several genes, including *sat* (secreted autotransporter toxin), *fimH* (type 1 fimbriae), *fyuA* (yersiniabactin receptor), *iha* (adhesin siderophore receptor), *ompT* (outer membrane receptor) *iucC* (aerobactin), *iutA* (aerobactin receptor), and *tratT* (serum resistance associated), that are frequently found in *E. coli* ST131 isolates. Studies have indicated the growing prevalence and spread of the ST131 clonal group in various African countries. In Malawi, a genomic epidemiology study at Queen Elizabeth Central Hospital, a tertiary care center and referral hospital in Lilongwe, found ST131 in 14.9% and 32.8% of the *E. coli* isolates sequenced, respectively [27,28]. Additionally, a study conducted in South Africa found that 18% of *E. coli* isolated from urinary tract infections in inpatients and outpatients belonged to ST131 [29]. Furthermore, a study found that ESBL-producing *E. coli* isolates from Tanzania and Uganda belonged to the ST131 strain, accounting for 8.4% and 2.9% of isolates, respectively [30]. Considering the increased prevalence and pandemic nature of the ST131 strain, which poses a significant threat to public health, due attention should be given to prevent and control its dissemination.

A high degree of diversity was observed in the *E. coli* population with 23 different serotypes. The predominant *E. coli* serotype was H30, which showed increased resistance to different classes of antibiotics. They harbored a plasmid encoding the $bla_{CTX-M-15}$ gene, one of the most widespread and predominant ESBL-encoding genes [31]. They also carried genes that determine resistance to aminoglycosides, quinolones, sulfonamides, trimethoprim, tetracyclines, macrolides, and amphenicols. The result also showed that this serotype harbors different groups of plasmids and MGEs containing genes for various VAGs encoding protectins, iron acquisition, cytotoxins, or adhesion factors [32]. The high diversity of ARGs and VAGs and their distribution among various serotypes and STs suggest multiple sources of resistant and virulent bacteria and the flow of these genes between different bacterial populations [33].

Antimicrobial resistance is one of the greatest global threats to public health. The fundamental reason why some bacteria are able to resist the effects of antibiotics is that they have specific genes that code for resistance mechanisms [34]. The resistance genes were detected in 96.6% of the *E. coli* isolates tested in this study. This rate is comparable to a previous report in which resistance genes were found in 84.9% of *E. coli* isolates from humans in the Central Zambia region [35]. However, a lower rate (38%) of ARGs was detected in *E. coli* isolates obtained from human samples in a systematic review and meta-analysis conducted in South Africa [36]. Bioinformatic analysis of the whole-genome sequence of the isolate (MBL protocol number 181) that was found to be phenotypically resistant to ampicillin, cefuroxime, cefotaxime, ceftriaxone, and trimethoprim-sulfamethoxazole did not reveal any ARGs, indicating that this isolate might possess a not yet identified ARG. Another possible explanation could be that the MGE carrying the ARG conferring this specific resistance phenotype was lost in sub-culturing the *E. coli* isolates.

E. coli has developed various mechanisms to resist the effects of antibiotics. In many cases, a single strain of *E. coli* can carry ARGs that confer resistance to different classes of antibiotics [37]. This ability significantly complicates the treatment of *E. coli* infections, as clinicians have fewer effective antibiotic options to choose from [38]. The current study found a high co-detection rate of multiple ARGs within the *E. coli* isolates tested. For example, all the isolates carrying aminoglycoside resistance genes were also confirmed to harbor resistance genes for β-lactams, quinolones, sulfonamides, and trimethoprim. Many of the detected resistance genes, including those encoding aminoglycoside-modifying enzymes (AMEs), β-lactamases, and determinants of quinolone resistance, sulfonamides, trimethoprim, and tetracyclines resistance, are often located on MGEs [39–41]. Furthermore, plasmids belonging to different incompatibility groups, and other MGEs, which have an important role in the transmission of ARGs, were simultaneously detected in the isolates in the current study. This observation suggests that the diverse MGEs may have the capacity to disseminate different antibiotic classes, further exacerbating the problem of multi-drug-resistant *E. coli* infections and the need to target plasmids to limit acquisition and transmission of antimicrobial resistance, as already reviewed by Vrancianu et al. [42].

Approximately 75.9% of plasmids were found to be self-transferable, possessing key genetic elements such as *oriT*s, relaxases, T4CPs, and T4SSs. This high proportion of self-transferable plasmids indicates their substantial potential for self-transfer and dissemination within bacterial populations. Additionally, it was observed that 95.4% of the conjugative plasmids examined were found to carry acquired ARGs. This suggests that these conjugative plasmids have the capacity to potentially spread and disseminate ARGs among different bacterial species, which poses a significant concern in the context of antimicrobial resistance.

The main limitation of this study was the inability to conclusively determine whether ARGs were plasmid-encoded or chromosomally located. This technical limitation restricts the understanding of the genetic mechanisms underlying the observed multi-drug resistance. Further investigations employing advanced sequencing techniques would be needed to address these limitations and provide a more comprehensive understanding of the antimicrobial resistance landscape in the region.

4. Materials and Methods

4.1. E. coli *Isolates*

The current study is a continuation of a previous cross-sectional study conducted in Addis Ababa and Hossana from October 2021 to September 2022, in which 260 *E. coli* strains were isolated from patients of all ages and sexes recruited according to patient flow from 13 randomly selected health centers from 4 sub-cities (40% of the total) in Addis Ababa and all 4 health centers in Hossana [43]. For the current study, all *E. coli* isolates that showed phenotypic resistance to third-generation cephalosporin antibiotics or were identified as ESBL-producing were selected from the previous collection, because *E. coli* isolates exhibiting these characteristics are critically important for human medicine. The sampling method used in the current study is shown in Figure 5.

4.2. DNA Extraction, Whole-Genome Sequencing, Raw Data Pre-Processing, De Novo Assembly, and Quality Control

Twelve selected isolates exhibiting phenotypic resistance to third-generation cephalosporins were subjected to whole-genome sequencing (WGS). Genomic DNA extraction, WGS, raw data pre-processing, de novo assembly and quality control was performed as previously described for the 19 ESBL-producing isolates possessing ESBL genes [43]. The whole-genome sequence data of the two sequenced third-generation cephalosporin-resistant *E. coli* isolates did not pass the preliminary quality assessment criteria and were excluded in further bioinformatic analysis.

4.3. Bioinformatics Analysis

The ClermonTyping tool available at http://clermontyping.iame-research.center/ (accessed on 7 June 2024) was used to determine the phylogenetic groups of *E. coli* isolates [44]. Sequence typing was performed based on the seven gene Achtman scheme using the website https://pubmlst.org/ (accessed on 7 June 2024) [45] and EnteroBase https://enterobase.warwick.ac.uk/ (accessed on 27 June 2024) [46]. The detection of *E. coli* virulence-associated genes and serotypes was performed using VirulenceFinder 2.0 and SerotypeFinder available at www.genomicepidemiology.org (accessed on 5 June 2024) with the default parameters [47,48]. In addition, the acquired antimicrobial resistance genes as well as chromosomal point mutations causing resistance were determined using ResFinder 4.0, available at http://genepi.food.dtu.dk/resfinder (accessed on 5 June 2024), with default parameters [49]. Plasmid replicon was identified to infer plasmid presence using PlasmidFinder 2.1, available at https://cge.food.dtu.dk/services/PlasmidFinder/ (accessed on 15 June 2024), with minimum identity of 95% and minimum coverage of 60% [50]. Integrated mobile genetic elements were predicted in the assembled genomes using MobileElementFinder version 1.0 with MGEdb v1.0.2, available at https://cge.food.dtu.dk/services/MobileElementFinder/ (accessed on 15 June 2024) [51].

4.4. Statistical Analysis

The data were captured in Microsoft Excel and analyzed using SPSS software (Version 25. IBM SPSS Inc., New York, NY, USA). Descriptive analyses using frequency and percentages were used to summarize the variables. The chi-square or Fisher exact test was used to assess the association between ARGs and MGEs in *E. coli* isolates. A p-value of <0.05 was considered statistically significant.

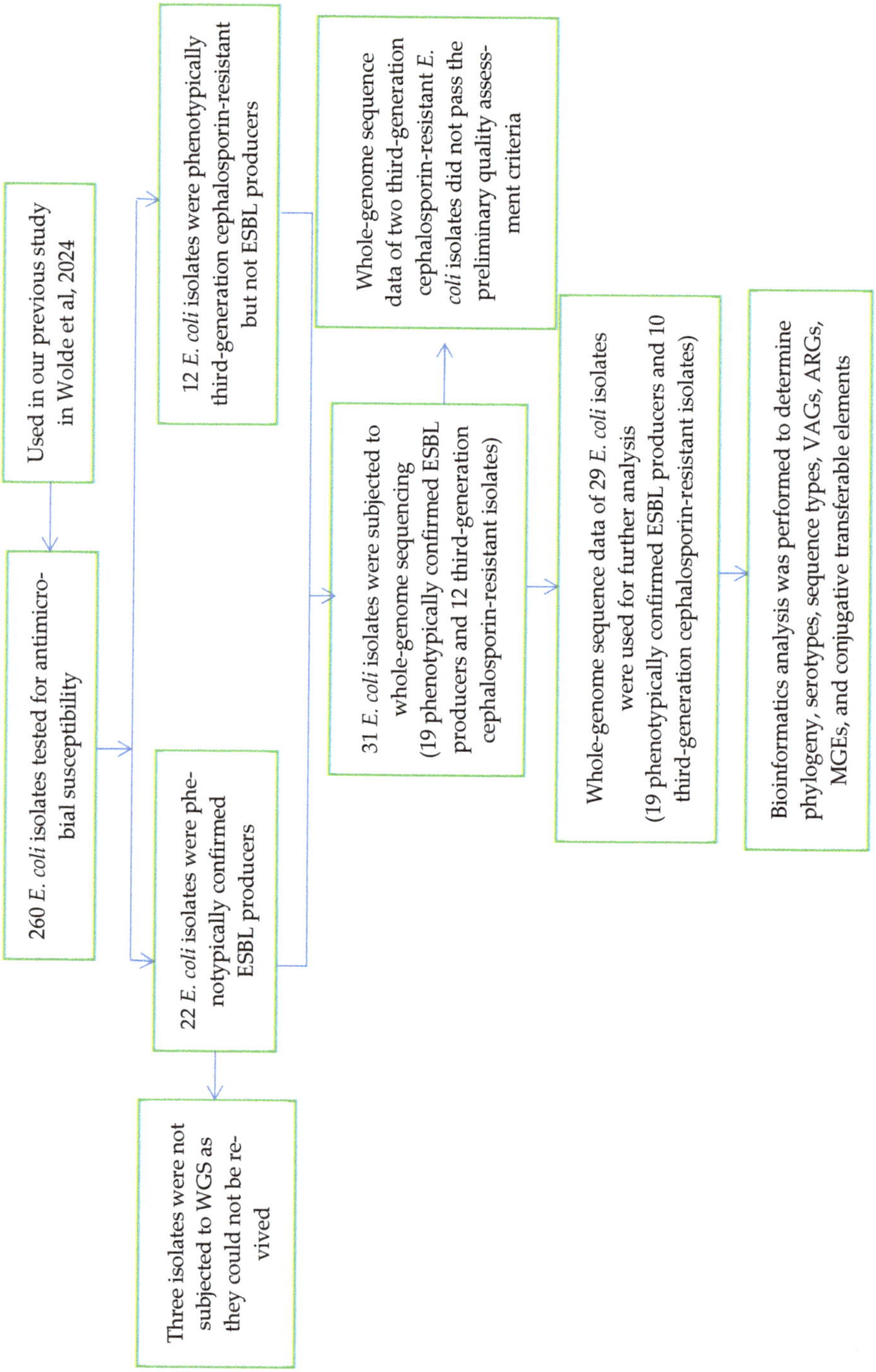

Figure 5. Flow diagram showing the selection of *E. coli* isolates for the study.

5. Conclusions

The overall finding of the current study showed the high genetic diversity of ESBL-producing and third-generation cephalosporin-resistant *E. coli* in patients from the two study areas. The observed diversity in the ARGs and MGEs among different STs and serotypes suggests the possible exchange of genes amongst different isolates, posing a significant challenge to infection control and to effectively treating infections. Additionally, a high proportion of self-transferable conjugative elements suggests a significant potential for horizontal gene transfer, with implications for the spread of clinically important traits, such as virulence and antimicrobial resistance. Therefore, the findings from this study highlight the pressing need for comprehensive strategies to address the challenge of AMR in the study area. In the future, based on the findings of this study, enhanced surveillance covering a large geographic area with a representative number of samples in collaboration with various stakeholders and sectors will be proposed to address this growing public health challenge. Implementation of strategies to reduce the spread of ARGs and VAGs is necessary to address this public health threat.

Supplementary Materials: The following supporting information can be downloaded at: https://www.mdpi.com/article/10.3390/antibiotics13090851/s1, Supplementary Table S1: VAGs detected in *E. coli* isolates; Supplementary Table S2: ARGs detected in *E. coli* isolates; Supplementary Table S3: MGEs detected in *E. coli* isolates.

Author Contributions: Conceptualization, D.W., M.P., J.A. and M.S.E.; methodology, D.W., K.S.S., J.A., M.P., T.C.K. and D.K.; software, D.W., G.M., T.C.K., T.J., A.S. and M.S.E.; validation, T.E. and M.S.E.; formal analysis, D.W., K.S.S., T.C.K. and M.S.E.; investigation, D.W., M.P., K.S.S., J.A., D.K., T.C.K., T.J. and A.S.; resources, M.P., J.A., A.S. and M.S.E.; data curation, D.W., A.F.H., H.A., A.M., K.S.S., J.A., D.K., T.C.K., T.J., A.S. and M.S.E.; writing—original draft preparation, D.W.; writing—review and editing, D.W., T.E., G.M., A.F.H., H.A., A.M., M.P., K.S.S., J.A., D.K., T.C.K., T.J., A.S. and M.S.E.; visualization, D.W., T.C.K. and M.S.E.; supervision, T.E., A.F.H., G.M, A.M. and M.S.E.; project administration, M.S.E.; funding acquisition, M.S.E. All authors have read and agreed to the published version of the manuscript.

Funding: This research was partially supported by Addis Ababa University through the thematic research project and by the Slovenian Research and Innovation Agency (research core funding no. P4-0092 'Animal health, environment and food safety'; no. P1-0198 'Molecular biology of microorganisms; no. P3-0083 'Host-parasite relationship'; and the APC was funded by P1-0198). D.W. was a grant recipient from the Swedish International Development Cooperation Agency.

Institutional Review Board Statement: This study was conducted in accordance with the Declaration of Helsinki and approved by the Institutional Review Board of ALIPB, AAU (protocol code ALIPB IRB/66/2013/21 and date of approval 19 August 2021).

Informed Consent Statement: Informed consent was obtained from all subjects involved in the study.

Data Availability Statement: The generated sequencing raw data and assembled genomes were submitted to SRA—Sequence Read Archive (accession number: PRJNA1105046, URL https://www.ncbi.nlm.nih.gov/bioproject/PRJNA1105046, accessed on 10 May 2024).

Acknowledgments: The authors would like to thank Marko Kolenc from the Institute of Microbiology and Immunology, University of Ljubljana and Tom Koritnik from the National Laboratory of Health, Environment, and Food, Ljubljana, Slovenia for their assistance in DNA extraction and sequencing isolates, respectively. The authors are also thankful to Nigel Dyer, research fellow, Bioinformatics Research Technology Platform, University of Warwick (EnteroBase), for his support in defining a new sequence type detected in our study.

Conflicts of Interest: The authors declare no conflicts of interest.

References

1. Walsh, T.R.; Gales, A.C.; Laxminarayan, R.; Dodd, P.C. Antimicrobial resistance: Addressing a global threat to humanity. *PLos Med.* **2023**, *20*, e1004264. [CrossRef]

2. Porooshat, D. Antimicrobial Resistance: Implications and costs. *Infect. Drug Resist.* **2019**, *12*, 3903–3910. [CrossRef]
3. Ohemu, G.P. Starved of ACTION: A critical look at the antimicrobial resistance action plans of African countries. *ACS Infect. Dis.* **2022**, *8*, 1377–1380. [CrossRef] [PubMed]
4. Poirel, L.; Madec, J.-Y.; Lupo, A.; Schink, A.-K.; Skieffer, N.; Nordmann, P.; Schwarz, S. Antimicrobial resistance in *Escherichia coli*. *Microbiol. Spectr.* **2018**, *6*, ARBA-0026-2017. [CrossRef]
5. WHO Africa. Addressing the challenge of antimicrobial resistance in Ethiopia. 2023. Available online: https://www.afro.who.int/photo-story/addressing-challenge-antimicrobial-resistance-ethiopia (accessed on 14 June 2024).
6. World Health Organization. *Critically Important Antimicrobials for Human Medicine: Ranking of Medically Important Antimicrobials for Risk Management of Antimicrobial Resistance due to Non-Human Use*, 6th ed.; World Health Organ: Geneva, Switzerland, 2019.
7. de Been, M.; Lanza, V.F.; de Toro, M.; Scharringa, J.; Dohmen, W.; Du, Y.; Hu, J.; Lei, Y.; Li, N.; Tooming-Klunderud, A.; et al. Dissemination of cephalosporin resistance genes between *Escherichia coli* Strains from farm animals and humans by specific plasmid lineages. *PLoS Genet.* **2014**, *10*, e1004776. [CrossRef] [PubMed]
8. Jariremombe, R.C. Mechanisms of antimicrobial resistance of *E. coli*. In *Escherichia coli—Old New Insights*; Starčič Erjavec, M., Ed.; IntechOpen: London, UK, 2022; pp. 1–9. [CrossRef]
9. Bush, K. Past and present perspectives on β-lactamases. *Antimicrob. Agents Chemother.* **2018**, *62*, e01076-18. [CrossRef]
10. Nørgaard, S.M.; Jensen, C.S.; Aalestrup, J.; Vandenbroucke-Grauls, C.M.J.E.; De Boer, M.G.J.; Pedersen, A.B. Choice of therapeutic interventions and outcomes for the treatment of infections caused by multidrug-resistant gram-negative pathogens: A systematic review. *Antimicrob. Resist. Infect. Control* **2019**, *8*, 1–13. [CrossRef]
11. Huddleston, J.R. Horizontal gene transfer in the human gastrointestinal tract: Potential spread of antibiotic resistance genes. *Infect. Drug Resist.* **2014**, *7*, 167–176. [CrossRef]
12. Partridge, S.R.; Kwong, S.M.; Firth, N.; Jensen, S.O. Mobile genetic elements associated with antimicrobial resistance. *Clin. Microbiol. Rev.* **2018**, *31*, e00088-17. [CrossRef]
13. Leekitcharoenphon, P.; Hans, M.; Johansson, K.; Munk, P.; Malorny, B.; Skarżyńska, M.; Wadepohl, K.; Moyano, G.; Hesp, A.; Veldman, K.T.; et al. Genomic evolution of antimicrobial resistance in *Escherichia coli*. *Sci. Rep.* **2021**, *11*, 15108. [CrossRef] [PubMed]
14. Wozniak, R.A.F.; Waldor, M.K. Integrative and conjugative elements: Mosaic mobile genetic elements enabling dynamic lateral gene flow. *Nat. Rev. Microbiol.* **2010**, *8*, 552–563. [CrossRef]
15. Virolle, C.; Goldlust, K.; Djermoun, S.; Bigot, S.; Lesterlin, C. Plasmid transfer by conjugation in gram-negative bacteria: From the cellular to the community level. *Genes* **2020**, *11*, 1239. [CrossRef]
16. Burrus, V. Mechanisms of stabilization of integrative and conjugative elements. *Curr. Opin. Microbiol.* **2017**, *38*, 44–50. [CrossRef]
17. Lukjancenko, O.; Wassenaar, T.M.; Ussery, D.W. Comparison of 61 sequenced *Escherichia coli* Genomes. *Microb. Ecol.* **2010**, *60*, 708–720. [CrossRef]
18. Picard, B.; Garcia, J.S.; Gouriou, S.; Duriez, P.; Brahimi, N.; Bingen, E.; Elion, J.; Denamur, E. The link between phylogeny and virulence in *Escherichia coli* extraintestinal infection. *Infect. Immun.* **1999**, *67*, 546–553. [CrossRef]
19. Olowe, O.A.; Adefioye, O.J.; Ajayeoba, T.A.; Schiebel, J.; Weinreich, J.; Ali, A.; Burdukiewicz, M.; Rödiger, S.; Schierack, P. Phylogenetic grouping and biofilm formation of multidrug resistant *Escherichia coli* isolates from humans, animals and food products in South-West Nigeria. *Sci. Afr.* **2019**, *6*, e00158. [CrossRef]
20. Alfinete, N.W.; Bolukaoto, J.Y.; Heine, L.; Potgieter, N.; Barnard, T.G. Virulence and phylogenetic analysis of enteric pathogenic *Escherichia coli* isolated from children with diarrhoea in South Africa. *Int. J. Infect. Dis.* **2022**, *114*, 226–232. [CrossRef]
21. Escobar-Páramo, P.; Grenet, K.; Le Menac'h, A.; Rode, L.; Salgado, E.; Amorin, C.; Gouriou, S.; Picard, B.; Rahimy, M.C.; Andremont, A.; et al. Large-scale population structure of human commensal *Escherichia coli* isolates. *Appl. Environ. Microbiol.* **2004**, *70*, 5698–5700. [CrossRef] [PubMed]
22. Clermont, O.; Dixit, O.V.A.; Vangchhia, B.; Condamine, B.; Dion, S.; Bridier-Nahmias, A.; Denamur, E.; Gordon, D. Characterization and rapid identification of phylogroup G in *Escherichia coli*, a lineage with high virulence and antibiotic resistance potential. *Environ. Microbiol.* **2019**, *21*, 3107–3117. [CrossRef] [PubMed]
23. Foley, S.L.; Kaldhone, P.R.; Ricke, S.C.; Han, J. Incompatibility Group I1 (*IncI1*) Plasmids: Their Genetics, Biology, and Public Health Relevance. *Microbiol. Mol. Biol. Rev.* **2021**, *85*, 10–1128. [CrossRef] [PubMed]
24. Murase, K.; Martin, P.; Porcheron, G.; Houle, S.; Helloin, E.; Pénary, M.; Nougayrède, J.-P.; Dozois, C.M.; Hayashi, T.; Oswald, E. HlyF produced by extraintestinal pathogenic *Escherichia coli* is a virulence factor that regulates outer membrane vesicle biogenesis. *J. Infect. Dis.* **2015**, *212*, 856–865. [CrossRef]
25. Valková, D.; Valkovičová, L.; Vávrová, S.; Kováčová, E.; Mravec, J.; Turňa, J. The contribution of tellurite resistance genes to the fitness of *Escherichia coli* uropathogenic strains. *Cent. Eur. J. Biol.* **2007**, *2*, 182–191. [CrossRef]
26. Pitout, J.D.D.; DeVinney, R. *Escherichia coli* ST131: A multidrug-resistant clone primed for global domination. *F1000Research* **2017**, *6*, 195. [CrossRef]
27. Musicha, P.; Feasey, N.A.; Cain, A.K.; Kallonen, T.; Chaguza, C.; Peno, C.; Khonga, M.; Thompson, S.; Gray, K.J.; Mather, A.E.; et al. Genomic landscape of extended-spectrum β-lactamase resistance in *Escherichia coli* from an urban African setting. *J. Antimicrob. Chemother.* **2017**, *72*, 1602–1609. [CrossRef]

28. Tegha, G.; Ciccone, E.J.; Krysiak, R.; Kaphatika, J.; Chikaonda, T.; Ndhlovu, I.; van Duin, D.; Hoffman, I.; Juliano, J.J.; Wang, J. Genomic epidemiology of *Escherichia coli* isolates from a tertiary referral center in lilongwe, Malawi. *Microb. Genomics* **2021**, *7*, 000490. [CrossRef]
29. Hoog, K.J. Molecular Epidemiology of *Escherichia coli* Isolated from Urinary Tract Infections in Inpatients and Outpatients in Gauteng, South Africa. Master's Thesis, University of Pretoria, Pretoria, South Africa, 2021.
30. Sserwadda, I.; Kidenya, B.R.; Kanyerezi, S.; Akaro, I.L.; Mkinze, B.; Mshana, S.E.; Hashim, S.O.; Isoe, E.; Seni, J.; Joloba, M.L.; et al. Unraveling virulence determinants in extended-spectrum beta-lactamase-producing *Escherichia coli* from East Africa using whole-genome sequencing. *BMC Infect. Dis.* **2023**, *23*, 587. [CrossRef] [PubMed]
31. Bevan, E.R.; Jones, A.M.; Hawkey, P.M. Global epidemiology of CTX-M β-lactamases: Temporal and geographical shifts in genotype. *J. Antimicrob. Chemother.* **2017**, *72*, 2145–2155. [CrossRef]
32. Pitout, J.D.D.; Chen, L. The significance of epidemic plasmids in the Success of multidrug-resistant drug pandemic extraintestinal pathogenic *Escherichia coli*. *Infect. Dis. Ther.* **2023**, *12*, 1029–1041. [CrossRef]
33. Katale, B.Z.; Misinzo, G.; Mshana, S.E.; Chiyangi, H.; Campino, S.; Clark, T.G.; Good, L.; Rweyemamu, M.M.; Matee, M.I. Genetic diversity and risk factors for the transmission of antimicrobial resistance across human, animals and environmental compartments in East Africa: A review. *Antimicrob. Resist. Infect. Control* **2020**, *9*, 127. [CrossRef]
34. Jian, Z.; Zeng, L.; Xu, T.; Sun, S.; Yan, S.; Yang, L.; Huang, Y.; Jia, J.; Dou, T. Antibiotic resistance genes in bacteria: Occurrence, spread, and control. *J. Basic. Microbiol.* **2021**, *61*, 1049–1070. [CrossRef]
35. Mainda, G.; Lupolova, N.; Sikakwa, L.; Richardson, E.; Bessell, P.R.; Malama, S.K.; Kwenda, G.; Stevens, M.P.; Bronsvoort, B.M.D.; Muma, J.B.; et al. Whole genome sequence analysis reveals lower diversity and frequency of acquired antimicrobial resistance (AMR) genes in *E. coli* from dairy herds compared with human isolates from the same region of central Zambia. *Front. Microbiol.* **2019**, *10*, 471–475. [CrossRef] [PubMed]
36. Ramatla, T.; Tawana, M.; Lekota, K.E.; Thekisoe, O. Antimicrobial resistance genes of *Escherichia coli*, a bacterium of "One Health" importance in South Africa: Systematic review and meta-analysis. *AIMS Microbiol.* **2023**, *9*, 75–89. [CrossRef] [PubMed]
37. Galindo-Méndez, M. Antimicrobial Resistance in *Escherichia coli*. In *E. coli Infections—Importance of Early Diagnosis and Effictive Treatment*; Rodrigo, L., Ed.; IntechOpen: London, UK, 2020; pp. 1–13. [CrossRef]
38. Barlow, G. Clinical challenges in antimicrobial resistance. *Nat. Microbiol.* **2018**, *3*, 258–260. [CrossRef]
39. Rozwandowicz, M.; Brouwer, M.S.M.; Fischer, J.; Wagenaar, J.A.; Guerra, B.; Mevius, D.J. Plasmids carrying antimicrobial resistance genes in *Enterobacteriaceae*. *J. Antimicrob. Chemother.* **2018**, *73*, 1121–1137. [CrossRef] [PubMed]
40. Tansirichaiya, S.; Goodman, R.N.; Guo, X.; Bulgasim, I.; Samuelsen, Ø.; Al-Haroni, M.; Roberts, A.P. Intracellular transposition and capture of mobile genetic elements following intercellular conjugation of multidrug resistance conjugative plasmids from clinical *Enterobacteriaceae* isolates. *Microbiol. Spectr.* **2022**, *10*, e0214021. [CrossRef]
41. Snaith, A.E.; Dunn, S.J.; Moran, R.A.; Newton, P.N.; Dance, D.A.B.; Davong, V.; Kuenzli, E.; Kantele, A.; Corander, J.; McNally, A. The highly diverse plasmid population found in *Escherichia coli* colonizing travellers to Laos and its role in antimicrobial resistance gene carriage. *Microb. Genomics* **2023**, *9*, 001000. [CrossRef]
42. Vrancianu, C.O.; Popa, L.I.; Bleotu, C.; Chifiriuc, M.C. Targeting plasmids to limit acquisition and transmission of antimicrobial resistance. *Front. Microbiol.* **2020**, *11*, 761. [CrossRef]
43. Wolde, D.; Eguale, T.; Alemayehu, H.; Medhin, G.; Haile, A.F.; Pirs, M.; Strašek Smrdel, K.; Avberšek, J.; Kušar, D.; Cerar Kišek, T.; et al. Antimicrobial susceptibility and characterization of extended-spectrum β-lactamase-producing *Escherichia coli* isolated from stools of primary healthcare patients in Ethiopia. *Antibiotics* **2024**, *13*, 93. [CrossRef] [PubMed]
44. Beghain, J.; Bridier-Nahmias, A.; Nagard HLe Denamur, E.; Clermont, O. ClermonTyping: An easy-to-use and accurate in silico method for *Escherichia* genus strain phylotyping. *Microb. Genomics* **2018**, *4*, e000192. [CrossRef]
45. Jolley, K.A.; Bray, J.E.; Maiden, M.C.J. Open-access bacterial population genomics: BIGSdb software, the PubMLST.org website and their applications. *Wellcome Open Res.* **2018**, *3*, 124. [CrossRef]
46. Zhou, Z.; Alikhan, N.; Mohamed, K.; Fan, Y.; Group, S.; Achtman, M. The EnteroBase user's guide, with case studies on *Salmonella* transmissions, *Yersinia pestis* phylogeny, and *Escherichia* core genomic diversity. *Genome Res.* **2020**, *30*, 138–152. [CrossRef] [PubMed]
47. Joensen, K.G.; Scheutz, F.; Lund, O.; Hasman, H.; Kaas, R.S.; Nielsen, E.M.; Aarestrup, F.M. Real-time whole-genome sequencing for routine typing, surveillance, and outbreak detection of verotoxigenic *Escherichia coli*. *J. Clin. Microbiol.* **2014**, *52*, 1501–1510. [CrossRef]
48. Joensen, K.G.; Tetzschner, A.M.M.; Iguchi, A.; Aarestrup, F.M.; Scheutz, F. Rapid and easy in silico serotyping of *Escherichia coli* isolates by use of whole-genome sequencing data. *J. Clin. Microbiol.* **2015**, *53*, 2410–2426. [CrossRef] [PubMed]
49. Bortolaia, V.; Kaas, R.S.; Ruppe, E.; Roberts, M.C.; Schwarz, S.; Cattoir, V.; Philippon, A.; Allesoe, R.L.; Rebelo, A.R.; Florensa, A.F.; et al. ResFinder 4.0 for predictions of phenotypes from genotypes. *J. Antimicrob. Chemother.* **2020**, *75*, 3491–3500. [CrossRef] [PubMed]

50. Carattoli, A.; Zankari, E.; Garciá-Fernández, A.; Larsen, M.V.; Lund, O.; Villa, L.; Møller Aarestrup, F.; Hasman, H. In Silico detection and typing of plasmids using plasmidfinder and plasmid multilocus sequence typing. *Antimicrob. Agents Chemother.* **2014**, *58*, 3895–3903. [CrossRef]
51. Johansson, M.H.K.; Bortolaia, V.; Tansirichaiya, S.; Aarestrup, F.M.; Roberts, A.P.; Petersen, T.N. Detection of mobile genetic elements associated with antibiotic resistance in *Salmonella enterica* using a newly developed web tool: MobileElementFinder. *J. Antimicrob. Chemother.* **2021**, *76*, 101–109. [CrossRef]

antibiotics

MDPI

Article

Genomic Characterization of Methicillin-Resistant and Methicillin-Susceptible *Staphylococcus aureus* Implicated in Bloodstream Infections, KwaZulu-Natal, South Africa: A Pilot Study

Bakoena A. Hetsa [1,*], Jonathan Asante [1,2], Joshua Mbanga [1,3], Arshad Ismail [4,5], Akebe L. K. Abia [1], Daniel G. Amoako [1,6] and Sabiha Y. Essack [1,7]

1 Antimicrobial Research Unit, College of Health Sciences, University of KwaZulu-Natal, Durban 4000, South Africa; jonathan.asante@ucc.edu.gh (J.A.); mbangaj@ukzn.ac.za (J.M.); abiaakebel@ukzn.ac.za (A.L.K.A.); amoakod@ukzn.ac.za (D.G.A.); essacks@ukzn.ac.za (S.Y.E.)
2 School of Pharmacy and Pharmaceutical Sciences, University of Cape Coast, PMB, Cape Coast, Ghana
3 Department of Applied Biology & Biochemistry, National University of Science and Technology, Corner Cecil Avenue & Gwanda Road, Bulawayo 263, Zimbabwe
4 Sequencing Core Facility, National Institute for Communicable Diseases, Division of the National Health Laboratory Service, Johannesburg 2193, South Africa; arshadi@nicd.ac.za
5 Department of Biochemistry and Microbiology, Faculty of Science, Engineering and Agriculture, University of Venda, Thohoyandou 0950, South Africa
6 Department of Pathobiology, University of Guelph, Guelph, ON N1G 2W1, Canada
7 School of Pharmacy, University of Jordan, Amman 11942, Jordan
* Correspondence: 218085934@stu.ukzn.ac.za

Abstract: *Staphylococcus aureus* is an opportunistic pathogen and a leading cause of bloodstream infections, with its capacity to acquire antibiotic resistance genes posing significant treatment challenges. This pilot study characterizes the genomic profiles of *S. aureus* isolates from patients with bloodstream infections in KwaZulu-Natal, South Africa, to gain insights into their resistance mechanisms, virulence factors, and clonal and phylogenetic relationships. Six multidrug-resistant (MDR) *S. aureus* isolates, comprising three methicillin-resistant *S. aureus* (MRSA) and three methicillin-susceptible *S. aureus* (MSSA), underwent whole genome sequencing and bioinformatics analysis. These isolates carried a range of resistance genes, including *blaZ*, *aac(6′)-aph(2″)*, *ant(9)-Ia*, *ant(6)-Ia*, and *fosB*. The *mecA* gene, which confers methicillin resistance, was detected only in MRSA strains. The isolates exhibited six distinct *spa* types (t9475, t355, t045, t1265, t1257, and t7888) and varied in virulence gene profiles. Panton–Valentine leukocidin (Luk-PV) was found in one MSSA isolate. Two SCC*mec* types, IVd(2B) and I(1B), were identified, and the isolates were classified into four multilocus sequence types (MLSTs), with ST5 (n = 3) being the most common. These sequence types clustered into two clonal complexes, CC5 and CC8. Notably, two MRSA clones were identified: ST5-CC5-t045-SCC*mec*_I(1B) and the human-associated endemic clone ST612-CC8-t1257-SCC*mec*_IVd(2B). Phylogenomic analysis revealed clustering by MLST, indicating strong genetic relationships within clonal complexes. These findings highlight the value of genomic surveillance in guiding targeted interventions to reduce treatment failures and mortality.

Keywords: *Staphylococcus aureus*; bloodstream infections; whole-genome sequencing; antibiotic resistance; virulence; bioinformatics

1. Introduction

Staphylococcus aureus is a Gram-positive bacterium inhabiting healthy individuals' nostrils and skin. However, it has become an important opportunistic pathogen in communities and hospitals [1]. It causes severe skin infections, pneumonia, endocarditis, and bloodstream infections (BSIs) [2]. BSIs caused by *S. aureus* infections have high morbidity and mortality if not treated timeously [3]. The most significant risk factors for *S. aureus* BSIs are intravascular devices, surgical procedures, and a debilitated immune system [4].

Methicillin-resistant *S. aureus* (MRSA) has become a significant cause of BSIs. MRSA poses a major public health threat because of multidrug resistance to different antibiotic classes that limit treatment options [5]. Methicillin resistance in MRSA strains is mediated by the *mecA* gene, found on a mobile genetic element (MGE) known as the staphylococcal cassette chromosome *mec* (SCC*mec*) [6]. Methicillin-susceptible *S. aureus* (MSSA) is also emerging as a causative agent of BSIs [7] and has been reported to display high virulence and multidrug resistance [8].

The pathogenicity of *S. aureus* depends on its ability to produce a wide array of virulence factors involved in adhesion, invasion of host tissues, immune system evasion, and biofilm formation [9,10]. Also, *S. aureus* produces metallophores that enable bacteria to scavenge metal ions such as iron and zinc essential for bacterial metabolism and pathogenicity [11]. Virulence factors and multiple resistance genes can be transmitted by horizontal gene transfer (HGT) [12] on diverse MGEs, amongst which plasmids are reported as the primary sources for dissemination [4].

The epidemiology of *S. aureus* strains indicates that its molecular characteristics continually change over time, resulting in new clones, which vary by region. In a study in the United States, ST5 and ST8 were the most prevalent sequence types [13]. In South Africa, ST612 is dominant in the hospital environment [14]. The ST612-IV [2B], belonging to *spa* type t1257, was identified as a typical clone in clinical settings [15] and sporadically in poultry settings [16]. The ST5 and ST8 clones are commonly associated with BSIs and the pandemic lineages of *S. aureus*, such as the clonal complex CC8 and CC5 [17]. Notably, the sequence types ST612, ST5, ST8, and ST72 have displayed high resistance to most antibiotic drug classes and are challenging to treat [17].

Multidrug-resistant (MDR) *S. aureus* infections pose a serious clinical concern. A high incidence of pathogenic MDR MRSA has been reported, and the data suggest that its prevalence is increasing in Africa [18]. A recent South African study investigating the genetic relatedness of hospital-acquired-associated MRSA isolates in two hospitals revealed that all isolates were resistant to aminoglycosides and β-lactams. All the isolates carried the *aacA-aphD* and *mecA*-resistant genes and clusters of virulence genes [19]. This pilot study aimed to comprehensively characterize the genomic profiles, resistance mechanisms, virulence factors, pathogenicity, phylogenomic relationships, and clonal diversity of *Staphylococcus aureus* clinical strains implicated in BSIs at a regional hospital in KwaZulu-Natal, South Africa.

2. Results

2.1. Patient Demographics and Characteristics

The 6 isolates investigated in this study were obtained from patients who visited a regional hospital in the uMgungundlovu District in the KwaZulu-Natal Province. Three of the six isolates were recovered from the neonatal ICU (n = 3, 50%), two from surgical wards, and one isolate from the pediatric ward. Four patients were male, while two were female. The age distribution of patients ranged from 0 to 33 years old, and the mean age was 8.83 years (Table 1). The demographic details of the source participants of the isolates that were selected for WGS are shown in Supplementary Table S1.

Table 1. Antibiotic susceptibility profiles, age, and demographic characteristics of patients with BSIs attributed to *S. aureus*.

Isolate ID	Species	Sex	Ward	Age	Antibiotics																			
					PEN	AMP	FOX	CIP	MXF	LEV	GEN	AMK	ERY	CLI	TET	DOX	TGC	CHL	NIT	SXT	VAN	RIF	LZD	TEC
S11	MRSA	F	Surgical ward	17 years	R	R	R	R	R	R	R	R	R	R	R	R	R	I	S	R	S	R	R	R
S29	MRSA	M	Pediatric ward	<1 year	R	R	R	R	R	R	R	R	I	R	R	R	S	I	S	R	S	I	R	R
S31	MRSA	F	Surgical ward	3 years	R	R	R	R	R	R	R	R	R	R	R	R	S	R	S	R	S	R	R	R
S24	MSSA	M	ICU	33 years	R	S	S	R	R	R	S	S	R	R	R	I	S	I	R	R	S	R	I	I
S13	MSSA	M	ICU	<1 year	R	S	S	R	R	R	I	R	I	R	I	R	S	S	S	I	S	S	I	R
S34	MSSA	M	NICU	<1 year	R	S	S	R	R	R	I	R	I	R	R	R	S	I	S	S	S	I	S	S

Key: PEN, penicillin; AMP, ampicillin; FOX, cefoxitin; CIP, ciprofloxacin; MXF, moxifloxacin; LEV, levofloxacin; GEN, gentamicin; AMK, amikacin; ERY, erythromycin; CLI, clindamycin; TET, tetracycline; DOX, doxycycline; TGC, tigecycline; CHL, chloramphenicol; NIT, nitrofurantoin; SXT, trimethoprim-sulfamethoxazole; VAN, vancomycin; RIF, rifampicin; LZD, linezolid; TEC, teicoplanin. R, resistant; I, intermediate; S, susceptible; M, male; F, female; NICU, neonatal intensive care unit; ICU, intensive care unit.

2.2. Antibiotic Susceptibility Test Results

The isolates displayed varying phenotypic resistance profiles, with most being resistant to penicillin G (n = 6), tetracycline (n = 5), doxycycline (n = 5), clindamycin (n = 5), moxifloxacin (n = 5), rifampicin (n = 4), and erythromycin (n = 3). The lowest resistance was against nitrofurantoin, tigecycline, and chloramphenicol (n = 1) (Table 1).

2.3. Phenotypic and Genotypic Identification of MRSA Isolates

MRSA isolates were confirmed by phenotypic resistance to cefoxitin (Table 1) and the detection of the *mecA* gene using polymerase chain reaction (PCR).

2.4. Genomic Features

The genome size of our draft genomes ranged from 2.7 Mb to 2.9 Mb. The genomic characteristics of the sequences in relation to G + C content (%), number of RNAs, number of coding sequences, size, N50, L50, and coverage are shown in Table S2.

2.5. In Silico ARGs Analysis

Isolates harbored various permutations and combinations of ARGs, which included ARGs against β-lactams [*blaZ*], aminoglycosides [*aac(6′)-aph(2″), aad(6′), ant(9)-la, ant(6)-Ia, aph(2″)-Ia, aph(3′)-IIa, sat-4*], trimethoprim [*dfrG, dfrC*], macrolides [*erm(C), erm(A)*], tetracycline [*tet(K), tet(M), mepR, mepA*], flouroquinolones [*parE, parC, grlA, gyrA, norA, norC* (multidrug efflux pumps)], rifampicin [*rpoB*] and fosfomycin [*fosB, murA*] (Table 2). Only the MRSA isolates harbored the *mecA* gene. There was good concordance between ARGs and phenotypic profiles for ARGs in all MRSA and MSSA isolates.

Table 2. Genotypic characteristics of *S. aureus* implicated in BSIs.

Isolate ID	MRSA/MSSA	MLST	*spa* Type	Resistome	Plasmid Replicon Type	Insertion Sequences	Confirmed CRISPRs (CAS)	Clonal Complex	* SCC*mec* Type	*agr* Type	Pathogenicity Score
S11	MRSA	ST8	t9475	*blaZ, mecA, aac(6′)-aph(2″), parC, dfrG, erm(C), grlA, tetK, mepR, mepA, norA, norC, fosB*	rep10, rep7a, rep7c	-	6 (0)	CC8	NT	Type I	0.982 (882)
S29	MRSA	ST5	t045	*blaZ, mecA, aph(3′)-III, aac(6′)-aph(2″), ant(6)-Ia, ant(9)-Ia, aad(6′), erm(C), erm(A), qacA, mepR, fosB, norA, norC, sat-4*	rep10, rep21	IS6, IS256	12 (0)	CC5	SCC*mec* type I(1B)	Type II	0.98 (914)
S31	MRSA	ST612	t1257	*blaZ, mecA, aac(6′)-aph(2″), aph(2″)-Ia, aad(6′), ant(6)-Ia, ant(9)-Ia, tet(M), mepR, mepA, dfrC, parC, erm(C), parE, gyrA, rpoB, fosB, norA, norC, murA*	rep7c, rep20	IS256, IS6	7 (0)	CC8	SCC*mec* type IVd(2B)	Type I	0.976 (978)
S24	MSSA	ST152	t355	*blaZ, dfrG, mepR, norC, murA*	rep16, rep5a	-	8 (0)	-	NA	Type IV	0.975(225)
S13	MSSA	ST5	t1265	*blaZ, norA, norC, fosB*	rep20	-	9 (0)	CC5	NA	Type II	0.985 (844)
S34	MSSA	ST5	t7888	*blaZ, norA, norC, mepR, fosB*	rep19, rep16, rep20, rep5a	1S6	7 (0)	CC5	NA	Type II	0.983 (871)

* SCC*mec* typing was predicted with the SCC*mec*Finder, MSSA—Methicillin-susceptible *Staphylococcus aureus*, MRSA—Methicillin-resistant *Staphylococcus aureus*—non-typeable (NT). The MSSA do not harbor SCC*mec* types, which is indicated as "None (NA)" to reflect their SCC*mec*-negative status.

2.5.1. MLST, spa Typing, and Clonal Complex

MLST revealed total four sequence types, i.e., ST5 (CC5, n = 3), ST152 (n = 1), ST612 (CC8, n = 1), and ST8 (CC8, n = 1). Two MRSA strains belonged to CC8 and one to CC5. Methicillin-susceptible (MSSA) isolates were identified as ST5 (n = 2) or ST152 (n = 1). The genetic diversity of the isolates was confirmed by *spa* typing, which revealed six different *spa* types: t9475, t1265, t355, t045, t1257, and t7888 (Table 2). CC and *spa* type combinations were CC8-t9475, CC8-t1257, and CC5-t045 among MRSA isolates, with CC5-t1265 belonging to one MSSA isolate. There was no association observed between STs, *spa* type, and CC. The grouping of the STs and *spa*-types yielded six genotypes, i.e., ST8-t9475, ST152-t355, ST5-t045, ST5-t1265, ST612-1257, and ST5-t7888, indicating that isolates were not clonally related.

The SCC*mec*Finder analysis identified two SCC*mec* types, i.e., IVd (2B) and I (1B), among the MRSA isolates (Table 2). One MRSA isolate was non-typeable (NT) for SCC*mec*. The combination of MLST, CC, *spa*, and SCC*mec* yielded the ST612-CC8-t1257-SCC*mec*_IVd (2B) and ST5-CC5-t045-SCCmec_I (1B), clones both of which have been reported in South Africa.

2.5.2. Mobilome (Plasmids, Insertion Sequences, Intact Prophages, and SCCmec Elements)

Analysis of the six isolate genomes identified various MGEs, including plasmid replicons, IS's, prophages, and SCC*mec* elements. A total of eight different plasmid replicons were detected, of which rep20 (n = 3) was the most prevalent (Table 2). There were no associations between plasmid replicons and STs. However, the rep7c was found in CC8 isolates in addition to other plasmid replicons, while rep16 and rep5a were found in isolates with the non-typeable CC. The rep20 plasmid replicon was associated with CC5 and CC8 isolates. The rep10 was carried in CC8 and CC5 isolates, while the re7a and rep21 were carried in CC8 and CC5 isolates, respectively. IS6 and IS256 were identified in three isolates, and their occurrence was not associated with any STs or CC (Table 2). The ddistribution of ISs and plasmid replicons detected among the isolates are shown in Supplementary Table S4. A total of six intact prophages were detected, of which the most identified were PHAGE_Staphy_phi2958PVL (n = 2) and PHAGE_Staphy_P282 (n = 2) (Table S5). PHAGE_Staphy_phiJB was associated with the *dfrG* gene.

2.5.3. Virulome and Pathogenicity of S. aureus Strains

A total of 82 virulence genes were detected across the isolates (Table S3). The virulence genes belonged to the five main virulence determinant classes of *S. aureus*: adherence factors, immune evasion, enzymes (exoenzymes), toxins, and the secretion system. It is noteworthy that the most prevalent toxins were hemolysins, i.e., gamma (*hlg*), delta (*hld*), alpha (*hly/hla*), staphylococcal enterotoxins (*se*, *set*, *sel*) genes, and leucocidin genes (*lukD/E*), while *lukS-PV* and *lukF-PV* genes were detected in two isolates (S24 and S29). The prediction of isolates pathogenicity towards humans yielded a high average probability score (Pscore $\approx$ 0.980).

2.6. Genetic Environment of the ARGs and Virulence Genes

The co-carriage of ARGs and virulence genes was evident across the isolates. Using NCBI annotation, we identified *blaZ* genes on five isolates in parallel with *cacD*, virulence genes, and the type 1 toxin–antitoxin system. Across the isolates, most *blaZ* genes were associated with regulator genes *blaR* and *blaI* and frequently found with either a recombinase, integrase, cadmium resistance (*cadD*) gene, or type I toxin–antitoxin system (Table 3). A similar genetic context was detected in the S13 isolate, where *blaZ*, *blaR*, and *blaI* were flanked by IS6, *cadD*, a type I toxin–antitoxin system, on a contig with the closest nucleotide homology to a plasmid from *S. aureus* pER10678.3A.1 (CP051928.1), suggesting that ARGs, heavy metal resistance genes (HMRGs), and virulence genes may be mobilized by plasmids (Table 3). It is noteworthy that IS1182 was associated with the *mecA*, *mecI*, and *mecR1* genes together with recombinases, while IS6 bracketed the *mecA* gene and its regulatory genes (*mecI* and *mecR*) in three MRSA isolates (Table 4). Most ARGs, including *erm(A)*, *ant(9)-Ia*, *dfrG*, and *tet(M)*, were associated with a recombinase and integrase. One isolate was found harboring the *dfrG* gene bracketed by ISL3 and recombinases (Table 4).

Regulatory Genes

The accessory gene regulator system (*agr*) involved in the regulation and expression of toxins, exoenzymes, and biofilm was detected in all isolates. Isolates carried *agr* type I and II. The distribution of the *agr* group in MRSA was: *agr* I (n = 1), *agr* II (n = 2), while in MSSA *agr* I (n = 2), and *agr* II (n = 1).

2.7. Phylogenomics

The phylogenetic analysis, integrated with metadata, reveals clear clustering patterns based on Multilocus Sequence Typing (MLST) and geographic origin. Isolates sharing the same MLST type generally clustered together, with additional grouping observed by country of isolation (Figure 1). For instance, the ST5 isolates from this study (S11-ST5, S34-ST5, and S29-ST5) are closely aligned with other South African isolates, indicating minimal genetic divergence within this MLST type in the region. This suggests a strong regional lineage for ST5 in South Africa. Additionally, this study isolates S24-ST152 clusters with other ST152 isolates from various African countries, including Kenya and Ghana, highlighting the broader geographic distribution and potential mobility of this MLST type across the continent (Figure 1). Notably, the tree analysis also reveals that ST8 and ST612 isolates are clustered together, suggesting close genetic relatedness despite being distinct MLST types (Figure 1). This finding could indicate a shared ancestry or recent genetic exchange between these groups, pointing to complex evolutionary dynamics within these populations.

The linkage between the *mecA* gene, SCC*mec* types, and clonal complexes is particularly notable, as it highlights the genetic mechanisms underlying methicillin resistance and the clustering of MRSA strains (Figure 2). The distinct clustering patterns observed for different isolates underscore the complex interplay between genetic background, resistance gene acquisition, and selective pressures in the evolution of these pathogens.

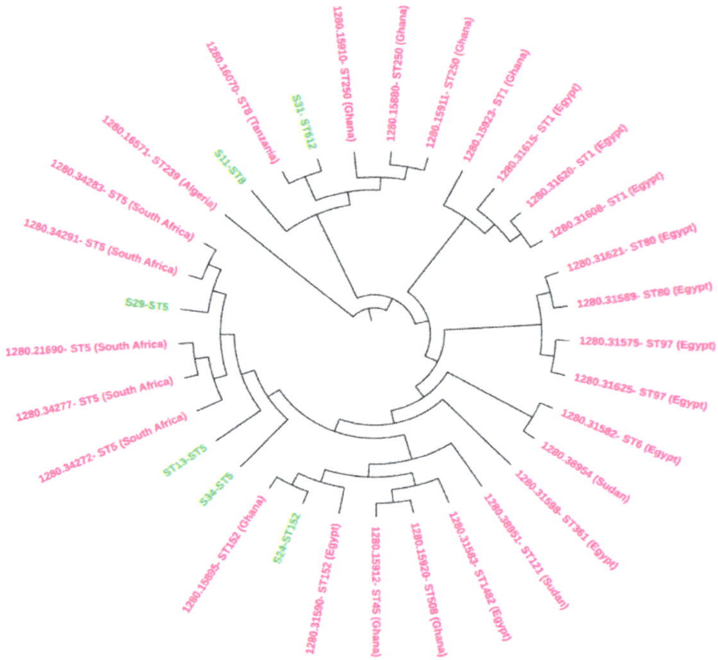

Figure 1. The circular phylogenetic tree provides a visual representation of the genetic relationships between *Staphylococcus aureus* isolates from this study (highlighted in green) and various African blood culture isolates (highlighted in purple). This tree illustrates how isolates cluster based on their Multilocus Sequence Typing (MLST) and geographic origin.

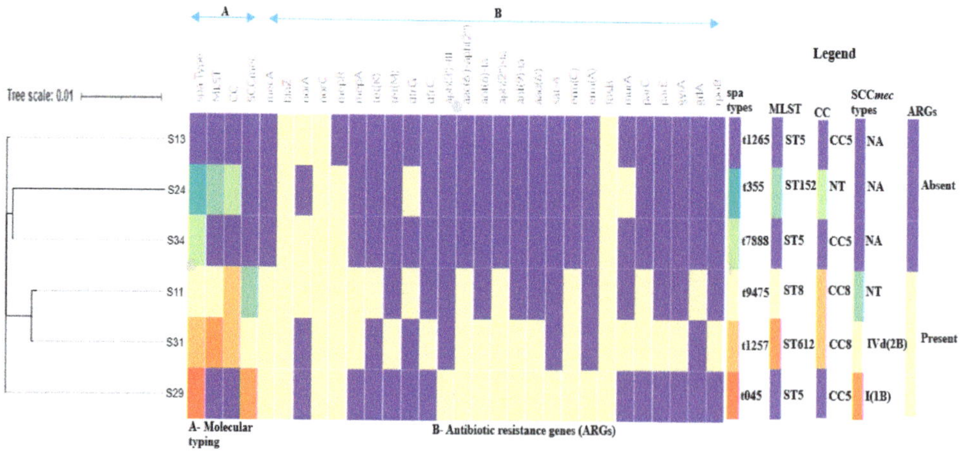

Figure 2. The phylogenetic branch and metadata, A—Molecular typing [*spa* type, sequence type (ST), clonal complex, SCC*mec* types], and B—Antibiotic resistance genes [ARGs], visualized using Phandango (https://github.com/jameshadfield/phandango/wiki accessed on 15 August 2024) in *S. aureus* isolates. The heat map in the middle indicates the presence (yellow) and absence (purple) of antibiotic resistance genes. NT indicates non-typeable. Isolates S13, S24, and S34 are MSSA and therefore do not harbor SCC*mec* types, which is indicated as "None (NA)" to reflect their SCC*mec*-negative status.

Table 3. Genetic context of virulence genes in *S. aureus* isolates.

Strain (MLST)	Strain	Contig	Synteny of Virulence Genes and MGEs	Plasmid/Chromosomal Sequence with Closest Nucleotide Homology (Accession Number)
S11 (ST8)	MRSA	4	pmtC:pmtB:pmtA:***eap***::***scn***::***sak***:::sph:: ***lukG***::***lukH***::intergrase:::agrB	*S. aureus* strain Laus385 chromosome (CP071350.1)
		6	***icaR***::***icaD***:***icaB***:***icaC***:vraD:vraE:vraH:: IS30:vraH::recombinase:IS6	*S. aureus* strain TF3198 chromosome, complete genome (CP023561.1)
		10	***lukE***:***lukD***::splA::epiE::***splA***:splB:splC: splD:splE:splF::pepA1:transposase	*S. aureus* strain 82 chromosome, complete genome (CP031661.1)
S29 (ST5)	MRSA	53	type I toxin–antitoxin system:IS6:cadD	*S. aureus* strain MIN-175 chromosome (CP086121.1)
		40	***clfA***:vwb:***emp***	*S. aureus* strain ER02693.3 chromosome, complete genome (CP030605.1)
S31 (ST612)	MRSA	11	pmtD:pmtC:pmtB:pmtA::***eap***:***scn***::***sak***	*S. aureus* strain 2395 USA500, complete genome (CP007499.1)
		15	***lukE***:***lukD***::::***splA***:***splB***:***splC***:***splF***::type I restriction-modification system	*S. aureus* strain NRL 02/947 chromosome, complete genome (CP103850.1)
		19	***lukG***:***lukH***:pathogenicity island:intergrase::phenol-soluble modulin:agrB	*S. aureus* strain 2395 USA500, complete genome (CP007499.1)
		22	***seq***:***sek***:integrase::::***emp***:***clfA***	*S. aureus* strain 2395 USA500, complete genome (CP007499.1)
		33	recombinase::universal stress protein:::cadD::***seq***:***sek***:integrase:::***emp***:***clfA***	*S. aureus* plasmid SAP017A, complete sequence (GQ900382.1)
		64	***sea***:putative holin-like toxin	*S. aureus* strain R50 chromosome, complete genome (CP039167.1)
S13 (ST5)	MSSA	4	***sbi***:***hlgA***:***hlgC***:***hlgB***	*S. aureus* strain AR462 chromosome, complete genome (CP029086.1)
		5	scpA:::**eap**::***scn***:***sak***::::::intergrase:sph: ***lukH***:***sbi***:***hlgA***:***hlgC***:***hlgB***	*S. aureus* strain pt239 chromosome, complete genome (CP049467.1)
		15	IS6::cadD:::***sed***:***sej***:***ser***::recombinases: cpA::eap::***scn***:**sak**::integrase:sph:***lukH***	*S. aureus* strain ER10678.3 plasmid pER10678.3A.1 (CP051928.1)

Table 3. *Cont.*

Strain (MLST)	Strain	Contig	Synteny of Virulence Genes and MGEs	Plasmid/Chromosomal Sequence with Closest Nucleotide Homology (Accession Number)
S24 (ST152)	MSSA	8	arsB::crcB::***scn***:***sak***:::recombinase::type II toxin–antitoxin system toxin:intergrase	*S. aureus* strain UMCG579 chromosome, complete genome (CP091066.1)
		21	cadD:type toxin–antitoxin::integrase	*S. aureus* strain GHA13 chromosome (CP043911.1)
		11	BrxA/BrxB:::msrA:msrB:::norD::cspA:cvfB	*S. aureus* strain NGA84b chromosome, complete genome (CP051165.2)
S34 (ST5)	MSSA	7	***eap/map***::***scn***:***sak***::::***sea***:::type II toxin–antitoxin:integrase:sph:***lukG***:***lukH***	*S. aureus* strain HPV107 chromosome, complete genome (CP026074.1)
		8	clfA:vwb:emp::thermonuclease protein:::**sek**:***seq***::pathogenicity island	*S. aureus* strain B4-59C chromosome, complete genome (CP042153.1)
		12	**sem**:**sei**:**seu**:**sen**:**seg**:::***lukE***:***lukD***::splA:splB:splC:splD:splF	*S. aureus* strain ER03588.3 chromosome, complete genome (CP030595.1)
		14	isdB:isdA:isdC:isdD:isdE:isdF::isdG::***ecb***::***efb***:***scb***	*S. aureus* strain B3-17D chromosome, complete genome (CP042157.1)
		20	SSL13:SSL12:**hyl**	*S. aureus* strain NAS_AN_239 chromosome, complete genome (CP062409.1)

Virulence gene(s) in bold.

Table 4. Genetic environment of antibiotic resistance genes in *S. aureus* isolates.

Isolate ID (MLST)	Strain	Contig	Synteny of Resistance Genes and MGEs	Plasmid/Chromosomal Sequence with Closest Nucleotide Homology (Accession Number)
S11 (ST8)	MRSA	4	*blaI*:*blaR1*:*blaZ*::recombinase/integrase	*S. aureus* strain ER02826.3 chromosome (CP030661.1)
		7	recombinase::dfrG:insertionelement:::ISL3::: recombinase:	*S. aureus* strain UP_403 chromosome (CP047849.1)
		59	IS6::*mecA*:*MecR1*:IS6::	*S. aureus* strain ER03868.3 chromosome (CP030403.1)
		123	Plasmid recombination:*tet(K)*	*S. epidermidis* isolate BPH0662 genome assembly, plasmid: 1 (LT614820.1)

Table 4. *Cont.*

Isolate ID (MLST)	Strain	Contig	Synteny of Resistance Genes and MGEs	Plasmid/Chromosomal Sequence with Closest Nucleotide Homology (Accession Number)
S29 (ST5)	MRSA	8	*erm(A):ant(9)-Ia*:transposase:recombinase:integrase	*S. aureus* strain 628 chromosome (CP022905.1)
		11	*gyrB:gyrA*:::ligase	*S. aureus* strain MIN-175 chromosome (CP086121.1)
		38	recombinase:IS1182::*mecR1:mecA*:::IS6	*S. aureus* subsp. aureus strain FDAARGOS_5 chromosome (CP007539.3)
		51 *	recombinase:*blaI:blaR1:blaZ:cadC:cadA*	*S. aureus* plasmid pSK57, partial sequence (GQ900493.1)
		56	*ant(6)-Ia:sat4:aph(3′)-IIIa*	*S. pseudintermedius* strain MAD627 chromosome (CP039743.1)
		64	*qacA/B*:qacR	*S. aureus* strain MIN-175 chromosome (CP086121.1)
		67	*ermCL:erm(C)*	*S. epidermidis* strain TMDU-137 plasmid p5, complete sequence (CP093178.1)
S31 (ST612)	MRSA	23	*mecA:mecR1*::IS1182::recombinase	*S. aureus* strain 2395 USA500 (CP007499.1)
		17	integrase::::::tet(M):::IS256	*S. aureus* strain NRS120 chromosome, complete genome (CP026072.1)
S13 (ST5)	MSSA	15 *	IS6 IS6::*cadD*:::type I toxin–antitoxin::recombinase::*blaI:blaR1:blaZ*	*S. aureus* strain ER10678.3 plasmid pER10678.3A.1 (CP051928.1)
S24 (ST152)	MSSA	21	cadD:typetoxinantitoxin::recombinase:*blaI:blaR1:blaZ*:recombinase	*S. aureus* strain GHA13 chromosome (CP043911.1)
S34 (ST5)	MSSA	19	recombinase:blaZ:blaR1:blaI:recombinase:integrase	*S. aureus* strain UP_678 plasmid unnamed (CP047840.1)

* Co-occurrence of a heavy metal resistance gene (HMRG) and antibiotic resistance genes (ARGs).

3. Discussion

We studied the genomic characteristics of six MDR *S. aureus* isolates implicated in BSIs. This study analyzed the resistome, virulome, mobilome, phylogeny, and genetic environment of the resistance genes using WGS and bioinformatics. The genomes analyzed herein were predominantly recovered from ≤1-year-old patients.

There was a diversity of ARGs encoding resistance to different antibiotics and good concordance between the observed phenotypic and genotypic resistance. The incidence of ARG's encoding resistance to β-lactams, aminoglycosides, macrolides, fosfomycin, trimethoprim, tetracycline, and genes coding multidrug resistance (MDR) efflux pumps (*norA*, *mepR*, and *mgrA*) was not dependent on the clonal type. The *erm(C)* and *erm(A)* genes that are commonly found in macrolide–lincosamide–streptogramin B (MLS_B)-resistant *S. aureus* were found in erythromycin and clindamycin-resistant isolates (Table 2), which was expected since resistance to erythromycin co-selects resistance to other antibiotics, such as streptogramin B (MLS_B) and lincosamides [20]. The *ermC* gene is among the primary *erm* types that facilitate ribosome methylation of the 23S rRNA, triggering conformational changes resulting in drug binding inhibition [21], and has been reported in clinical *S. aureus* isolates from South Africa [22]. In this study, the *ermC* encoding macrolide resistance was carried on a plasmid, on a contig that had the closest nucleotide homology to plasmids from *S. epidermidis* strain TMDU-137 plasmid p5, complete sequence (CP093178.1), implying the likelihood of horizontal transfer of *ermC* genes in clinical *S. aureus* isolates. The *ermC* are often plasmid-mediated, resulting in high resistance to macrolides in *S. aureus* [23].

The *blaZ* gene, which inactivates penicillin through hydrolysis of the beta-lactam ring, was observed in all six isolates that were phenotypically resistant to penicillin. The *blaZ* genes have also been isolated in clinical isolates of Staphylococci in South Africa [24]. In this study, the *blaZ* genes were found on contigs with closest homology to either chromosomes or plasmids. This agrees with a study conducted in Spain that analyzed ARGs presence in chromosomes and plasmids from the genomes of *S. aureus*. WGS analysis of *S. aureus* revealed that *blaZ* (n = 2) was located on chromosomic contigs, while *blaZ* was found in plasmid contigs in three isolates [25]. It is important to note that most *blaZ* and associated MGEs from isolates belonging to ST5 (S13, S34) isolated from the intensive care unit (ICU) and pediatric ward (S29) were located on contigs that had the closest homology to plasmids, implying that plasmids play a crucial role in mobilizing the *blaZ* gene in clinical *S. aureus* isolates. The S29 isolate, belonging to the t045-CC5 lineage, carried an assortment of ARGs encoding resistance to different antibiotics (Table 4). Similar ARGs in MRSA lineage t045-CC5-MRSA were also reported in a study conducted in South Africa, where t045-CC5 MRSA lineages obtained from different clinical samples from South Africa and Nigeria reported that t045 lineages were MDR, suggesting that this lineage is hospital-associated and their multidrug resistance nature may compromise treatment [26].

Also, the *blaZ* genes, heavy metal genes, and associated MGEs were carried on either plasmid or chromosome. The *blaZ* and *cadAC* genes were found on the genetic element recombinase *blaI:blaR1:blaZ:cadC:cadA* for isolates S24 (MSSA) that was from the ICU and S29 (MRSA) from the pediatric ward, suggesting co-selection of heavy metal resistance dissemination and adaptation in different wards. The *cadA* gene confers a high resistance to cadmium and other heavy metals like zinc and lead in *S. aureus* isolates [27]. The *cadA* was associated with a plasmid, similar to the findings of a study that was conducted by Al-Trad et al. (2023) in Malaysia, who used WGS to analyze the plasmid content of clinical MRSA isolates and reported that heavy metal resistance plasmids harbored cadmium resistance genes, with the majority being *cadAC* [28]. The HMRGs have been reported to trigger a co-selection mechanism with antibiotics, which may complicate treatment [29]. This may pose a challenge, especially among patients in the ICU, where broad-spectrum antibiotics are often used.

Tetracycline resistance genes (*tetK* and *tetM*) were observed in two isolates. Isolate S11 carried *tet(K)* associated with the following genetic context: plasmid recombination *tet(K)* that had a high similarity to *S. epidermidis* BPH0662, and plasmid 1 (LT614820.1), which could be significant in mobilizing TET-resistant genes. Also, the *tet(M)* was bracketed by integrase and IS256 in isolate S31. The IS*256* is a retrotransposon that can mobilize the resistance genes through a copy-and-paste mechanism and has been shown to confer a robust genomic plasticity in MRSA strains [30].

We found that ARGs and virulence genes were associated with MGEs, which may enable their transfer within and between plasmids and chromosomes [31]. In this study, the *mecA* gene was located on IS1182 in two MRSA isolates, surrounded by recombinase in the genetic context *mecA:mecR1*::IS1182::recombinase. The insertion sequence IS1182 was present in 2/3 MRSA strains that contained *mecA*. IS1182 has been shown to occur close to the SCC*mec* element and increase resistance through inactivating the *lytH* gene encoding a putative lytic enzyme in pathogenic MRSA isolates [32].

MLST, clonal complex typing, *spa* typing, and SCC*mec* typing were used to analyze the molecular characteristics of the *S. aureus* isolates. Four ST types and two clonal clusters (CCs) were found among the six clinical isolates in this study, with ST5 as the most predominant complex clonal CC5 and CC8. Generally, clonal lineages ST5, ST8, ST152, and ST612 are among the most commonly reported in hospital environments, along with other sequence types of *S. aureus* [33]. *S. aureus* ST5, belonging to CC5, was predominant in this study and was previously reported among patients with bloodstream infections at Ruijin Hospital in Shanghai [3]. The detection of clonal complexes CC5 and CC8 agrees with a study by Smith et al. [17], which also found CC8 and CC5 were predominant in a study that analyzed the genomic epidemiology of MRSA and MSSA from bloodstream infections in the USA. Their results revealed that the MDR phenotype observed in strains belonging to CC5 and CC8 was responsible for the occurrence of multidrug and methicillin resistance in the *S. aureus* population. MRSA strains belonging to CC8 and CC5 are frequently associated with global outbreaks and have been identified in Africa [34].

The *spa* typing revealed six different *spa* types, suggesting a non-clonal MRSA and MSSA distribution. The detection of *spa* types t1257, t045, and t355 agrees with a study conducted in South Africa, which analyzed the diversity of SCC*mec* elements and *spa* types in *S. aureus* isolates from blood culture in the Gauteng, KwaZulu-Natal, Free State, and Western Cape provinces [15], in which t037 and t1257 were the most common and predominated throughout the seven-year study period. In this study, some antibiotic resistance genes were associated with specific MRSA clones belonging to *spa* types t1257, t045, and t9475. Shittu et al. (2021) found the *spa* types t045 and t1257 to be the most prevalent and associated with genes conferring resistance to aminoglycosides, trimethoprim, macrolides, and tetracycline in clinical isolates of *S. aureus* from South Africa and Nigeria [26].

The analysis of SCC*mec* types revealed the presence of SCC*mec* type IVd (2B) and SCC*mec* type I (B) carrying the *mecA* gene, which occurred in tandem with *mecR1* in both MRSA isolates. However, one MRSA (S11) isolate had a non-typeable SCC*mec* element cassette due to the missing cassette chromosome recombinase (*ccr*) gene complex [35]. The *ccr* gene complex is an essential component required to facilitate the integration or excision of the SCC*mec* element in the staphylococcal chromosome, and their loss has also been reported [36]. The SCC*mec* IV detected in our study is associated with the *spa* type t1257, previously reported in South Africa in *S. aureus* obtained from poultry isolates [16], implying its possible transfer between humans and animals.

We found different MRSA genotypes, ST612-t1257-CC8, ST8-t9475-CC8, and ST5-t045-CC5, suggesting that MRSA isolates were not clonally and epidemiologically related. The ST612-t1257-CC8 identified in this study is an endemic MRSA clone that has been reported in animal and clinical settings [15,16]. The ST5-I-MRSA, known as the pandemic British EMRSA-3 clone, was detected in the pediatric ward. This is similar to a study conducted in South Africa, where the t045-MRSA strain occurred in pediatric patients [19]. The isolation of the t045-ST5-MRSA strain could confirm its successful persistence in the hospital and its capacity to cause infections in neonatal and pediatric wards [37].

Several virulence factors, including adherence, immune invasion, toxins, and exoenzymes associated with invasive infections, were detected in our isolates. The virulence genes encoding clumping factor proteins (*clfA* and *clfB*) are involved in the pathogenesis of *S. aureus*, including bacteremia [9]. Consistent with pathogenic *S. aureus* strains isolated in various environments globally, our isolates were characterized by the *icaADBC* operon and *sdrC*, *sdrD*, and *sdrE* involved in biofilm-forming genes [38]. Most strains harbored genes, including the alpha and gamma-hemolysin genes (*hlgA*, *hlgB*, *hlgC*, *hly/hla*, and *hlb*), and the *ica* operon associated with pathogenicity and adhesion. Additionally, our isolates were characterized by various toxins, including *lukE/D* genes and Panton–Valentine leukocidin (PVL) *lukS-PV*/*lukF-PV* genes in one MSSA and MRSA strain. The expression of these PVL toxin genes in *S. aureus* isolates lyses host cells and promotes virulence of the bacteria [39], which might worsen the outcomes of *S. aureus* infection. Consistent with clinical *S. aureus* strains, our isolates were characterized by a capsular polysaccharide (CP) serotype 8, which shields the bacterial pathogen from host immune defense mechanisms associated with increased virulence in BSIs [40].

Most virulence genes, including those encoding SEs, *sak*, *hlg*, *luk*, *scn*, *clfA*, *sbi*, and associated MGEs, were carried on chromosomes in the majority of isolates. The *ica* gene operon and *vra* genes were found to be associated with ISs (IS30, IS6) and recombinase for S11 (ST8) isolate from the surgical ward. The *ica* genes *vraDEH* genes have been shown to play an important role in biofilm formation [41] and daptomycin resistance in *S. aureus* [42], which could enhance antibiotic resistance traits and chronic infection. The occurrence of ST8-t9475 MRSA strains co-harboring *ica* genes and genes encoding daptomycin resistance in ST8 MRSA could be advantageous to the ST8-t9475 colonization, invasion, and survival in the surgical ward. The virulence genes encoding SEs, *eap*, *scn*, *sak*, *sph*, *lukH*, and *cadA*, were found on a contig that had high sequence similarity to *S. aureus* strain ER10678.3 plasmid pER10678.3A.1 (CP051928.1), implying that they are mobilized by plasmids. Virulence genes, including those encoding *hla*/*hld*, toxin production, and biofilm formation, are plasmid-mediated [43], thus could easily facilitate their transfer, resulting in highly pathogenic strains that may be difficult to treat.

Phylogenomic analyses revealed that the clinical isolates in this study clustered mainly with clinical isolates from hospital patients (Figure 2). ST5 study isolates were closely related to clinical isolates from South Africa, suggesting possible dissemination of ST5 strains and adaptation in hospital environments. Furthermore, the ST152 isolate was closely related to ST152 strains from Egypt and Ghana, implying a possible spread and epidemiological linkage between these isolates. ST152-PVL-producing *S. aureus* isolates are particularly frequent and widespread in West and Central Africa [44] and livestock [45]. The ST152-PVL-positive MSSA has also been reported from cutaneous abscesses among mine workers at a gold mine in Gauteng, South Africa [46]. Identifying ST152 in livestock and humans suggests animal–human transmission, which requires further investigation. ST8 and ST612 isolates were closely related to ST8 isolated from Tanzania, indicating that ST612 is a double-locus variant of ST8. ST8 and ST612 isolates are potentially multidrug-resistant and highly virulent strains associated with hospital outbreaks [47]. The integration of phylogenetic data with resistance profiles offers valuable insights into the epidemiology and evolutionary dynamics of these isolates, highlighting potential patterns of transmission and resistance development [48].

In light of the increasing resistance observed in *S. aureus* strains, innovative strategies are being explored to combat antimicrobial resistance. One promising approach focuses on targeting microbial metallophores, molecules that bacteria use to scavenge essential metals from their environment. By inhibiting metallophore function, it is possible to disrupt bacterial metabolism and enhance the effectiveness of existing antibiotics [49]. Additionally, alternative strategies, such as the use of bacteriophages, antimicrobial peptides, and immune system modulation, offer potential avenues for treating resistant infections [50]. Finally, combination therapy, which involves using multiple antibiotics or combining antibiotics with adjuvants that inhibit resistance mechanisms, is gaining attention as a way to overcome multi-drug resistance and reduce the likelihood of treatment failure [51]. These emerging strategies represent critical avenues for future research and clinical application, aiming to curb the growing threat of antimicrobial resistance.

This study analyzed a limited number of *Staphylococcus aureus* isolates (n = 6), which may not fully represent the epidemiology of MRSA/MSSA in South Africa or bloodstream infections more broadly. The small sample size limits the generalizability of the findings to the wider population. Therefore, while this study offers valuable insights into the genomic characteristics and resistance profiles of these isolates, it should be viewed as a pilot study that lays the groundwork for larger, more comprehensive investigations. The small sample size also limits the ability to validate the findings, particularly concerning virulence factors and other genomic features. Future studies with larger sample sizes and experimental validation are necessary to confirm and extend these observations.

4. Materials and Methods

4.1. Ethical Consideration

The ethical approval for this study was issued by the Biomedical Research Ethics Committee of the University of KwaZulu-Natal under the following reference number: BCA444/16.

4.2. Sample Collection and Bacterial Identification

A total of forty-five presumptive *Staphylococcus* isolates from blood cultures sourced from patients with BSIs at two hospitals in the uMgungundlovu district in the KwaZulu-Natal province from November 2017 to December 2018. All isolates were confirmed as *S. aureus* using the automated VITEK 2 system (BioMérieux, MarcyL'Etoile, France). We selected a subset of 10 MDR isolates for WGS based on their antibiotic-resistant profiles/patterns, but 4 isolates were excluded during the quality control process.

4.3. Antimicrobial Susceptibility and MRSA Detection

Isolates were tested for antibiotic susceptibility by disk-diffusion method on Mueller–Hinton agar as recommended by the European Committee on Antimicrobial Susceptibility Testing (EUCAST) [52] or Clinical and Laboratory Standards Institute (CLSI) [53]. The antibiotics tested and interpreted according to the EUCAST breakpoints (EUCAST, 2017) were penicillin G (10 µg), ampicillin (10 µg), cefoxitin (30 µg), tigecycline (15 µg), and nitrofurantoin (300 µg). The CLSI guidelines (CLSI, 2017) were used for the following antibiotics: ciprofloxacin (5 µg), levofloxacin (5 µg), moxifloxacin (5 µg), erythromycin (15 µg), gentamicin (10 µg), amikacin (30 µg), chloramphenicol (30 µg), tetracycline (30 µg), doxycycline (30 µg), sulphamethoxazole/trimethoprim (1.25 µg + 23.75 µg), teicoplanin (30 µg), linezolid (30 µg), clindamycin (2 µg), and rifampicin (5 µg). MRSA isolates were identified using a cefoxitin disk (30 µg). The antibiotic disks were obtained from Oxoid (Oxoid, Basingstoke, UK). *S. aureus* ATCC 29213, was used as the quality control strain. Multidrug resistance (MDR) was defined as resistance to three or more antibiotic classes [54].

4.4. Whole-Genome Sequencing (WGS) and Bioinformatic Analysis

Genomic DNA extraction was performed using the GenElute Bacterial Genomic DNA kit (Sigma Aldrich, St. Louis, MO, USA) following the manufacturer's instructions. The quality of the DNA was assessed using NanoDrop 8000 (Thermo Fisher Scientific Waltham, MA, USA). Genome libraries were constructed using the Nextera XT DNA Library Preparation Kit (Illumina, San Diego, CA, USA) and sequenced on the Illumina NextSeq Machine (Illumina, San Diego, CA, USA). The raw reads were trimmed using Sickle v1.33 (https://github.com/najoshi/sickle accessed on 15 August 2020) and assembled using the SPAdes v3.6.2 assembler (https://cab.spbu.ru/software/spades/ accessed on 15 August 2020). Assembled genome sequences were submitted to Genebank and assigned accession numbers under the BioProject number PRJNA400143.

4.5. Genomic Analysis

Genotyping of the assembled genomes was performed using the MLST 2.0, *spa* Typer 1.0, and SCC*mec*Finder 1.2 available at the Centre for Genomic Epidemiology (CGE) (https://www.genomicepidemiology.org/services/ accessed on 1 October 2023). The detection of antibiotic resistance genes (ARGs) was examined by ResFinder 4.1 (https://cge.cbs.dtu.dk/servic es/ResFinder/ accessed on 1 October 2023) and the comprehensive antibiotic resistance database (https://card.mcmaster.ca/analyze/rgi accessed on 1 October 2023). Virulence determinants were identified with default settings using the virulence factor database (VFDB: http://www.mgc.ac.cn/VFs/main.htm accessed on 15 December 2023) and VirulenceFinder 2.0 https://cge.food.dtu.dk/services/VirulenceFinder/ accessed on 1 October 2023). Pathogenicity of isolates was found out using PathogenFinder 1.1 (https://cge.food.dtu.dk/services/PathogenFinder/ accessed on 1 October 2023). PHASTER was used to identify prophage elements (https://phaster.ca/ accessed on 1 October 2023). Mobile genetic elements (MGEs) in relation to ARGs, virulent factors, and plasmid replicons were identified using MobileElementFinder (https://cge.food.dtu.dk/services/MobileElementFinder/ accessed on 1 October 2023) and Plasmid Finder 2.1 (https://cge.cbs.dtu.dk/services/PlasmidFinder/ accessed on 1 October 2023). The genetic environment of ARGs, virulence factors, and associated MGEs was examined using GenBank's general feature format (GFF3) files and imported into Geneious Prime 2020.2 (https://www.geneious.com accessed on 10 December 2023) for analysis [55]. The accessory gene regulator (*agr*) typing was performed by employing nucleotide BLAST, and the following GenBank accession numbers: AFS50129.1, AFS50128.1, AFS50130.1, and AFS50131.1 were used as reference sequences for *agr* types I–IV [56].

4.6. Phylogenomic Analysis

For phylogeny analysis, whole-genome sequences of *S. aureus* isolates from blood culture were selected from Northern Africa (Egypt, Algeria, and Sudan), Western Africa (Ghana), and Eastern Africa (Tanzania) and downloaded from the bacterial and viral bioinformatics resource center's (BV-BRC) online platform (https://www.bv-brc.org/ accessed on 15 May 2024) and used together with our study's isolates. We constructed a phylogenomic tree using the online Phylogenetic Tree Building tool available on the BV-BRC website (https://www.bv-brc.org/ accessed on 15 May 2024). The generated phylogenetic tree was visualized, annotated, and edited using iTOL (https://itol.embl.de/ accessed on 15 May 2024) and Figtree (http://tree.bio.ed.ac.uk/software/figtree/ accessed on 15 May 2024).

4.7. Nucleotide Sequence Accession Number

The nucleotide sequences of MRSA (S29, S11, S31) and MSSA (S13, S24, S34) isolates were submitted to the NCBI GenBank database under the following accession numbers: JADQTH000000000, JADIXB000000000, JADIXC000000000, JADIXA000000000, JADIXE000000000, and JADIXD000000000.

5. Conclusions

This study presents an insight into ARGs, virulence genes, MGEs, and genetic diversity of *S. aureus* collected from a public hospital in uMgungundlovu. We observed high diversity of *spa* types, STs, and a predominance of CC8 and CC5, indicating the genetic variability of *S. aureus* in hospital settings. The occurrence of pathogenic and MDR strains in the hospital setting, especially in the ICU, can pose a serious threat that limits the therapeutic options available. Here, we demonstrate that while MRSA displayed multidrug resistance, MSSA reflected potentially increasing resistance to the antibiotics used for treatment. Continuous surveillance and monitoring of MRSA and MSSA strains circulating in hospital environments is needed.

Supplementary Materials: The following supporting information can be downloaded at: https://www.mdpi.com/article/10.3390/antibiotics13090796/s1. Table S1: Patient demographics. Table S2: Genomic characteristics of *S. aureus* strains. Table S3: Virulence genes identified in MSSA and

MRSA isolates in this study. Table S4: Distribution of insertion sequences and plasmid replicon among the *Staphylococcus aureus* strains. Table S5: Distribution of intact prophage region among the *Staphylococcus aureus* strains.

Author Contributions: Conceptualization, B.A.H., D.G.A. and S.Y.E.; methodology, B.A.H., A.I. and A.L.K.A.; formal analysis, B.A.H., D.G.A., J.M., J.A. and A.L.K.A.; investigation, B.A.H.; resources, A.I., D.G.A. and S.Y.E.; writing—original draft preparation, B.A.H.; writing—review and editing, B.A.H., J.A., J.M., A.I., A.L.K.A., D.G.A. and S.Y.E.; supervision, S.Y.E., A.L.K.A. and D.G.A. All authors have read and agreed to the published version of the manuscript.

Funding: This work was supported by the South African Research Chairs Initiative of the Department of Science and Technology and the National Research Foundation of South Africa (Grant No. 98342), the SAMRC and UK MRC Newton Fund, the SAMRC Self-Initiated Research Grant, and the College of Health Sciences University of KwaZulu-Natal, South Africa.

Institutional Review Board Statement: Ethical approval for this study was obtained from the Biomedical Research Ethics Committee of the University of KwaZulu-Natal under the following reference number BCA444/16. The study isolates were part of a larger surveillance study using the Global Antimicrobial Resistance and Use Surveillance System (GLASS) guidelines.

Informed Consent Statement: Not applicable.

Data Availability Statement: The data presented in this study are openly available in GenBank and assigned accession numbers under the BioProject PRJNA400143. [NCBI Genebank] [https://www.ncbi.nlm.nih.gov/bioproject/?term=PRJNA400143 accessed on 15 May 2024] [PRJNA400143].

Acknowledgments: We are grateful to Sumayya Haffejee of the National Health Laboratory Services for her assistance during sample collection and obtaining demographic data.

Conflicts of Interest: S.Y.E. is a chairperson of the Global Respiratory Infection Partnership and member of the Global Hygiene Council, both funded by unrestricted educational grants from Reckitt and Benckiser (Pty.), UK. The remaining authors declare that research was conducted in the absence of any commercial or financial relationships that could be construed as a potential conflict of interest.

References

1. Asadollahi, P.; Farahani, N.N.; Mirzaii, M.; Khoramrooz, S.S.; van Belkum, A.; Asadollahi, K.; Dadashi, M.; Darban-Sarokhalil, D. Distribution of the Most Prevalent Spa Types among Clinical Isolates of Methicillin-Resistant and-Susceptible Staphylococcus Aureus Around the World: A Review. *Front. Microbiol.* **2018**, *9*, 163. [CrossRef] [PubMed]
2. Liang, Y.; Tu, C.; Tan, C.; El-Sayed Ahmed, M.A.E.G.; Dai, M.; Xia, Y.; Liu, Y.; Zhong, L.L.; Shen, C.; Chen, G.; et al. Antimicrobial Resistance, Virulence Genes Profiling and Molecular Relatedness of Methicillin-Resistant Staphylococcus Aureus Strains Isolated from Hospitalized Patients in Guangdong Province, China. *Infect. Drug Resist.* **2019**, *12*, 447–459. [CrossRef] [PubMed]
3. Gu, F.; He, W.; Xiao, S.; Wang, S.; Li, X.; Zeng, Q.; Ni, Y.; Han, L. Antimicrobial Resistance and Molecular Epidemiology of Staphylococcus Aureus Causing Bloodstream Infections at Ruijin Hospital in Shanghai from 2013 to 2018. *Sci. Rep.* **2020**, *10*, 6019. [CrossRef]
4. Diekema, D.J.; Hsueh, P.R.; Mendes, R.E.; Pfaller, M.A.; Rolston, K.V.; Sader, H.S.; Jones, R.N. The Microbiology of Bloodstream Infection: 20-Year Trends from The SENTRY Antimicrobial Surveillance Program. *Antimicrob. Agents Chemother.* **2019**, *63*, e00355-19. [CrossRef]
5. Turner, N.A.; Sharma-Kuinkel, B.K.; Maskarinec, S.A.; Eichenberger, E.M.; Shah, P.P.; Carugati, M.; Holland, T.L.; Fowler, V.G. Methicillin-Resistant Staphylococcus Aureus: An Overview of Basic and Clinical Research. *Nat. Rev. Microbiol.* **2019**, *17*, 203–218. [CrossRef]
6. Hadyeh, E.; Azmi, K.; Seir, R.A.; Abdellatief, I.; Abdeen, Z. Molecular Characterization of Methicillin Resistant Staphylococcus Aureus in West Bank-Palestine. *Front. Public Health* **2019**, *7*, 130. [CrossRef]
7. Jin, Y.; Zhou, W.; Zhan, Q.; Chen, Y.; Luo, Q.; Shen, P.; Xiao, Y. Genomic Epidemiology and Characterisation of Penicillin-Sensitive Staphylococcus Aureus Isolates from Invasive Bloodstream Infections in China: An Increasing Prevalence and Higher Diversity in Genetic Typing Be Revealed. *Emerg. Microbes Infect.* **2022**, *11*, 326–336. [CrossRef] [PubMed]
8. Yuan, W.; Liu, J.; Zhan, Y.; Wang, L.; Jiang, Y.; Zhang, Y.; Sun, N.; Hou, N. Molecular Typing Revealed the Emergence of PVL-Positive Sequence Type 22 Methicillin-Susceptible Staphylococcus Aureus in Urumqi, North Western China. *Infect. Drug Resist.* **2019**, *12*, 1719–1728. [CrossRef]
9. Foster, T.J. The MSCRAMM Family of Cell-Wall-Anchored Surface Proteins of Gram-Positive Cocci. *Trends Microbiol.* **2019**, *27*, 927–941. [CrossRef]

10. Foster, T.J.; Geoghegan, J.A.; Ganesh, V.K.; Hook, M. Adhesion, Invasion and Evasion: The Many Functions of the Surface Proteins of Staphylococcus Aureus. *Nat. Rev. Microbiol.* **2014**, *12*, 46–62. [CrossRef]
11. Ghssein, G.; Ezzeddine, Z. The Key Element Role of Metallophores in the Pathogenicity and Virulence of Staphylococcus Aureus: A Review. *Biology* **2022**, *11*, 1525. [CrossRef] [PubMed]
12. Warnes, S.L.; Highmore, C.J.; Keevil, C.W. Horizontal Transfer of Antibiotic Resistance Genes on Abiotic Touch Surfaces: Implications for Public Health. *MBio* **2012**, *3*, e00489-12. [CrossRef] [PubMed]
13. Park, K.H.; Greenwood-Quaintance, K.E.; Uhl, J.R.; Cunningham, S.A.; Chia, N.; Jeraldo, P.R.; Sampathkumar, P.; Nelson, H.; Patel, R. Molecular Epidemiology of Staphylococcus Aureus Bacteremia in a Single Large Minnesota Medical Center in 2015 As Assessed Using MLST, Core Genome MLST and Spa Typing. *PLoS ONE* **2017**, *12*, e0179003. [CrossRef] [PubMed]
14. Perovic, O.; Iyaloo, S.; Kularatne, R.; Lowman, W.; Bosman, N.; Wadula, J.; Seetharam, S.; Duse, A.; Mbelle, N.; Bamford, C.; et al. Prevalence and Trends of Staphylococcus Aureus Bacteraemia in Hospitalized Patients in South Africa, 2010 to 2012: Laboratory-Based Surveillance Mapping of Antimicrobial Resistance and Molecular Epidemiology. *PLoS ONE* **2015**, *10*, e0145429. [CrossRef] [PubMed]
15. Singh-Moodley, A.; Lowe, M.; Mogokotleng, R.; Perovic, O. Diversity of SCC*mec* Elements and Spa Types in South African Staphylococcus Aureus MecA-Positive Blood Culture Isolates. *BMC Infect. Dis.* **2020**, *20*, 816. [CrossRef]
16. Amoako, D.G.; Somboro, A.M.; Abia, A.L.K.; Allam, M.; Ismail, A.; Bester, L.; Essack, S.Y. Genomic Analysis of Methicillin-Resistant Staphylococcus Aureus Isolated from Poultry and Occupational Farm Workers in Umgungundlovu District, South Africa. *Sci. Total Environ.* **2019**, *670*, 704–716. [CrossRef] [PubMed]
17. Smith, J.T.; Eckhardt, E.M.; Hansel, N.B.; Eliato, T.R.; Martin, I.W.; Andam, C.P. Genomic Epidemiology of Methicillin-Resistant and-Susceptible Staphylococcus Aureus from Bloodstream Infections. *BMC Infect. Dis.* **2021**, *21*, 589. [CrossRef] [PubMed]
18. Schaumburg, F.; Alabi, A.S.; Peters, G.; Becker, K. New Epidemiology of Staphylococcus Aureus Infection in Africa. *Clin. Microbiol. Infect.* **2014**, *20*, 589–596. [CrossRef] [PubMed]
19. Strasheim, W.; Perovic, O.; Singh-Moodley, A.; Kwanda, S.; Ismail, A.; Lowe, M. Ward-Specific Clustering of Methicillin-Resistant Staphylococcus Aureus Spa-Type T037 and T045 in Two Hospitals in South Africa: 2013 to 2017. *PLoS ONE* **2021**, *16*, e0253883. [CrossRef]
20. Silva, V.; Hermenegildo, S.; Ferreira, C.; Manaia, C.M.; Capita, R.; Alonso-Calleja, C.; Carvalho, I.; Pereira, J.E.; Maltez, L.; Capelo, J.L.; et al. Genetic Characterization of Methicillin-Resistant Staphylococcus Aureus Isolates from Human Bloodstream Infections: Detection of MLSB Resistance. *Antibiotics* **2020**, *9*, 375. [CrossRef]
21. Asante, J.; Govinden, U.; Owusu-Ofori, A.; Bester, L.A.; Essack, S.Y. Molecular Characterization of Methicillin-Resistant Staphylococcus Aureus Isolates from a Hospital in Ghana. *African J. Clin. Exp. Microbiol.* **2019**, *20*, 164. [CrossRef]
22. Mkhize, S.; Amoako, D.G.; Shobo, C.O.; Zishiri, O.T.; Bester, L.A. Genotypic and Phenotypic Characterizations of Methicillin-Resistant Staphylococcus Aureus (MRSA) on Frequently Touched Sites from Public Hospitals in South Africa. *Int. J. Microbiol.* **2021**, *2021*, 6011045. [CrossRef] [PubMed]
23. McCarthy, A.J.; Lindsay, J.A. The Distribution of Plasmids That Carry Virulence and Resistance Genes in Staphylococcus Aureus Is Lineage Associated. *BMC Microbiol.* **2012**, *12*, 104. [CrossRef] [PubMed]
24. Ocloo, R.; Newton-Foot, M.; Ziebuhr, W.; Whitelaw, A.C. Molecular Epidemiology and Antibiotic Resistance of Staphylococci Other than Staphylococcus Aureus in Children in Cape Town, South Africa. *Front. Microbiol.* **2023**, *14*, 1239666. [CrossRef] [PubMed]
25. Pennone, V.; Prieto, M.; Avelino, Á.; Cobo-diaz, J.F. Antimicrobial Resistance Genes Analysis of Publicly Available Staphylococcus Aureus Genomes. *Antibiotics* **2022**, *11*, 1632. [CrossRef] [PubMed]
26. Shittu, A.O.; Adesoji, T.; Udo, E.E. DNA Microarray Analysis of Staphylococcus Aureus from Nigeria and South Africa. *PLoS ONE* **2021**, *16*, e0237124. [CrossRef]
27. Parsons, C.; Lee, S.; Kathariou, S. Dissemination and Conservation of Cadmium and Arsenic Resistance Determinants in Listeria and Other Gram-Positive Bacteria. *Mol. Microbiol.* **2020**, *113*, 560–569. [CrossRef]
28. Al-Trad, E.I.; Chew, C.H.; Che Hamzah, A.M.; Suhaili, Z.; Rahman, N.I.A.; Ismail, S.; Puah, S.M.; Chua, K.H.; Kwong, S.M.; Yeo, C.C. The Plasmidomic Landscape of Clinical Methicillin-Resistant Staphylococcus Aureus Isolates from Malaysia. *Antibiotics* **2023**, *12*, 733. [CrossRef]
29. Zhong, Q.; Cruz-Paredes, C.; Zhang, S.; Rousk, J. Can Heavy Metal Pollution Induce Bacterial Resistance to Heavy Metals and Antibiotics in Soils from An Ancient Land-Mine? *J. Hazard. Mater.* **2021**, *411*, 124962. [CrossRef]
30. Kleinert, F.; Kallies, R.; Hort, M.; Zweynert, A.; Szekat, C.; Nagel, M.; Bierbaum, G. Influence of IS256 on Genome Variability and Formation of Small-Colony Variants in Staphylococcus Aureus. *Antimicrob. Agents Chemother.* **2017**, *61*, e00144-17. [CrossRef]
31. Mbelle, N.M.; Feldman, C.; Osei Sekyere, J.; Maningi, N.E.; Modipane, L.; Essack, S.Y. Publisher Correction: The Resistome, Mobilome, Virulome and Phylogenomics of Multidrug-Resistant Escherichia Coli Clinical Isolates from Pretoria, South Africa. *Sci. Rep.* **2020**, *10*, 1270. [CrossRef]
32. Wang, W.; Baker, M.; Hu, Y.; Xu, J.; Yang, D.; Maciel-Guerra, A.; Xue, N.; Li, H.; Yan, S.; Li, M.; et al. Whole-Genome Sequencing and Machine Learning Analysis of Staphylococcus Aureus from Multiple Heterogeneous Sources in China Reveals Common Genetic Traits of Antimicrobial Resistance. *mSystems* **2021**, *6*, e0118520. [CrossRef]

33. Sands, K.; Carvalho, M.J.; Spiller, O.B.; Portal, E.A.R.; Thomson, K.; Watkins, W.J.; Mathias, J.; Dyer, C.; Akpulu, C.; Andrews, R.; et al. Characterisation of Staphylococci Species from Neonatal Blood Cultures in Low- and Middle-Income Countries. *BMC Infect. Dis.* **2022**, *22*, 593. [CrossRef]
34. Lawal, O.U.; Ayobami, O.; Abouelfetouh, A.; Mourabit, N.; Kaba, M.; Egyir, B.; Abdulgader, S.M.; Shittu, A.O. A 6-Year Update on the Diversity of Methicillin-Resistant Staphylococcus Aureus Clones in Africa: A Systematic Review. *Front. Microbiol.* **2022**, *13*, 860436. [CrossRef] [PubMed]
35. Elshabrawy, W.; Elsayed Zaki, M.; Farag Kamel, M. Genetic and Phenotypic Study of Methicillin-Resistant Staphylococcus Aureus Among Patients and Health Care Workers in Mansoura University Hospital, Egypt. *Iran. J. Microbiol.* **2017**, *9*, 82–88. [PubMed]
36. Soliman, M.S.; Soliman, N.S.; El-Manakhly, A.R.; Elbanna, S.A.; Aziz, R.K.; El-Kholy, A.A. Genomic Characterization of Methicillin-Resistant Staphylococcus Aureus (MRSA) by High-Throughput Sequencing in A Tertiary Care Hospital. *Genes* **2020**, *11*, 1219. [CrossRef]
37. Abdulgader, S.M.; van Rijswijk, A.; Whitelaw, A.; Newton-Foot, M. The Association Between Pathogen Factors and Clinical Outcomes in Patients With Staphylococcus Aureus Bacteraemia in a Tertiary Hospital, Cape Town. *Int. J. Infect. Dis.* **2020**, *91*, 111–118. [CrossRef]
38. Peng, Q.; Tang, X.; Dong, W.; Sun, N.; Yuan, W. A Review of Biofilm Formation of Staphylococcus Aureus and Its Regulation Mechanism. *Antibiotics* **2023**, *12*, 12. [CrossRef]
39. Vasquez, M.T.; Lubkin, A.; Reyes-Robles, T.; Day, C.J.; Lacey, K.A.; Jennings, M.P.; Torres, V.J. Identification of a Domain Critical for Staphylococcus Aureus LuKED Receptor Targeting and Lysis of Erythrocytes. *J. Biol. Chem.* **2020**, *295*, 17241–17250. [CrossRef] [PubMed]
40. Mohamed, N.; Timofeyeva, Y.; Jamrozy, D.; Rojas, E.; Hao, L.; Silmon de Monerri, N.C.; Hawkins, J.; Singh, G.; Cai, B.; Liberator, P.; et al. Molecular Epidemiology and Expression of Capsular Polysaccharides in Staphylococcus Aureus Clinical Isolates in the United States. *PLoS ONE* **2019**, *14*, e0208356. [CrossRef]
41. Singh, S.; Singh, S.K.; Chowdhury, I.; Singh, R. Understanding the Mechanism of Bacterial Biofilms Resistance to Antimicrobial Agents. *Open Microbiol. J.* **2017**, *11*, 53–62. [CrossRef]
42. Popella, P.; Krauss, S.; Ebner, P.; Nega, M.; Deibert, J.; Götz, F. VraH Is the Third Component of the Staphylococcus Aureus VraDEH System Involved in Gallidermin and Daptomycin Resistance and Pathogenicity. *Antimicrob. Agents Chemother.* **2016**, *60*, 2391–2401. [CrossRef] [PubMed]
43. Bukowski, M.; Piwowarczyk, R.; Madry, A.; Zagorski-Przybylo, R.; Hydzik, M.; Wladyka, B. Prevalence of Antibiotic and Heavy Metal Resistance Determinants and Virulence-Related Genetic Elements in Plasmids of Staphylococcus Aureus. *Front. Microbiol.* **2019**, *10*, 805. [CrossRef] [PubMed]
44. Abdulgader, S.M.; Shittu, A.O.; Nicol, M.P.; Kaba, M. Molecular Epidemiology of Methicillin-Resistant Staphylococcus Aureus in Africa: A Systematic Review. *Front. Microbiol.* **2015**, *6*, 348. [CrossRef] [PubMed]
45. Agabou, A.; Ouchenane, Z.; Essebe, C.N.; Khemissi, S.; Chehboub, M.T.E.; Chehboub, I.B.; Sotto, A.; Dunyach-Remy, C.; Lavigne, J.P. Emergence of Nasal Carriage of ST80 and ST152 PVL+ Staphylococcus Aureus Isolates from Livestock in Algeria. *Toxins* **2017**, *9*, 303. [CrossRef] [PubMed]
46. Ismail, H.; Govender, N.P.; Singh-Moodley, A.; Van Schalkwyk, E.; Shuping, L.; Moema, I.; Feller, G.; Mogokotleng, R.; Strasheim, W.; Lowe, M.; et al. An Outbreak of Cutaneous Abscesses Caused by Panton-Valentine Leukocidin-Producing Methicillin-Susceptible Staphylococcus Aureus Among Gold Mine Workers, South Africa, November 2017 to March 2018. *BMC Infect. Dis.* **2020**, *20*, 621. [CrossRef] [PubMed]
47. Wang, X.; Zhao, H.; Wang, B.; Zhou, Y.; Xu, Y.; Rao, L.; Ai, W.; Guo, Y.; Wu, X.; Yu, J.; et al. Identification of Methicillin-Resistant Staphylococcus Aureus ST8 Isolates in China with Potential High Virulence. *Emerg. Microbes Infect.* **2022**, *11*, 507–518. [CrossRef]
48. Agyepong, N.; Govinden, U.; Owusu-Ofori, A.; Amoako, D.G.; Allam, M.; Janice, J.; Pedersen, T.; Sundsfjord, A.; Essack, S. Genomic characterization of multidrug-resistant ESBL-producing Klebsiella pneumoniae isolated from a Ghanaian teaching hospital. *Int. J. Infect. Dis.* **2019**, *85*, 117–123. [CrossRef]
49. Ezzeddine, Z.; Ghssein, G. Towards new antibiotics classes targeting bacterial metallophores. *Microb. Pathog.* **2023**, *29*, 06221. [CrossRef]
50. Murugaiyan, J.; Kumar, P.A.; Rao, G.S.; Iskandar, K.; Hawser, S.; Hays, J.P.; Mohsen, Y.; Adukkadukkam, S.; Awuah, W.A.; Jose, R.A.M.; et al. Progress in Alternative Strategies to Combat Antimicrobial Resistance: Focus on Antibiotics. *Antibiotics* **2022**, *11*, 200. [CrossRef]
51. Fischbach, M.A. Combination Therapies for Combating Antimicrobial Resistance. *Curr. Opin. Microbiol.* **2011**, *14*, 519–523. [CrossRef] [PubMed]
52. EUCAST European Committee on Antimicrobial Susceptibility Testing, Breakpoint Tables for Interpretation of MICs and Zone Diameters. Version 8 2017. Available online: www.eucast.org/clinical_breakpoints (accessed on 12 November 2022).
53. *CLSI Document M100-S27*; Performance Standards for Antimicrobial Susceptibility Testing; Twenty-Seventh Informational Supplement. CLSI: Malvern, PA, USA, 2017; ISBN 1562387855.
54. Magiorakos, A.P.; Srinivasan, A.; Carey, R.B.; Carmeli, Y.; Falagas, M.E.; Giske, C.G.; Harbarth, S.; Hindler, J.F.; Kahlmeter, G.; Olsson-Liljequist, B.; et al. Multidrug-Resistant, Extensively Drug-Resistant and Pandrug-Resistant Bacteria: An International Expert Proposal for Interim Standard Definitions for Acquired Resistance. *Clin. Microbiol. Infect.* **2012**, *18*, 268–281. [CrossRef] [PubMed]

55. Mbanga, J.; Amoako, D.G.; Abia, A.L.K.; Allam, M.; Ismail, A.; Essack, S.Y. Genomic Insights of Multidrug-Resistant Escherichia Coli From Wastewater Sources and Their Association With Clinical Pathogens in South Africa. *Front. Vet. Sci.* **2021**, *8*, 636715. [CrossRef]
56. Francois, P.; Koessler, T.; Huyghe, A.; Harbarth, S.; Bento, M.; Lew, D.; Pittet, D.; Schrenzel, J. Rapid Staphylococcus Aureus Agr Type Determination by a Novel Multiplex Real-Time Quantitative PCR Assay. *J. Clin. Microbiol.* **2006**, *44*, 1892–1895. [CrossRef] [PubMed]

antibiotics

MDPI

Article

Genomic Characterization of Multidrug-Resistant Enterobacteriaceae Clinical Isolates from Southern Thailand Hospitals: Unraveling Antimicrobial Resistance and Virulence Mechanisms

Thunchanok Yaikhan [1], Sirikan Suwannasin [1], Kamonnut Singkhamanan [1], Sarunyou Chusri [2], Rattanaruji Pomwised [3], Monwadee Wonglapsuwan [3] and Komwit Surachat [1,4,*]

1 Department of Biomedical Sciences and Biomedical Engineering, Faculty of Medicine, Prince of Songkla University, Songkhla 90110, Thailand; ythuncha@medicine.psu.ac.th (T.Y.); sirikan4036@gmail.com (S.S.); skamonnu@medicine.psu.ac.th (K.S.)

2 Division of Infectious Diseases, Department of Internal Medicine, Faculty of Medicine, Prince of Songkla University, Songkhla 90110, Thailand

3 Division of Biological Science, Faculty of Science, Prince of Songkla University, Songkhla 90110, Thailand; rattanaruji.p@psu.ac.th (R.P.); monwadee.wo@psu.ac.th (M.W.)

4 Translational Medicine Research Center, Faculty of Medicine, Prince of Songkla University, Songkhla 90110, Thailand

* Correspondence: komwit.s@psu.ac.th

Abstract: The emergence and spread of antimicrobial resistance (AMR) among Enterobacteriaceae pose significant threats to global public health. In this study, we conducted a short-term surveillance effort in Southern Thailand hospitals to characterize the genomic diversity, AMR profiles, and virulence factors of Enterobacteriaceae strains. We identified 241 carbapenem-resistant Enterobacteriaceae, of which 12 were selected for whole-genome sequencing (WGS) and genome analysis. The strains included *Proteus mirabilis*, *Serratia nevei*, *Klebsiella variicola*, *Klebsiella aerogenes*, *Klebsiella indica*, *Klebsiella grimontii*, *Phytobacter ursingii*, *Phytobacter palmae*, *Kosakonia* spp., and *Citrobacter freundii*. The strains exhibited high levels of multidrug resistance, including resistance to carbapenem antibiotics. Whole-genome sequencing revealed a diverse array of antimicrobial resistance genes (ARGs), with strains carrying genes for ß-lactamase, efflux pumps, and resistance to other antibiotic classes. Additionally, stress response, metal tolerance, and virulence-associated genes were identified, highlighting the adaptability and pathogenic potential of these strains. A plasmid analysis identified several plasmid replicons, including IncA/C2, IncFIB(K), and Col440I, as well as several plasmids identical to those found globally, indicating the potential for the horizontal gene transfer of ARGs. Importantly, this study also identified a novel species of *Kosakonia* spp. PSU27, adding to the understanding of the genetic diversity and resistance mechanisms of Enterobacteriaceae in Southern Thailand. The results reported in this study highlight the critical importance of implementing effective antimicrobial management programs and developing innovative treatment approaches to urgently tackle AMR.

Keywords: Enterobacteriaceae; antimicrobial resistance gene; virulence factor; plasmid

1. Introduction

The emergence and spread of antimicrobial resistance (AMR) among Enterobacteriaceae play significant roles to global public health. These bacteria are responsible for a wide range of infections, including urinary tract infections (UTIs), hospital-acquired and ventilator-associated pneumonia (HAP/VAP), complicated intra-abdominal infections (cIAIs), and bloodstream infections (BSIs) [1]. AMR is a critical issue that contributes to significant morbidity and mortality worldwide. According to the World Health Organization (WHO), AMR causes an estimated 700,000 deaths annually, and this number could rise to 10 million by 2050 if no effective measures are taken [2]. The economic burden is also substantial, with healthcare costs rising due to prolonged hospital stays, the need for more expensive treatments, and increased mortality rates. Infections caused by multidrug-resistant (MDR) Enterobacteriaceae are particularly challenging to manage and control, leading to significant treatment costs and posing a threat to public health [3].

In Southern Thailand, where infectious diseases are a major public health concern, understanding the genomic diversity, AMR profiles, and virulence factors of Enterobacteriaceae is crucial for guiding local treatment strategies and infection control measures. Despite the importance of this issue, there is a scarcity of studies focusing on the genomics of Enterobacteriaceae in this region. The ability of this bacterial family to rapidly acquire and disseminate resistance genes has made them difficult challengers in the clinical setting, often resulting in treatment failures and increased healthcare costs.

We obtained a diverse family of bacteria that includes several clinically significant genera, such as *Proteus*, *Serratia*, *Klebsiella*, *Phytobacter*, *Kosakonia*, and *Citrobacter*, from short-term survey of Enterobacteriaceae in Southern Thailand hospitals. Therefore, this study aims to address this gap by providing genomic insights into Enterobacteriaceae diversity, AMR, and virulence in Southern Thailand. By employing whole-genome sequencing and bioinformatic analyses, we seek to characterize the genetic diversity of Enterobacteriaceae isolates, identify AMR genes and mechanisms, and elucidate virulence factors contributing to their pathogenicity. The findings of this study are expected to have significant implications for clinical practice and public health in Southern Thailand. By providing a comprehensive understanding of the genomic landscape of Enterobacteriaceae in this region, we might be able to suggest local treatment guidelines, enhance antimicrobial deployment efforts, and contribute to global AMR surveillance efforts.

2. Results and Discussions

2.1. Clinical Data and Antimicrobial Susceptibility Testing Results

The clinical data and antimicrobial resistance susceptibility results are exhibited in Table 1, providing a detailed overview of Enterobacteriaceae strains isolated from various hospital sources and their resistance patterns to a range of antibiotics. The included strains are *P. mirabilis*, *S. nevei*, two isolates of *K. variicola*, *K. aerogenes*, *K. indica*, *K. grimontii*, *P. ursingii*, *P. palmae*, *Kosakonia* spp., and two isolates of *C. freundii* from different source of isolations, such as rectal, throat, endotracheal tube, and nasopharynx samples, highlighting the diverse nature of these pathogens and their potential to cause infections in different clinical contexts. All strains in this study present a resistance to carbapenem antibiotics (meropenem, imipenem, and ertapenem) and some types of third-generation cephalosporin (ceftriaxone and ceftazidime), indicating a high level of multidrug resistance among these pathogens. Previous studies have also reported the problem of carbapenem-resistant Enterobacteriaceae (CRE). For example, in 2017, Logan and Weinstein reported on the evolution and epidemiology of CRE globally, highlighting the rapid global dissemination of carbapenemase-producing Enterobacteriaceae [4]. Another study in 2021 emphasized the global challenge of multidrug-resistant (MDR) Enterobacteriaceae and the need for new treatment options to combat CRE infections [5]. In addition to resistance to ß-lactam drugs, the resistance profiles reveal variability among the strains. For instance, *K. variicola* PSU7 shows resistance to most antibiotics except amikacin, while *K. variicola* PSU16 exhibits resistance to all antibiotics tested. This bacterium is recognized as an emerging pathogen in humans and has been identified in various environments. It is a member of the *K. pneumoniae* complex and has been found to be a more serious pathogen, containing a broad range of resistance genes that confer the resistance phenotype. This indicates a particularly worrisome level of MDR, especially the ß-lactam class [5].

Table 1. Antimicrobial susceptibility testing in Enterobacteriaceae strains in this study.

Strain	Hospital	Source	Antimicrobial Susceptibility Test (AST)										
			MEM	IPM	ETP	GEN	AMK	TZP	CIP	LVX	CRO	CAZ	SAM
Proteus mirabilis PSU2	PSU	R	R	R	R	S	R	R	R	R	R	R	R
Serratia nevei PSU6	PSU	Th	R	R	R	R	R	R	I	I	R	R	R
Klebsiella variicola PSU7	PSU	R	R	R	R	R	S	R	R	R	R	R	R
Klebsiella variicola PSU16	PT	R	R	R	R	R	R	R	R	R	R	R	R
Klebsiella aerogenes PSU22	ST	Tu	R	R	R	R	S	R	R	R	R	R	R
Klebsiella indica PSU33	PA	Th	R	R	R	S	R	R	R	R	R	R	S
Klebsiella grimontii PSU35	PA	Ng	R	R	R	R	R	R	R	R	R	R	R
Phytobacter ursingii PSU26	ST	Ng	R	R	R	R	R	R	I	R	R	R	R
Phytobacter palmae PSU29	ST	Ng	R	R	R	S	R	R	R	I	R	R	R
Kosakonia spp. PSU27	ST	Ng	R	R	R	R	R	R	R	R	R	R	R
Citrobacter freundii PSU41	YL	R	R	R	R	R	S	R	I	R	R	R	S
Citrobacter freundii PSU42	YL	R	R	R	R	R	R	R	R	R	R	R	R

Hospital	Source of isolation
PSU: Songklanagarind Hospital	Ng: nasopharynx
PT: Patthalung Hospital	Tu: endotracheal tube
ST: Satun Hospital	Th: throat
PA: Pattani Hospital	R: rectum
YL: Yala Hospital	

Meropenem (MEM), Imipenem (IPM), Ertapenem (ETP), Gentamicin (GEN), Amikacin (AMK), Tazocin (TZP), Ciprofloxacin (CIP), Levofloxacin (LVX), Ceftriaxone (CRO), Ceftazidime (CAZ), Sulperazon (SAM). Resistance (R), Susceptible (S), and Intermediate (I). The results were analyzed according to the Clinical and Laboratory Standards Institute 2023 (CLSI 2023).

Surprisingly, we identified a novel species of *Kosakonia* spp. PSU27, isolated from the nasopharynx of a patient at Satun Hospital. The average nucleotide identity (ANI) shows a 94.55% similarity to the closest species, *Kosakonia sacchari* BO-1. We also collected data from the RefSeq database, which contains 99 isolates of *Kosakonia* genomes (Figure 1). Among these isolates, only five are clinical samples: one *Kosakonia* spp. from an oral metagenome, three *Kosakonia radicincitans* from blood, and one *Kosakonia cawanii* from an oral swab. Our *Kosakonia* spp. PSU27 is the first novel species reported in Thailand.

This discovery of a novel species, *Kosakonia* spp. PSU27, isolated from a patient's nasopharynx at Satun Hospital, is significant for several reasons. Firstly, it expands the known diversity of the *Kosakonia* genus, which can have implications for understanding its ecological niche, pathogenicity, and potential biotechnological applications. The identification of this novel species was supported by an average nucleotide identity (ANI) analysis, which showed a 94.55% similarity to the closest known species, *Kosakonia sacchari* BO-1. This level of genetic divergence suggests that *Kosakonia* spp. PSU27 represents a distinct species within the genus. Furthermore, the rarity of clinical isolates of *Kosakonia* in public databases, as evidenced by the limited number of clinical samples among the 99 isolates of *Kosakonia* genomes in RefSeq NCBI [6], underscores the novelty of this finding. Among these isolates, only five were clinical samples, which indicates the importance of this discovery in expanding our understanding of the clinical relevance of *Kosakonia* species.

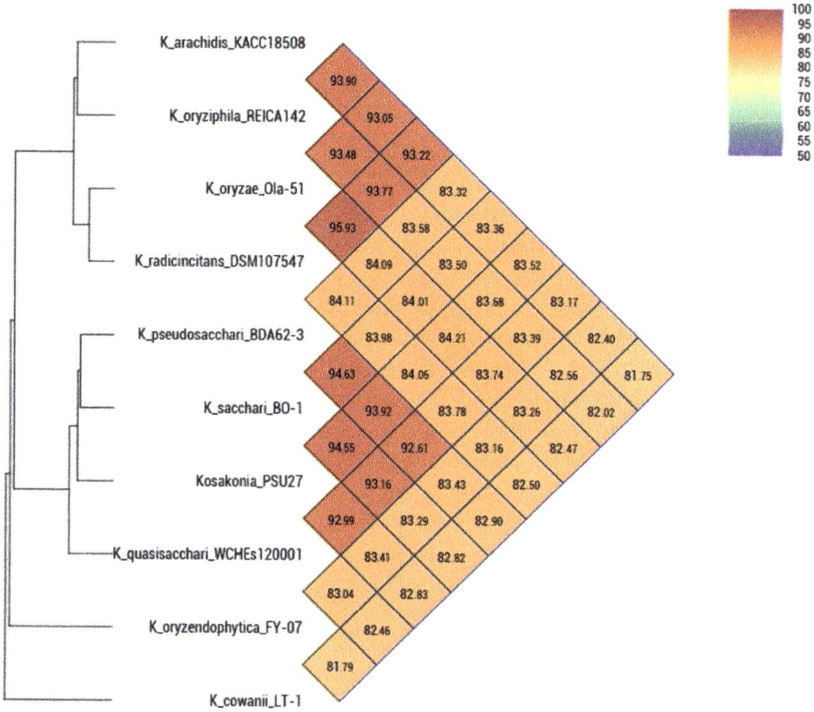

Figure 1. Average nucleotide identity of nine *Kosakonia* isolates collected from RefSeq NCBI database and *Kosakonia* spp. PSU27 in this study. Percent nucleotide identity lower than 95% is considered as a novel species.

These antimicrobial profiles and the ANI provide valuable insights into the diversity and antimicrobial resistance patterns of Enterobacteriaceae strains in clinical settings. The inclusion of various strains from different hospital sources highlights the diverse nature of these pathogens and their potential to cause infections in different clinical contexts. The high level of multidrug resistance, including resistance to carbapenem antibiotics, highlights the urgent need for new treatment options and effective antimicrobial management programs. The variability in resistance profiles among strains further emphasizes the complexity of antimicrobial resistance mechanisms within Enterobacteriaceae. The discovery of a novel species, *Kosakonia* spp. PSU27, expands our understanding of the genus and raises awareness among clinical staff and researchers about the importance of ongoing surveillance and research to identify emerging pathogens.

2.2. Antimicrobial Resistance Genes in Enterobacteriaceae

The Enterobacteriaceae isolates in this study exhibit a wide range of antimicrobial resistance genes, with a total of 45 ARGs, which were categorized into 14 classes based on the antibiotics they confer resistance to (Figure 2). Among the isolates, *Klebsiella* and *Citrobacter* strains were found to carry several ß-lactamase genes (e.g., bla_{LEN}, bla_{CMY}, bla_{DHA}, bla_{NDM}, bla_{OXA}, bla_{TEM}, and bla_{VEB}), indicating resistance to ß-lactam antibiotics. The presence of these genes suggests the enzymatic inactivation of beta-lactams. A study by Chen et al. found that bla_{TEM}, $bla_{CTX\text{-}M}$, and bla_{OXA} were the most common ß-lactamase genes detected in Enterobacteriaceae isolates [7], aligning with our findings. Furthermore, our study identified additional resistance genes, including those for efflux pumps (e.g., *kdeA*, *emrD*, *sdeB*, *sdeY*, and *smfY*) in strains such as *S. nevei* PSU6 and some isolates of

Klebsiella. Efflux pumps, regulated by these genes, play a crucial role in bacterial survival by expelling toxic substances, including antibiotics. Wang et al. similarly identified efflux pump genes, such as *acrAB* and *oqxAB*, in Enterobacteriaceae isolates, which contribute significantly to multidrug resistance [8].

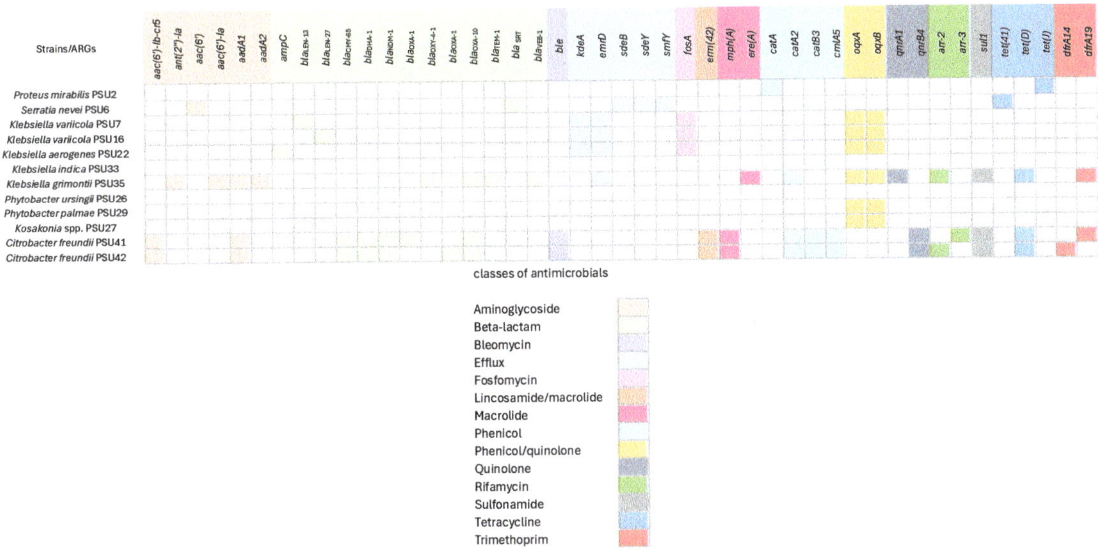

Figure 2. Antimicrobial resistance genes (ARGs) in 13 Enterobacteriaceae isolates. Forty-five genes conferring antimicrobial resistance genes which were categorized into 14 classes. The presence of the genes is represented by different colors. Classes of related antimicrobial resistance genes are represented below.

For Fosfomycin resistance (*fosA*) gene detection, studies from Hong Kong and Japan have examined the prevalence of fosfomycin resistance in clinical isolates of Enterobacteriaceae, including *Klebsiella* species. These studies revealed that the *fosA* gene is commonly found among resistant isolates, indicating a widespread distribution of this resistance determinant. For example, a study in Hong Kong reported a high prevalence of *fosA* among fosfomycin-resistant clinical isolates, emphasizing the significance of this gene in conferring resistance [9]. Likewise, research from Japan identified *fosA* as a frequent contributor to fosfomycin resistance in Enterobacteriaceae [9]. The presence of *fosA* in our *Klebsiella* isolates aligns with these findings and suggests the gene's role in resistance mechanisms and its potential impact on the efficacy of fosfomycin as a treatment option. This highlights the importance of monitoring the prevalence of *fosA* and other resistance genes to inform treatment strategies and combat antibiotic resistance. Additionally, genes associated with phenicol resistance (*catA* and *catB*), quinolone resistance (*oqxA* and *oqxB*), sulfonamide resistance (*sul1*), and tetracycline resistance (*tet(41)*, *tet(D)* and *tet(J)*) were found in various strains. These identified ARGs are mostly related to AST results, indicating the presence of these genes correlates with observed resistance phenotypes.

There are some discrepancies observed in *P. ursingii* PSU26, which shows no ARGs, and *P. palmae* PSU29 and *Kosakonia* spp. PSU27, which contain only *oqxA* and *oqxB*. However, this species was shown to resist all tested antimicrobials. This could be explained by the capability of the efflux system regulated by *oqxAB* [10]. The absence of ARGs in *P. ursingii* PSU26 and *K. indica* PSU33 is particularly important and could suggest alternative mechanisms of resistance. One such mechanism is the presence of multidrug efflux systems, which are known to confer resistance by actively pumping out antibiotics from the bacterial cell, thus reducing their intracellular concentrations to sub-lethal levels [11]. Another possible mechanism could be modifications in the bacterial cell membrane permeability. Bacteria can alter their outer membrane porins, reducing antibiotic uptake, and, thereby, conferring resistance. Additionally, the enzymatic inactivation of antibiotics through less well-characterized mechanisms or the presence of chromosomal mutations that confer resistance could also play roles [12]. Furthermore, short-read sequencing tends to produce errors and limitations that may impact the accuracy of results, especially in complex genomic regions or when dealing with highly similar sequences [13]. In addition, the presence of only *oqxA* and *oqxB* in *P. palmae* PSU29 and *Kosakonia* spp. PSU27 raises questions about the mechanisms underlying their MDR. These isolates might rely on specific efflux pumps, encoded by these genes, to expel multiple antibiotics from their cells [14]. The results suggest the need for effective antibiotic management programs and the development of novel antimicrobial agents to combat infections caused by MDR bacteria.

2.3. Stress Response, Metal Tolerance, and Virulence-Associated Genes in Enterobacteriaceae

A comprehensive examination of stress response, metal tolerance, and virulence genes in Enterobacteriaceae was presented in Figure 3. This study targeted these specific genes because they play crucial roles in the survival and pathogenicity of bacteria, impacting both human health and environmental management. Stress response genes help bacteria adapt to various environmental challenges, including oxidative stress, osmotic stress, and heat shock, which are common in hospital and natural settings. These genes are involved in the cellular response to environmental stressors and are essential for bacterial survival in challenging conditions [15]. Metal tolerance genes are vital due to the increasing environmental contamination with metals, which not only affects bacterial survival but can also be linked to antibiotic resistance mechanisms, posing significant treatment challenges. Virulence genes are essential for understanding the pathogenic potential, as they contribute to the ability of bacteria to cause disease in hosts.

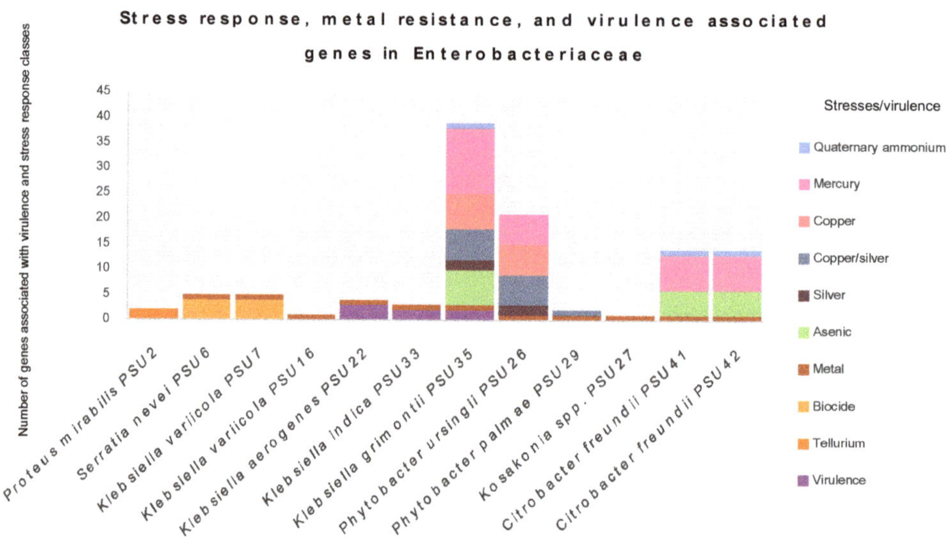

Figure 3. The graph depicts the determination of stress response, metal tolerance, and virulence-associated genes in 12 Enterobacteriaceae isolates from this study. Different colors represent classes of stress or virulence.

The results reveal that all Enterobacteriaceae in this study contain at least one type of virulence or stress gene. *P. mirabilis* PSU2 possesses the *terD* and *terZ* genes, conferring tellurium resistance. *S. nevei* PSU6 exhibits biocide-regulating genes, such as *smdA*, *smdB*, *sdeA*, and *ssmE*, associated with multidrug efflux systems. These systems play a crucial role in bacterial resistance to antimicrobial compounds, including antibiotics, by pumping them out of the cell [16–18]. The strain PSU6 also carries *fieF*, which confers metal tolerance. This gene is identified in almost all isolates in this study, except for *P. mirabilis* PSU2. Generally, Enterobacteriaceae can exhibit intrinsic or acquired resistance mechanisms, often involving efflux pumps or metal-binding proteins [19]. Metal tolerance is concerning, as it can be related to antibiotic resistance, potentially leading to treatment challenges [20]. Moreover, environmental contamination with metals can further contribute to the development and spread of metal resistance in this family.

Other interesting virulence-associated genes detected were the *iroB*, *iroC*, and *iroN* genes, carried by *K. aerogenes* PSU22. These genes are important components of the iron acquisition system in *Escherichia coli*, related to the production, export, and uptake of the siderophore salmochelin, crucial for scavenging iron, an essential nutrient, particularly in environments with limited iron [21]. For *ybtP*, *ybtQ*, and the yersiniabactin ABC Transporter ATP-Binding/Permease Protein, detected in *K. indica* PSU33 and *K. grimontii* PSU35, they are key components of the yersiniabactin gene cluster in Enterobacteriaceae, contributing to the biosynthesis and uptake of the siderophore yersiniabactin, crucial for acquiring iron from the host environment [22]. The details of stress response, metal tolerance, and virulence-associated genes are displayed in Table S2. The presence of multidrug efflux genes in *S. nevei* PSU6 highlights its ability to resist various antimicrobial agents, which might explain the lack of sensitivity to all tested antimicrobials in this isolate. *K. variicola* PSU7 and *P. mirabilis* PSU2 exhibit genes for tellurite resistance, which is particularly interesting given the scarceness of tellurium resistance mechanisms in bacteria [23]. *K. aerogenes* PSU22 and *K. indica* PSU33 display genes associated with virulence, indicating their potential pathogenicity. *K. grimontii* PSU35 and PSU36 are notable for their extensive repertoire of stress response genes, suggesting their adaptability to diverse environmental stresses, including metals, arsenic, and biocides [24,25]. *P. ursingii* PSU26 and *P. palmae* PSU29 also exhibit stress response genes, highlighting their ability to survive in metal-contaminated environments. *Kosakonia* spp. PSU27 demonstrates a similar trend, indicating its adaptation to metal stress. *C. freundii* PSU41 and PSU42 show a combination of stress response genes related to metals, arsenic, and mercury, along with genes conferring resistance to quaternary ammonium compounds, reflecting their flexibility against multiple stressors [26]. Understanding these genes emphasizes the genetic diversity and adaptive capacity of bacteria, offering insights into their survival strategies and potential implications for human health and environmental management.

2.4. Plasmid Identification in Enterobacteriaceae

According to PlasmidFinder, various plasmid replicons were found in bacterial strains, along with their lengths. In this study, plasmid replicons were identified in five isolates: *P. ursingii* PSU26, *P. palmae* PSU29, *K. grimontii* PSU35, and *C. freundii* PSU42 and PSU41. The identified plasmids included IncFIB(K), Col440I, IncA/C2, IncFIB(pB171), IncFII(K), and RepA, ranging in length from 4332 to 70,146 base pairs (Figure 4). The most abundant plasmid replicon type found was IncA/C2, identified in *C. freundii* PSU41 and PSU42, and *K. grimontii* PSU35 (Figure 4). IncA/C2 is commonly found in Enterobacteriaceae and other Gram-negative bacteria, known for carrying multiple ARGs and other mobile genetic elements [27,28].

A further analysis using the PLSDB tool confirmed that these isolates carry plasmids with a high sequence identity to those found in a variety of bacterial species globally. A PLSDB analysis identified several plasmids in Enterobacteriaceae isolates. These plasmids range from 1657 to 71,960 base pairs and are associated with strains from diverse geographic locations, including China, the USA, Canada, the UK, India, Japan, and Hong Kong. The same set of plasmids was identified in two isolates of *C. freundii*. These identified plasmids are associated with a variety of bacterial species, such as *Enterobacter hormaechei*, *Citrobacter* sp., *Salmonella enterica*, *Klebsiella pneumoniae*, *Raoultella* sp., *Klebsiella quasipneumoniae*, and *Enterobacter kobei* (Table 2). The presence of high-sequence-identity plasmids in these isolates probably suggests a potential for horizontal gene transfer and the global dissemination of plasmid-borne genes [29]. This is particularly significant for genes related to antimicrobial resistance and virulence, highlighting the role of plasmids in spreading resistance traits across different bacterial species and geographic regions. Moreover, the identical set of plasmids present in *C. freundii* collected from different samples might indicate a clonal spread or a common source of infection, suggesting that these plasmids confer a selective advantage that allows for their persistence and proliferation within and across different hosts [30]. The varying lengths of plasmid sequences underscore the diverse nature of plasmid genomes, which can carry different sets of genes that contribute to bacterial adaptability and survival. Taxonomically, the plasmids identified in this study are associated with several clinically significant bacterial species. Specially, species such as *K. pneumoniae* and *S. enterica* are known for their roles in hospital-acquired infections and their ability to acquire multidrug resistance [31].

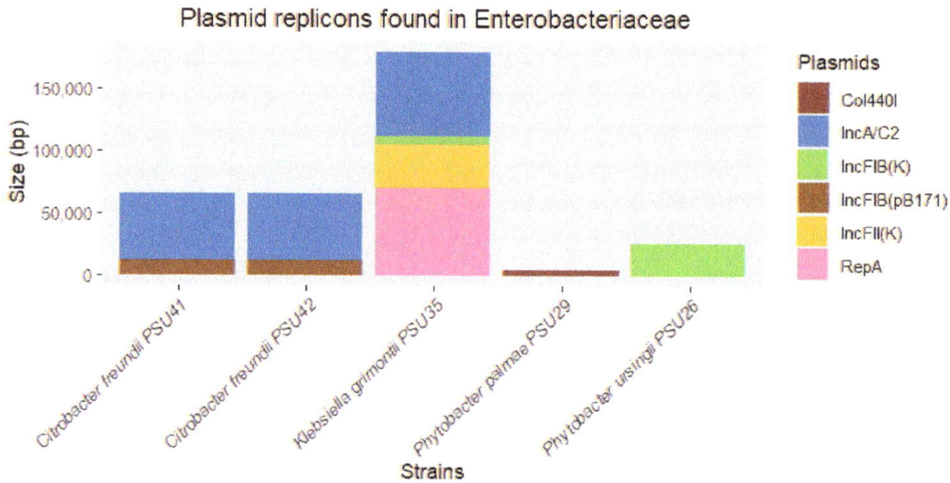

Figure 4. Determination of plasmid replicons identified from five out of twelve Enterobacteriaceae in this study. The thickness of each stack indicates the size of sequences, and different colors represent the types of replicons.

Table 2. Plasmid identification using PLSDB.

Isolate	Accession No.	Identity	Length	Location	Taxonomy
Phytobacter ursingii PSU26	NZ_CP083851.1	0.998551	3115	China	*Enterobacter hormaechei*
Phytobacter palmae PSU29	NZ_CP056253.1	0.98551	71,960	United Kingdom	*Citrobacter* sp. RHBSTW-00903
	CP101347.1	0.984609	1702	China	*Salmonella enterica*
	CP093099.1	0.983292	4448	USA	*Salmonella enterica*
	NZ_CP038598.1	0.983159	4448	Canada	*Salmonella enterica*
	CP093106.1	0.982958	4448	USA	*Salmonella enterica*
	CP093079.1	0.982958	4448	USA	*Salmonella enterica*
	NZ_CP113909.1	0.982282	4448	United Kingdom	*Klebsiella michiganensis*
Klebsiella grimontii PSU35	NZ_CP079817.1	1	1657	India	*Klebsiella pneumoniae*
	NZ_CP093279.1	0.999377	20,796	China	*Raoultella* sp. HC6
	NZ_AP026417.1	0.999038	2889	Japan	*Klebsiella quasipneumoniae*

Table 2. *Cont.*

Isolate	Accession No.	Identity	Length	Location	Taxonomy
Citrobacter freundii PSU41	NZ_MN370929.1	1	3579	missing	*Klebsiella pneumoniae*
	NZ_CP079817.1	1	1657	India	*Klebsiella pneumoniae*
	NZ_CP079667.1	0.999522	2186	India	*Klebsiella pneumoniae*
	NZ_CP083865.1	0.999473	17,041	China	*Enterobacter kobei*
	NZ_CP069298.1	0.999425	17,017	China	*Salmonella enterica*
	NZ_CP055215.1	0.999281	13,247	Hong Kong	*Klebsiella quasipneumoniae*
Citrobacter freundii PSU42	NZ_CP079817.1	1	1657	India	*Klebsiella pneumoniae*
	NZ_MN370929.1	0.999618	3579	missing	*Klebsiella pneumoniae*
	NZ_CP079667.1	0.999522	2186	India	*Klebsiella pneumoniae*
	NZ_CP083865.1	0.999425	17,041	China	*Enterobacter kobei*
	NZ_CP069298.1	0.999377	17,017	China	*Salmonella enterica*
	NZ_CP055215.1	0.999184	13,247	Hong Kong	*Klebsiella quasipneumoniae*

Despite identifying several ARGs and plasmid, no ARGs were found to be mediated by the plasmids. This could be because ARGs may be located on the bacterial chromosome rather than on plasmids. Furthermore, the ARGs may have integrated into the bacterial chromosome from plasmids through mechanisms such as transposition or recombination. The short-read sequencing used in this study could not obtain the complete genome of the bacteria, leading to limitations in identifying plasmid–ARG associations, such as sensitivity issues or incomplete plasmid databases [32,33]. The detection of these plasmids in our isolates might suggest that this study may uncover critical information regarding the spread and characteristics of resistance plasmids in different bacterial hosts.

3. Materials and Methods

3.1. Sample Collection

This study is part of a short-term surveillance effort conducted in ICU patients from five different hospitals in Southern Thailand over a 6-month period in 2019 [34–36]. The project identified 241 Enterobacteriaceae suspected of being resistant to carbapenem antimicrobials, all screened for resistance using MacConkey agar supplemented with 2 μg/mL of imipenem. Twelve out of 241 Enterobacteriaceae isolates were selected for this study. These included one *Proteus mirabilis*, one *Serratia nevei*, two *Klebsiella variicola*, one *Klebsiella aerogenes*, one *Klebsiella indica*, one *Klebsiella grimontii*, one *Phytobacter ursingii*, one *Phytobacter palmae*, one *Kosakonia* spp., and two *Citrobacter freundii*. They were chosen because they are uncommon in clinical samples but are found in outside communities or contaminated surfaces in the environment. These twelve isolates were collected from patients admitted to intensive care units (ICUs) in hospitals in southern Thailand, including Songklanagarind Hospital, Patthalung Hospital, Satun Hospital, Pattani Hospital, and Yala Hospital. The bacteria were isolated from various specimens, such as nasopharynx, swabs, endotracheal tubes, throats, and rectums. All patients had received prior antibiotics before their specimens were collected (Table S1).

Initial species identification of the twelve strains was initially conducted using biochemical tests based on Bergey's Manual of Systematic Bacteriology [37], with further confirmation accomplished using Matrix-Assisted Laser Desorption/Ionization–Time of Flight (MALDI-TOF) mass spectrometry (MS). However, recognizing the limitations of MALDI-TOF MS in distinguishing between species within the *Klebsiella pneumoniae* complex, we confirmed the identification of our selected. All of them were stored at −20 °C for further identification and analysis using WGS.

3.2. Antimicrobial Susceptibility Testing

All isolates underwent antimicrobial susceptibility testing (AST) using the disk diffusion method, according to Kirby–Bauer disk diffusion susceptibility test protocol [38]. The study utilized antimicrobial disks, including ciprofloxacin (5 μg), levofloxacin (5 μg), amikacin (10 μg), gentamicin (10 μg), imipenem (10 μg), meropenem (10 μg), tazocin (100/10 μg), ertapenem (30 μg), ceftriaxone (30 μg), ceftazidime (30 μg), and sulperazon (100/10 μg). *Escherichia coli* ATCC® 25922 (for co-trimoxazole) and *Pseudomonas aeruginosa* ATCC® 27853 were used as quality controls. AST results were interpreted according to the Clinical & Laboratory Standards Institute (CLSI) standard [39].

3.3. DNA Extraction and Sequencing

The genomic DNA from all Enterobacteriaceae isolates were obtained by the TIANamp Bacterial DNA Kit (Tiangen, Beijing, China), followed the manufacturer's guidelines [40]. The DNA concentrations were assessed using a NanoDrop™ 2000/2000c Spectrophotometer (Thermo Scientific, Norristown, PA, USA), while the integrity and purity of the DNA were verified through Agarose Gel Electrophoresis. Subsequently, the DNA samples were sent to the Beijing Genomics Institute (BGI) for short-read WGS with 150 bp paired-end reads using MGISEG-2000 platform.

3.4. Bioinformatics and Sequence Analysis

The BacSeq pipeline (Accession date: 18 February 2024) [41], a program-based bioinformatic tool for analyzing bacterial genomes, was used for processing Enterobacteriaceae genomes. All analyses and programs used in this study were performed using default parameters. Firstly, FastQC was utilized for quality control of the sequences [42]. The sequences that passed quality control were then assembled using SPAdes [43]. The assembly quality and completeness were checked using Quast and BUSCO [44,45], respectively. Genome annotation was performed using Prokka [46], and all annotated results were employed for downstream analysis. For the downstream analysis, AMRFinderPlus (Accession date: 1 April 2024) [47] was utilized for the detection of antimicrobial resistance, stress response, and virulence genes, while plasmid replicon and plasmid were identified using PlasmidFinder (Accession date: 6 April 2024) and PLSDB (Accession date: 20 May 2024) [48,49]. RStudio version 4.3.2 (Accession date: 22 April 2024) was used for generating figures [50]. To confirm the novel species of *Kosakonia* spp. PSU27, OrthoANI (Accession date: 12 April 2024) [51] was used to evaluate average nucleotide identity (ANI) of *Kosakonia* spp. PSU27 and other nine representative clinical isolates of *Kosakonia* obtained from RefSeq NCBI. An identity value lower than 95% is considered indicative of a novel species. The complete plasmid was selected for comparison with highly similar plasmids available in the NCBI database and was visualized using Proksee (Accession date: 21 April 2024) [52].

4. Conclusions

This study provides valuable insights into the genomic characteristics, antimicrobial resistance profiles, and virulence factors of Enterobacteriaceae strains isolated from hospitals in Southern Thailand. The findings highlight the diversity and complexity of these pathogens, emphasizing their potential to cause a wide range of infections and the challenges they pose to clinical management. The high prevalence of multidrug resistance, particularly to carbapenem antibiotics, is a major concern and underscores the urgent need for effective antimicrobial management programs and the development of new treatment options. The variability in resistance profiles among strains further emphasizes the need for tailored treatment strategies based on accurate identification and susceptibility testing. This necessitates the regular monitoring of resistance patterns and the implementation of precise diagnostic methods, such as whole-genome sequencing, to guide the appropriate therapy. The discovery of a novel species, *Kosakonia* spp. PSU27, expands our understanding of the genus and highlights the importance of ongoing surveillance for emerging pathogens.

Moreover, the presence of various stress response, metal tolerance, and virulence genes underscores the adaptability and pathogenic potential of these bacteria. These genetic traits contribute to their survival in harsh environments and their ability to evade host immune responses, posing significant challenges in both clinical and environmental settings.

Furthermore, we identified plasmids with a high sequence identity to those found in various bacterial species globally, highlighting the extensive horizontal gene transfer occurring across diverse geographic locations and bacterial species. The identification of multiple plasmids in isolates from this study, which match those detected in other regions, emphasizes the global challenge posed by plasmid-mediated resistance. This finding points to the need for robust monitoring systems and effective strategies to limit the dissemination of resistance genes within microbial populations. Our study underscores the critical need for continuous surveillance and increased awareness among healthcare staff regarding these rare, high-virulence, and multidrug-resistant pathogens. Understanding the local epidemiology of these pathogens is vital for informing regional public health strategies and improving infection control practices. The data collected from hospitals in Southern Thailand provide a representative snapshot of the current challenges faced in this area, highlighting the necessity for an enhanced healthcare infrastructure to better manage and mitigate the risks posed by these dangerous pathogens.

Further research is needed to elucidate the mechanisms underlying resistance and virulence in these strains. This includes studying the regulation and expression of resistance and virulence genes, as well as investigating the ecological and evolutionary pressures that drive their emergence and persistence. Understanding these mechanisms will be crucial for developing innovative strategies to combat the spread of multidrug-resistant Enterobacteriaceae.

Supplementary Materials: The following supporting information can be downloaded at: https://www.mdpi.com/article/10.3390/antibiotics13060531/s1, Table S1: Demographic data of isolates in this study. Table S2: Stress response, metal resistance, and virulence-associated genes.

Author Contributions: Conceptualization, T.Y., K.S. (Komwit Surachat) and K.S. (Kamonnut Singkhamanan); methodology, T.Y., S.S. and M.W.; software, T.Y., S.S. and K.S. (Komwit Surachat); validation, K.S. (Komwit Surachat), T.Y. and R.P.; formal analysis, T.Y. and K.S. (Komwit Surachat); investigation, T.Y. and K.S. (Komwit Surachat); resources, S.C. and R.P.; data curation, T.Y. and K.S. (Komwit Surachat); writing—original draft preparation, T.Y., S.S., K.S. (Komwit Surachat) and K.S. (Kamonnut Singkhamanan); writing—review and editing, T.Y., K.S. (Komwit Surachat) and K.S. (Kamonnut Singkhamanan); visualization, T.Y. and K.S. (Komwit Surachat); supervision, K.S. (Komwit Surachat) and K.S. (Kamonnut Singkhamanan); project administration, K.S. (Komwit Surachat); funding acquisition, K.S. (Komwit Surachat) and S.C. All authors have read and agreed to the published version of the manuscript.

Funding: This research has received funding support from the NSRF via the Program Management Unit for Human Resources & Institutional Development, Research Innovation (Grant No. B13F670075). In addition, this research was supported by the Postdoctoral Fellowship from Prince of Songkla University, Thailand, and has also received funding support from the NSRF via the Program Management Unit for Human Resources & Institutional Development, Research and Innovation, grant number B13F660074.

Institutional Review Board Statement: This study was conducted in accordance with the Declaration of Helsinki and approved by the Human Research Ethics Committee (HREC) of Prince of Songkla University (protocol code: 64-284-14-1, date of approval: 9 June 2021).

Informed Consent Statement: According to retrospective reviews, the ethical committee allowed the waiver of consent forms.

Data Availability Statement: The assembled genomes of all Enterobacteriaceae isolates in this study have been deposited in the NCBI GenBank under BioProject number PRJNA1080727, with BioSample numbers SAMN40146458 to SAMN40146472.

Conflicts of Interest: The authors declare no conflicts of interest.

References

1. Bologna, E.; Licari, L.C.; Manfredi, C.; Ditonno, F.; Cirillo, L.; Fusco, G.M.; Abate, M.; Passaro, F.; Di Mauro, E.; Crocetto, F.; et al. Carbapenem-Resistant Enterobacteriaceae in Urinary Tract Infections: From Biological Insights to Emerging Therapeutic Alternatives. *Medicina* **2024**, *60*, 214. [CrossRef] [PubMed]
2. Tang, K.W.K.; Millar, B.C.; Moore, J.E. Antimicrobial Resistance (AMR). *Br. J. Biomed. Sci.* **2023**, *80*, 11387. [CrossRef] [PubMed]
3. Poudel, A.N.; Zhu, S.; Cooper, N.; Little, P.; Tarrant, C.; Hickman, M.; Yao, G. The economic burden of antibiotic resistance: A systematic review and meta-analysis. *PLoS ONE* **2023**, *18*, e0285170. [CrossRef] [PubMed]
4. Logan, L.K.; Weinstein, R.A. The Epidemiology of Carbapenem-Resistant Enterobacteriaceae: The Impact and Evolution of a Global Menace. *J. Infect. Dis.* **2017**, *215*, S28–S36. [CrossRef] [PubMed]
5. Tilahun, M.; Kassa, Y.; Gedefie, A.; Ashagire, M. Emerging Carbapenem-Resistant Enterobacteriaceae Infection, Its Epidemiology and Novel Treatment Options: A Review. *Infect. Drug Resist.* **2021**, *14*, 4363–4374. [CrossRef] [PubMed]
6. National Center for Biotechnology Information. Kosakonia Assemblies. RefSeq NCBI. Available online: https://www.ncbi.nlm.nih.gov/assembly/?term=Kosakonia (accessed on 12 April 2024).
7. Chen, L.; Todd, R.; Kiehlbauch, J.; Walters, M.; Kallen, A. Notes from the Field: Pan-Resistant New Delhi Metallo-Beta-Lactamase-Producing *Klebsiella pneumoniae*—Washoe County, Nevada, 2016. *MMWR. Morb. Mortal. Wkly. Rep.* **2017**, *66*, 33. [CrossRef] [PubMed]
8. Wang, X.; Chen, H.; Zhang, Y.; Wang, Q.; Zhao, C.; Li, H.; He, W.; Zhang, F.; Wang, Z.; Li, S.; et al. Genetic characterisation of clinical *Klebsiella pneumoniae* isolates with reduced susceptibility to tigecycline: Role of the global regulator RamA and its local repressor RamR. *Int. J. Antimicrob. Agents* **2015**, *45*, 635–640. [CrossRef] [PubMed]
9. Ho, P.L.; Chan, J.; Lo, W.U.; Lai, E.L.; Cheung, Y.Y.; Lau, T.C.K.; Chow, K.H. Prevalence and molecular epidemiology of plasmid-mediated fosfomycin resistance genes among blood and urinary *Escherichia coli* isolates. *J. Med. Microbiol.* **2013**, *62*, 1707–1713. [CrossRef]
10. Moosavian, M.; Khoshkholgh Sima, M.; Ahmad Khosravi, N.; Abbasi Montazeri, E. Detection of OqxAB Efflux Pumps, a Multidrug-Resistant Agent in Bacterial Infection in Patients Referring to Teaching Hospitals in Ahvaz, Southwest of Iran. *Int. J. Microbiol.* **2021**, *2021*, 2145176. [CrossRef] [PubMed]
11. Kerluku, M.; Ratkova Manovska, M.; Prodanov, M.; Stojanovska-Dimzoska, B.; Hajrulai-Musliu, Z.; Jankuloski, D.; Blagoevska, K. Phenotypic and Genotypic Analysis of Antimicrobial Resistance of Commensal Escherichia coli from Dairy Cows' Feces. *Processes* **2023**, *11*, 1929. [CrossRef]
12. Bobate, S.; Mahalle, S.; Dafale, N.A.; Bajaj, A. Emergence of environmental antibiotic resistance: Mechanism, monitoring and management. *Environ. Adv.* **2023**, *13*, 100409. [CrossRef]
13. Li, K.; Xu, P.; Wang, J.; Yi, X.; Jiao, Y. Identification of errors in draft genome assemblies at single-nucleotide resolution for quality assessment and improvement. *Nat. Commun.* **2023**, *14*, 6556. [CrossRef]
14. Amereh, F.; Arabestani, M.R.; Shokoohizadeh, L. Relationship of OqxAB efflux pump to antibiotic resistance, mainly fluoroquinolones in Klebsiella pneumoniae, isolated from hospitalized patients. *Iran. J. Basic Med. Sci.* **2023**, *26*, 93–98. [CrossRef]
15. da Cruz Nizer, W.S.; Inkovskiy, V.; Versey, Z.; Strempel, N.; Cassol, E.; Overhage, J. Oxidative Stress Response in *Pseudomonas aeruginosa*. *Pathogens* **2021**, *10*, 1187. [CrossRef]
16. Matsuo, T.; Chen, J.; Minato, Y.; Ogawa, W.; Mizushima, T.; Kuroda, T.; Tsuchiya, T. SmdAB, a heterodimeric ABC-Type multidrug efflux pump, in Serratia marcescens. *J. Bacteriol.* **2008**, *190*, 648–654. [CrossRef]
17. Dalvi, S.D.; Worobec, E.A. Gene expression analysis of the SdeAB multidrug efflux pump in antibiotic-resistant clinical isolates of Serratia marcescens. *Indian J. Med. Microbiol.* **2012**, *30*, 302–307. [CrossRef]
18. Minato, Y.; Shahcheraghi, F.; Ogawa, W.; Kuroda, T.; Tsuchiya, T. Functional gene cloning and characterization of the SsmE multidrug efflux pump from Serratia marcescens. *Biol. Pharm. Bull.* **2008**, *31*, 516–519. [CrossRef]
19. Nguyen, T.H.T.; Nguyen, H.D.; Le, M.H.; Nguyen, T.T.H.; Nguyen, T.D.; Nguyen, D.L.; Nguyen, Q.H.; Nguyen, T.K.O.; Michalet, S.; Dijoux-Franca, M.G.; et al. Efflux Pump Inhibitors in Controlling Antibiotic Resistance: Outlook under a Heavy Metal Contamination Context. *Molecules* **2023**, *28*, 2912. [CrossRef]
20. Vats, P.; Kaur, U.J.; Rishi, P. Heavy metal-induced selection and proliferation of antibiotic resistance: A review. *J. Appl. Microbiol.* **2022**, *132*, 4058–4076. [CrossRef]
21. Mohsen, Y.; Tarchichi, N.; Barakat, R.; Kawtharani, I.; Ghandour, R.; Ezzeddine, Z.; Ghssein, G. The Different Types of Metallophores Produced by Salmonella enterica: A Review. *Microbiol. Res.* **2023**, *14*, 1457–1469. [CrossRef]
22. Koh, E.I.; Hung, C.S.; Henderson, J.P. The Yersiniabactin-Associated ATP Binding Cassette Proteins YbtP and YbtQ Enhance Escherichia coli Fitness during High-Titer Cystitis. *Infect. Immun.* **2016**, *84*, 1312–1319. [CrossRef] [PubMed]
23. Peng, W.; Wang, Y.; Fu, Y.; Deng, Z.; Lin, S.; Liang, R. Characterization of the Tellurite-Resistance Properties and Identification of the Core Function Genes for Tellurite Resistance in Pseudomonas citronellolis SJTE-3. *Microorganisms* **2022**, *10*, 95. [CrossRef] [PubMed]
24. Kovács, J.K.; Felső, P.; Horváth, G.; Schmidt, J.; Dorn, Á.; Ábrahám, H.; Cox, A.; Márk, L.; Emődy, L.; Kovács, T.; et al. Stress Response and Virulence Potential Modulating Effect of Peppermint Essential Oil in Campylobacter jejuni. *BioMed Res. Int.* **2019**, *2019*, 2971741. [CrossRef] [PubMed]

25. Mourão, J.; Magalhães, M.; Ribeiro-Almeida, M.; Rebelo, A.; Novais, C.; Peixe, L.; Novais, Â.; Antunes, P. Decoding Klebsiella pneumoniae in poultry chain: Unveiling genetic landscape, antibiotic resistance, and biocide tolerance in non-clinical reservoirs. *Front. Microbiol.* **2024**, *15*, 1365011. [CrossRef] [PubMed]
26. Abd Elnabi, M.K.; Elkaliny, N.E.; Elyazied, M.M.; Azab, S.H.; Elkhalifa, S.A.; Elmasry, S.; Mouhamed, M.S.; Shalamesh, E.M.; Alhorieny, N.A.; Abd Elaty, A.E.; et al. Toxicity of Heavy Metals and Recent Advances in Their Removal: A Review. *Toxics* **2023**, *11*, 580. [CrossRef] [PubMed]
27. Papagiannitsis, C.C.; Dolejska, M.; Izdebski, R.; Giakkoupi, P.; Skálová, A.; Chudějová, K.; Dobiasova, H.; Vatopoulos, A.C.; Derde, L.P.; Bonten, M.J.; et al. Characterisation of IncA/C2 plasmids carrying an In416-like integron with the blaVIM-19 gene from Klebsiella pneumoniae ST383 of Greek origin. *Int. J. Antimicrob. Agents* **2016**, *47*, 158–162. [CrossRef] [PubMed]
28. Pauly, N.; Hammerl, J.A.; Grobbel, M.; Käsbohrer, A.; Tenhagen, B.A.; Malorny, B.; Schwarz, S.; Meemken, D.; Irrgang, A. Identification of a bla(VIM-1)-Carrying IncA/C(2) Multiresistance Plasmid in an Escherichia coli Isolate Recovered from the German Food Chain. *Microorganisms* **2020**, *9*, 29. [CrossRef] [PubMed]
29. von Wintersdorff, C.J.; Penders, J.; van Niekerk, J.M.; Mills, N.D.; Majumder, S.; van Alphen, L.B.; Savelkoul, P.H.; Wolffs, P.F. Dissemination of Antimicrobial Resistance in Microbial Ecosystems through Horizontal Gene Transfer. *Front. Microbiol.* **2016**, *7*, 173. [CrossRef] [PubMed]
30. Marí-Almirall, M.; Ferrando, N.; Fernández, M.J.; Cosgaya, C.; Viñes, J.; Rubio, E.; Cuscó, A.; Muñoz, L.; Pellice, M.; Vergara, A.; et al. Clonal Spread and Intra- and Inter-Species Plasmid Dissemination Associated with *Klebsiella pneumoniae* Carbapenemase-Producing Enterobacterales During a Hospital Outbreak in Barcelona, Spain. *Front. Microbiol.* **2021**, *12*, 781127. [CrossRef]
31. Peleg, A.Y.; Hooper, D.C. Hospital-acquired infections due to gram-negative bacteria. *N. Engl. J. Med.* **2010**, *362*, 1804–1813. [CrossRef]
32. Wang, X.; Zhang, H.; Yu, S.; Li, D.; Gillings, M.R.; Ren, H.; Mao, D.; Guo, J.; Luo, Y. Inter-plasmid transfer of antibiotic resistance genes accelerates antibiotic resistance in bacterial pathogens. *ISME J.* **2024**, *18*, wrad032. [CrossRef] [PubMed]
33. Zhang, J.; Xu, Y.; Wang, M.; Li, X.; Liu, Z.; Kuang, D.; Deng, Z.; Ou, H.-Y.; Qu, J. Mobilizable plasmids drive the spread of antimicrobial resistance genes and virulence genes in *Klebsiella pneumoniae*. *Genome Med.* **2023**, *15*, 106. [CrossRef]
34. Yaikhan, T.; Chukamnerd, A.; Singkhamanan, K.; Nokchan, N.; Chintakovid, N.; Chusri, S.; Pomwised, R.; Wonglapsuwan, M.; Surachat, K. Genomic Characterization of Mobile Genetic Elements Associated with Multidrug-Resistant Acinetobacter Non-baumannii Species from Southern Thailand. *Antibiotics* **2024**, *13*, 149. [CrossRef] [PubMed]
35. Chukamnerd, A.; Pomwised, R.; Chusri, S.; Singkhamanan, K.; Chumtong, S.; Jeenkeawpiam, K.; Sakunrang, C.; Saroeng, K.; Saengsuwan, P.; Wonglapsuwan, M.; et al. Antimicrobial Susceptibility and Molecular Features of Colonizing Isolates of Pseudomonas aeruginosa and the Report of a Novel Sequence Type (ST) 3910 from Thailand. *Antibiot. Antibiot.* **2023**, *12*, 165. [CrossRef] [PubMed]
36. Chukamnerd, A.; Singkhamanan, K.; Chongsuvivatwong, V.; Palittapongarnpim, P.; Doi, Y.; Pomwised, R.; Sakunrang, C.; Jeenkeawpiam, K.; Yingkajorn, M.; Chusri, S.; et al. Whole-genome analysis of carbapenem-resistant Acinetobacter baumannii from clinical isolates in Southern Thailand. *Comput. Struct. Biotechnol. J.* **2022**, *20*, 545–558. [CrossRef] [PubMed]
37. Paul De Vos, E. *Bergey's Manual of Systematic Bacteriology. Volume Three, The Firmicutes*, 2nd ed.; Springer: Dordrecht, The Netherlands; New York, NY, USA, 2009.
38. Hudzicki, J. *Kirby-Bauer Disk Diffusion Susceptibility Test Protocol*; American Society for Microbiology: Washington, DC, USA, 2009.
39. Weinstein Melvin, P.; Lewis James, S. The Clinical and Laboratory Standards Institute Subcommittee on Antimicrobial Susceptibility Testing: Background, Organization, Functions, and Processes. *J. Clin. Microbiol.* **2020**, *58*, 3. [CrossRef] [PubMed]
40. *TIANamp Bacterial DNA Kit: User Manual*; Tiangen Biotech: Bejing, China, 2021.
41. Chukamnerd, A.; Jeenkeawpiam, K.; Chusri, S.; Pomwised, R.; Singkhamanan, K.; Surachat, K. BacSeq: A User-Friendly Automated Pipeline for Whole-Genome Sequence Analysis of Bacterial Genomes. *Microorganisms* **2023**, *11*, 1769. [CrossRef] [PubMed]
42. Andrews, S. FastQC: A Quality Control Tool for High Throughput Sequence Data. Available online: https://www.bioinformatics.babraham.ac.uk/projects/fastqc/ (accessed on 18 February 2024).
43. Bankevich, A.; Nurk, S.; Antipov, D.; Gurevich, A.A.; Dvorkin, M.; Kulikov, A.S.; Lesin, V.M.; Nikolenko, S.I.; Pham, S.; Prjibelski, A.D.; et al. SPAdes: A new genome assembly algorithm and its applications to single-cell sequencing. *J. Comput. Biol. J. Comput. Mol. Cell Biol.* **2012**, *19*, 455–477. [CrossRef] [PubMed]
44. Gurevich, A.; Saveliev, V.; Vyahhi, N.; Tesler, G. QUAST: Quality assessment tool for genome assemblies. *Bioinformatics* **2013**, *29*, 1072–1075. [CrossRef]
45. Simão, F.A.; Waterhouse, R.M.; Ioannidis, P.; Kriventseva, E.V.; Zdobnov, E.M. BUSCO: Assessing genome assembly and annotation completeness with single-copy orthologs. *Bioinformatics* **2015**, *31*, 3210–3212. [CrossRef]
46. Seemann, T. Prokka: Rapid prokaryotic genome annotation. *Bioinformatics* **2014**, *30*, 2068–2069. [CrossRef] [PubMed]
47. Feldgarden, M.; Brover, V.; Gonzalez-Escalona, N.; Frye, J.G.; Haendiges, J.; Haft, D.H.; Hoffmann, M.; Pettengill, J.B.; Prasad, A.B.; Tillman, G.E.; et al. AMRFinderPlus and the Reference Gene Catalog facilitate examination of the genomic links among antimicrobial resistance, stress response, and virulence. *Sci. Rep.* **2021**, *11*, 12728. [CrossRef] [PubMed]
48. Carattoli, A.; Zankari, E.; García-Fernández, A.; Voldby Larsen, M.; Lund, O.; Villa, L.; Møller Aarestrup, F.; Hasman, H. In silico detection and typing of plasmids using PlasmidFinder and plasmid multilocus sequence typing. *Antimicrob. Agents Chemother.* **2014**, *58*, 3895–3903. [CrossRef]

49. Galata, V.; Fehlmann, T.; Backes, C.; Keller, A. PLSDB: A resource of complete bacterial plasmids. *Nucleic Acids Res.* **2019**, *47*, D195–D202. [CrossRef] [PubMed]
50. R Studio Team. *RStudio: Integrated Development for R*; RStudio, PBC: Boston, MA, USA, 2022.
51. Lee, I.; Ouk Kim, Y.; Park, S.C.; Chun, J. OrthoANI: An improved algorithm and software for calculating average nucleotide identity. *Int. J. Syst. Evol. Microbiol.* **2016**, *66*, 1100–1103. [CrossRef]
52. Grant, J.R.; Enns, E.; Marinier, E.; Mandal, A.; Herman, E.K.; Chen, C.-Y.; Graham, M.; Van Domselaar, G.; Stothard, P. Proksee: In-depth characterization and visualization of bacterial genomes. *Nucleic Acids Res.* **2023**, *51*, W484–W492. [CrossRef]

antibiotics

MDPI

Article

Genomic Epidemiology of C2/H30Rx and C1-M27 Subclades of *Escherichia coli* ST131 Isolates from Clinical Blood Samples in Hungary

Kinga Tóth [1,2,*], Ivelina Damjanova [2], Levente Laczkó [3,4], Lilla Buzgó [2], Virág Lesinszki [2], Erika Ungvári [2], Laura Jánvári [2], Adrienn Hanczvikkel [2], Ákos Tóth [2,†] and Dóra Szabó [1,5,6,†]

1 Institute of Medical Microbiology, Faculty of Medicine, Semmelweis University, 1089 Budapest, Hungary
2 Department of Bacteriology, Parasitology and Mycology, National Center for Public Health and Pharmacy, 1097 Budapest, Hungary; buzgo.lilla@med.unideb.hu (L.B.); janvari.laura@nngyk.gov.hu (L.J.); toth.akos@nngyk.gov.hu (Á.T.)
3 One Health Institute, Faculty of Health Sciences, University of Debrecen, 4032 Debrecen, Hungary
4 HUN-REN-DE Conservation Biology Research Group, University of Debrecen, 4032 Debrecen, Hungary
5 HUN-REN-SE Human Microbiota Research Group, 1052 Budapest, Hungary
6 Neurosurgical and Neurointervention Clinic, Semmelweis University, 1083 Budapest, Hungary
* Correspondence: toth.kinga@phd.semmelweis.hu
† These authors contributed equally to this work.

Abstract: Extended-spectrum β-lactamase-producing *Escherichia coli* ST131 has become widespread worldwide. This study aims to characterize the virulome, resistome, and population structure of *E. coli* ST131 isolates from clinical blood samples in Hungary. A total of 30 C2/H30Rx and 33 C1-M27 ST131 isolates were selected for Illumina MiSeq sequencing and 30 isolates for MinION sequencing, followed by hybrid de novo assembly. Five C2/H30Rx and one C1-M27 cluster were identified. C1-M27 isolates harbored the F1:A2:B20 plasmid in 93.9% of cases. Long-read sequencing revealed that $bla_{\text{CTX-M-27}}$ was on plasmids. Among the C2/H30Rx isolates, only six isolates carried the C2-associated F2:A1:B- plasmid type. Of 19 hybrid-assembled C2/H30Rx genomes, the $bla_{\text{CTX-M-15}}$ gene was located on plasmid only in one isolate, while in the other isolates, IS*Ecp1* or IS*26*-mediated chromosomal integration of $bla_{\text{CTX-M-15}}$ was detected in unique variations. In one isolate a part of F2:A1:B- plasmid integrated into the chromosome. These results suggest that CTX-M-15-producing C2/H30Rx and CTX-M-27-producing C1-M27 subclades may have emerged and spread in different ways in Hungary. While $bla_{\text{CTX-M-27}}$ was carried mainly on the C1/H30R-associated F1:A2:B20 plasmid, the IncF-like plasmids of C2/H30Rx or its composite transposons have been incorporated into the chromosome through convergent evolutionary processes.

Keywords: whole genome sequencing (WGS); ST131; *Escherichia coli*; C1-M27; C2/H30RX; $bla_{\text{CTX-M-27}}$; $bla_{\text{CTX-M-15}}$; long-read sequencing

1. Introduction

The dissemination of 3rd generation cephalosporin-resistant (3GCR) *Escherichia coli* (*E. coli*) strains has been driven by a few specific clones that have rapidly emerged in hospitals globally. The overuse and misuse of broad-spectrum antibiotics in healthcare institutions have led to the expansion of clonal variants showing resistance to three or more antibiotic classes (termed multidrug-resistant—MDR) [1]. These MDR clones are capable of causing severe infections, typically associated with limited treatment options, high morbidity, and mortality [2]. The highest disease burden in Europe is attributed to 3GCR *E. coli* [3], of which one of the most prominent clones is a group of *E. coli* strains with sequence type 131 (ST131). The high-risk *E. coli* clone ST131 includes mostly extraintestinal pathogenic *E. coli* (ExPEC) strains, which mainly cause bloodstream infections and urinary tract infections [3]. The main lineages of the ST131 clone belong to clade C, associated with fluoroquinolone resistance and extended-spectrum β-lactamase (ESBL) production. Clade C comprises two subclades: C1 and C2. While the C2 isolates defined as C2/H30Rx carry mainly the ESBL gene $bla_{CTX\text{-}M\text{-}15}$, the C1 isolates defined as C1/H30R carry $bla_{CTX\text{-}M\text{-}14}$ or $bla_{CTX\text{-}M\text{-}27}$ genes [4–6]. Recently, a new sublineage within the C1, termed C1-M27, carrying $bla_{CTX\text{-}M\text{-}27}$, has emerged as a common cause of infection next to the C2/H30Rx isolates [7]. The selective advantage of the spread of some ExPEC clones, such as ST131, was the acquisition of epidemic plasmids that allow the spread of CTX-M-type ESBLs and other antibiotic resistance genes within and even between species through horizontal gene transfer. Meanwhile, the plasmids have mostly co-evolved with bacterial genomes, allowing their hosts to spread clonally. In most cases, typical IncF-type plasmids co-evolved with the C1 and C2 subclades of ST131 [4,8,9]. Strains belonging to the C2/H30Rx subclade often carry the F2:A1:B- plasmid, while those belonging to the C1-M27 subclade carry the F1:A2:B20 plasmid carrying ESBL genes [4,8,10].

The results of a recent prospective cohort study showed that 24% (6/25) of invasive *E. coli* isolated from blood cultures at a major tertiary-care hospital in Budapest, Hungary, collected between October and November 2018, proved to be ESBL-producers. Whole genome sequence analysis showed that five *E. coli* isolates belonged to the ST131 clone. The study highlighted the dominance of the ST131, but due to the small sample size and short time period, it was unable to determine the population structure of the clone in Hungary [11].

This study aims to characterize *E. coli* ST131 isolates belonging to the C2/H30Rx or C1-M27 subclades collected from blood culture with spatiotemporal distribution from two time periods (2015–2018 and 2021) from Hungary through comparison and characterization of their virulome, resistome, major resistance plasmids and determination of their population structure. These two time periods were used to investigate possible changes in the population structure of ESBL-producing, invasive *E. coli* ST131 before and during the COVID-19 pandemic in Hungary.

2. Results

2.1. Selection of Isolates for the Study

Between 2015 and 2018 and in 2021, 59.6% (130/218) and 67.7% (157/232) of invasive ESBL-producing *E. coli* isolates investigated at the National Center for Public Health and Pharmacy (NCPHP) belonged to the ST131 clone where the ratio of C2/H30Rx and C1-M27 was 1 to 0.8 and 1 to 0.5, respectively. Based on the inclusion criteria, a total of 30 C2/H30Rx and 33 C1-M27 ESBL-producing *E. coli* ST131 isolates originated from 21 healthcare institutions in Hungary were selected for short read sequencing and 30 isolates (19 C2/H30Rx and 11 C1-M27) were selected for the long-read sequencing and for further analysis.

2.2. Antimicrobial Susceptibility

All isolates were found to be resistant to ceftriaxone and ciprofloxacin but susceptible to ceftazidime/avibactam, ertapenem, tigecycline, fosfomycin, imipenem, and meropenem. The C2/H30Rx isolates showed significantly lower susceptibility rates to ceftazidime, amikacin, tobramycin, and gentamicin than the C1-M27 ones. Two C1-M27 isolates showed resistance against colistin (Table 1).

Table 1. Antimicrobial susceptibility of the *E. coli ST131* isolates.

Antimicrobial Agent	C2/H30Rx			C1-M27			*p*-Value
	R% (*n* = 30)	MIC_{50} (*n* = 30)	MIC_{90} (*n* = 30)	R% (*n* = 33)	MIC_{50} (*n* = 33)	MIC_{90} (*n* = 33)	
FOS	0	2	4	0	2	4	1
ETP	0	0.0625	0.125	0	0.016	0.0625	1
CZA	0	0.25	0.5	0	0.125	0.25	1
CAZ	83.9	16	64	15.6	4	8	<0.001 *

Table 1. *Cont.*

Antimicrobial Agent	C2/H30Rx			C1-M27			p-Value
	R% (n = 30)	MIC_{50} (n = 30)	MIC_{90} (n = 30)	R% (n = 33)	MIC_{50} (n = 33)	MIC_{90} (n = 33)	
CRO	100	256	256	100	128	256	1
COL	0	0.25	0.5	6.3	0.25	1	0.5
TGC	0	ND	ND	0	ND	ND	1
AK	40	ND	ND	6.1	ND	ND	<0.01 *
TM	66.7	ND	ND	3.0	ND	ND	<0.001 *
GM	40	ND	ND	0	ND	ND	<0.001 *
MEM	0	ND	ND	0	ND	ND	1
IMI	0	ND	ND	0	ND	ND	1
CIP	100	ND	ND	100	ND	ND	1

Legend: Comparison of the susceptibility of 33 C1-M27 and 30 C2/H30Rx ST131 *E. coli* to different antibacterial agents. The antibiotics tested were fosfomycin (FOS), ertapenem (ETP), ceftazidime/avibactam (CZA), ceftazidime (CAZ), ceftriaxone (CRO), colistin (COL), tigecycline (TGC), amikacin (AK), tobramycin (TM), gentamicin (GM), meropenem (MEM), imipenem (IMI), and ciprofloxacin (CIP). The R% corresponds to the resistance rate; MIC corresponds to the minimum inhibitory concentrations. $MIC_{50/90}$ is the MIC value at which ≥50% and ≥90% of isolates are inhibited. ND refers to not conducted. * Statistic value was revealed using Fisher's exact test: $p < 0.05$.

2.3. Molecular Epidemiology

The phylogenetic tree supported the separation (i.e., aLRT > 0.8 and ultrafast bootstrap > 0.95) of the 33 C1-M27 isolates from all C2/H30Rx isolates (Figure 1). In the latter group, five clusters could be identified with similarly high statistical support (Figure 1 and Figure S2).

A total of 63 virulence factors and 29 antibiotic-resistance genes were detected in the two subclades. The median number of virulence genes was 25 (range 22 to 29), and of antibiotic resistance genes was 15 (range 6 to 16) among C1-M27 isolates versus 29 (range 24 to 31) and 14 (range 6 to 18) among C2/H30Rx isolates. (Figure 1 and Figure S1).

All C1-M27 isolates harbored $bla_{CTX-M-27}$ and showed virotype C. The *traT* serum-resistance gene, *sen*B, *mcb*A, *pic* toxin genes, and some antibiotic resistance genes ($bla_{CTX-M-27}$, *sul2*, *aph(3'')-Ib*, *aph(6)-Id*, *mph(A)*) were significantly more frequent in the C1-M27 isolates than in the C2/H30Rx isolates ($p < 0.05$). All C2/H30Rx isolates harbored $bla_{CTX-M-15}$. Out of 30 isolates, six exhibited virotype A, one virotype B, and 23 virotype C. Some adhesins (*afa* operon, *nfa*E, *pap* operon), toxins (*cnf1*, *hly* operon, *aslA*), invasins (*daaA-F*, *draA-D*), and the *hra* protectin gene, as well as some resistance genes ($bla_{CTX-M-15}$, bla_{OXA-1}, *aac(6')-Ib-cr*, *qnrB19*, *catB3*), were present only in C2/H30Rx isolates ($p < 0.01$). The *aac(3)-IIa* occurred with higher frequency among C2/H30Rx subclade isolates (26.7% vs. 6.1%, $p < 0.05$).

Regarding the whole collection of isolates tested, an association was found between resistance to gentamicin and the presence of *aac(3)-IIa* ($p < 0.001$), between resistance to amikacin and the presence of *aac(6')-Ib-cr* ($p < 0.001$), and between resistance to tobramycin and the presence of one of *aac(3)-IIa* or *aac(6')-Ib-cr* ($p < 0.001$).

2.3.1. Phylogenetic Analysis of *E. coli* ST131 Isolates

Within the C1-M27 subclade, one large cluster with 33 isolates could be identified, referred to as Cluster A. Cluster A isolates were collected from 18 healthcare institutions: three isolates from 2015, two from 2016, eight from 2017, four from 2018, and 16 isolates from 2021. The main plasmid replicon type was IncF (n = 33). Thirty-one out of 33 isolates had F1:A2:B20, one had F2:A2:B20, and another isolate had the F1:A2:B- FAB formula. The isolates carried other plasmid groups such as IncI (n = 2), IncX1 (n = 4), IncX3 (n = 1), IncX4 (n = 2), IncN (n = 1), and IncB/O/K/Z plasmid (n = 1). Col-like plasmid replicons were also present among Cluster A isolates: Col(MG828) (n = 29), Col156 (n = 30), Col(BS512) (n = 15), ColRNAI (n = 8), Col8282 (4), Col(pHAD28) (n = 2), and Col(KPHS6) (n = 1).

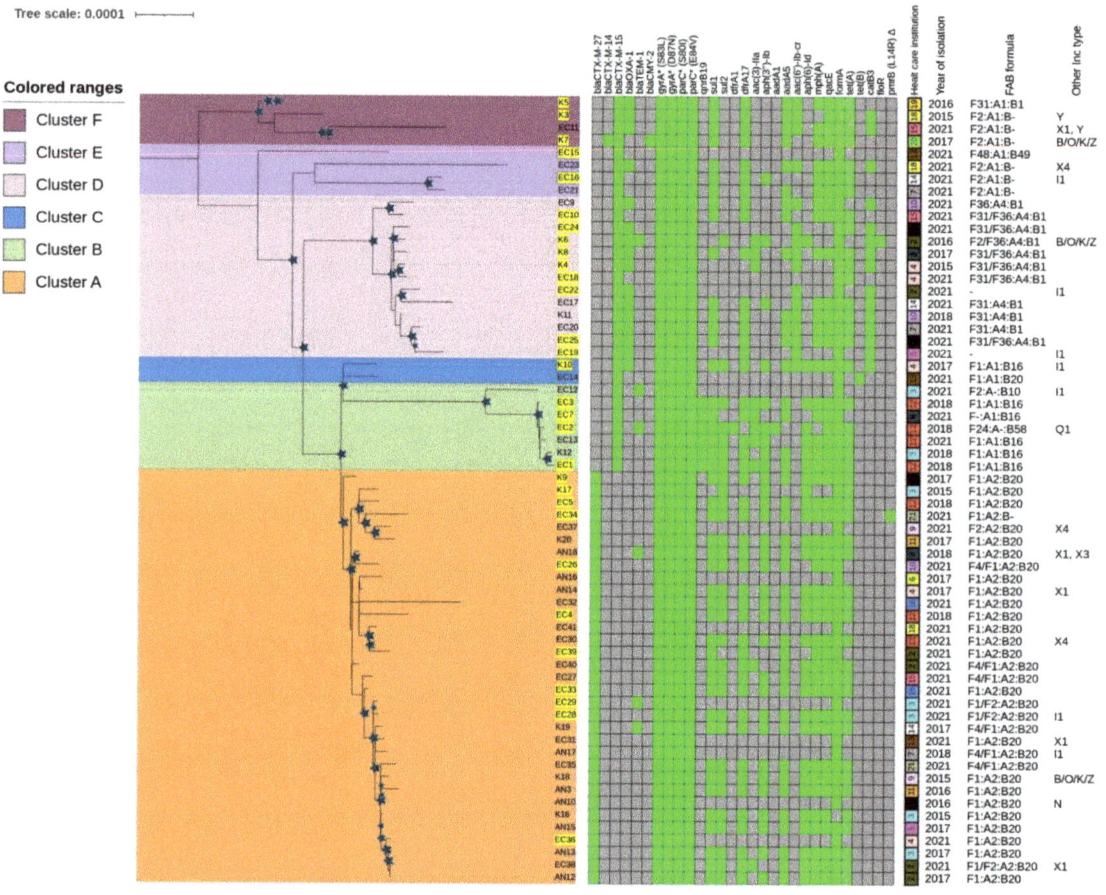

Figure 1. Maximum likelihood phylogeny of 63 ESBL-producing *E. coli* ST131 isolates and their genetic characteristic reconstructed by IQtree. Legend: Rectangles of different colors indicate clusters in the phylogenetic tree, and the star symbol indicates bootstrap values (LRT: >0.8 and UF bootstrap > 0.95). Hybrid genome assembly is indicated by a yellow background. In the table, the cells indicate the absence (grey) or presence of certain antibiotic-resistance genes (green). The symbol Δ indicates the mutation in chromosomally mediated colistin resistance, * the mutations in quinolone resistance determining region. The features show the profile of C1-M27 and C2/H30Rx isolates by health care institution, year of isolation, FAB formula, and other Inc types.

Within the C2/H30Rx subclade, five clusters could be identified (Cluster B–F). The isolates were collected from 15 healthcare institutions: two isolates from 2015, two from 2016, three from 2017, five from 2018, and 18 isolates from 2021. The main plasmid replicon type was IncF ($n = 28$), and the following FAB formulas were found: F31/F36:A4:B1 ($n = 10$), F2:A1:B- ($n = 6$), F1:A1:B16 ($n = 5$), F31:A1:B1 ($n = 1$), F-:A1:B16 ($n = 1$), F1:A1:B20 ($n = 1$), F2/F36:A4:B1 ($n = 1$), F2:A-:B10 ($n = 1$), F24:A-:B1 ($n = 1$), and F48:A1:B49 ($n = 1$), while two isolates did not carry IncF-type plasmid. Other Inc-types were found among the C2/H30Rx isolates, such as IncI ($n = 5$), IncX1 ($n = 1$), IncX4 ($n = 1$), IncB/O/K/Z plasmid ($n = 1$), IncY ($n = 1$), and IncQ1 ($n = 1$). Col-like plasmid replicons were also present in the following subclades: Col(MG828) ($n = 9$), Col156 ($n = 16$), Col(BS512) ($n = 2$), ColRNAI ($n = 8$), Col8282 ($n = 6$), and Col(pHAD28) ($n = 8$).

Seven isolates formed Cluster B. The isolates were collected in 2018 (n = 4 isolates) and 2021 (*n* = 3) from three healthcare institutions. Compared to the characteristic resistance and virulence genes of the C2 subclade, isolates of this cluster additionally carried *qnrB19* (*n* = 6), *sul2* (*n* = 5), *aph(6)-Id* (*n* = 5), and *aph(3″)-Ib* (*n* = 5) genes. Except for EC12, all isolates carried *astA* and *east1* virulence genes. All isolates exhibited virotype C except EC2, which had virotype B and additionally carried *iroBCDEN* genes. Five out of seven isolates carried plasmid with the F1:A1:B16 FAB formula.

Two isolates, which formed Cluster C, originated from different years and healthcare institutions and also possessed different FAB formulas (F1:A1:B20 and F1:A1:B16). Both isolates exhibited virotype C.

The thirteen isolates forming Cluster D were collected from nine different healthcare institutions and different years: one from each year except in 2021, when nine isolates were investigated. Compared to the characteristic resistance and virulence genes of the C2 subclade, isolates of this cluster additionally carried *hyl* operon and *cnf*1 genes. They showed virotype C and eleven isolates also carried bla_{OXA-1} and *aac(6′)-Ib-cr*. Ten of the 13 isolates possessed plasmid with the F31/F36:A4:B1 FAB formula.

Four isolates formed Cluster E, all collected in 2021 from four different healthcare institutions. In addition to the characteristic virulome and resistome for C2/H30Rx, EC15 and EC23 were identified by the presence of *nfaE*, *daaA-F*, *afa* operon, and *draA-F* virulence genes, the lack of the *pap* operon and exhibited virotype A. EC16 and EC21 exhibited virotype C. Three of the four isolates carried plasmid with the F2:A1:B- FAB formula.

The four isolates forming Cluster F were obtained from various years and different healthcare institutions. Each isolate harbored *nfaE*, *daaA-F*, and *afa* virulence genes, lacked *pap* operon, and exhibited virotype A. Three out of four isolates possessed the F2:A1:B- FAB formula.

2.3.2. Localization and Genetic Environment of $bla_{CTX-M-15}$ in the C2/H30Rx Isolates (Group A–I)

The localization and genetic environment of $bla_{CTX-M-15}$ genes were investigated in detail through 19 selected hybrid assembled genomes of the 30 C2/H30Rx isolates. Only one isolate from 2015 carried the $bla_{CTX-M-15}$ located on a plasmid (F2:A1:B-). In the other 18 isolates, chromosomal integration of the $bla_{CTX-M-15}$ linked to IS*Ecp1* (*n* = 7) or IS*26* (*n* = 12) translocable elements were detected. The group profile A-I was classified based on the structure of the genetic environment of each $bla_{CTX-M-15}$ copy and its association with the upstream IS element (Figure 2).

In Group A, the IS*Ecp1*-linked orf-$bla_{CTX-M-15}$ gene was present, and no other antibiotic-resistance genes were detected in its environment.

In Group B, the genetic environment of $bla_{CTX-M-15}$ was similar to Group A, but it was followed by the 273 bp sequence of the gene encoding the cupin fold metalloprotein *wbu*C.

The genetic context observed in Group C consists of two distinct segments. The first segment corresponded to the Group B sequence (IS*Ecp1*-orf-$bla_{CTX-M-15}$-*wbuC*), followed by a second segment. The second segment was a Tn*2* and IS*26*-linked composite transposon carrying Δ*catB3*, bla_{oxa-1} and *aac(6′)-Ib-cr5* resistance genes.

In Group D, the genetic environment of $bla_{CTX-M-15}$ consisted of two different segments of transposable units. The first segment was identical to the Group B sequence. The second segment was an IS*26*-linked composite transposon carrying ΔIS*Kpn11*, *tmrB*, and *aac(3)-IIa*. The two segments were linked by Tn*2*.

In Group E, the IS*26*-linked composite transposon carrying orf-$bla_{CTX-M-15}$-*wbuC* genes were present in the first segment and followed by two other different transposable unit segments. The second segment corresponded to the second segment of Group D (ΔIS*Kpn11*-*tmrB*-*aac(3)-IIa*), and the third segment corresponded to the second segment of Group C (Δ*catB3*-bla_{oxa-1}-*aac(6′)-Ib-cr5*).

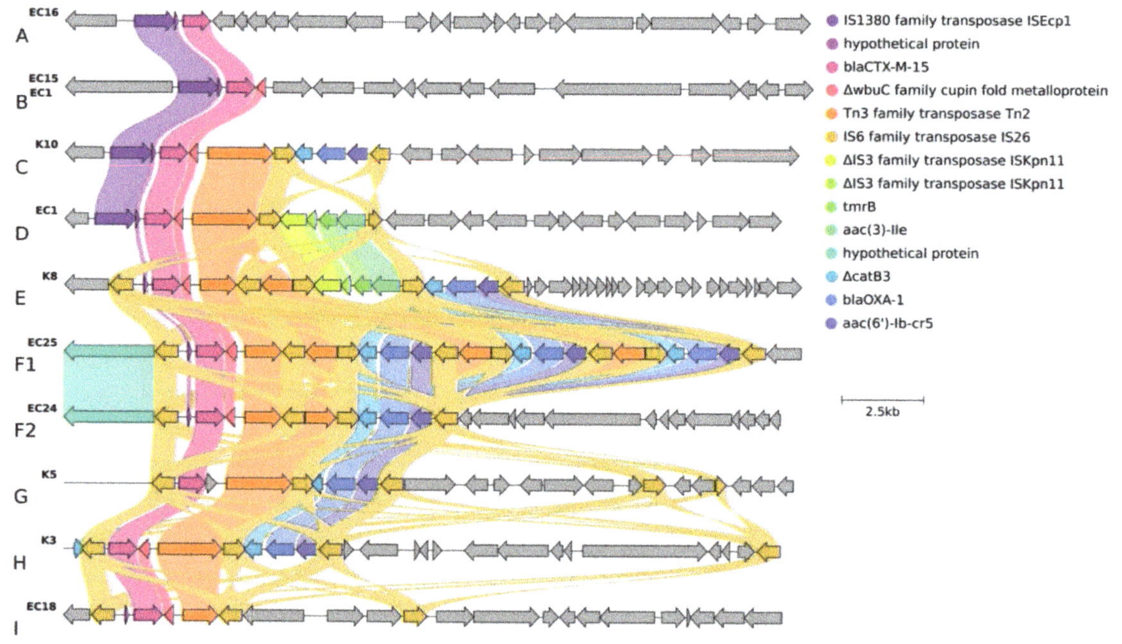

Figure 2. Linear sequence comparison of the genetic context (range 22,117 to 23,084 bp) of $bla_{\text{CTX-M-15}}$ on chromosome and plasmid. Legend: Genes are grouped and color-coded according to function. The color-coded arrows indicated genes correspond to the circles in the top right corner of the figure. Grey arrows indicate genes with no similarity. Colorful wavy links represent the sequence identity of homologous gene groups identified by Clinker. Groups are indicated as "A–I" next to the designated isolate name and genome sequence.

In Group F, the genetic environment of $bla_{\text{CTX-M-15}}$ genes consisted of two different segments of transposable units. The first segment corresponded to the IS*26*-linked $bla_{\text{CTX-M-15}}$. The IS*26*-linked composite transposon carried Δ*catB3*-$bla_{\text{oxa-1}}$-*aac(6′)-Ib-cr5* was the second segment. The two segments were linked by Tn2. Group F2 had one repeat, while Group F1 had three repeats of the second segment.

In Group G, the genetic environment of $bla_{\text{CTX-M-15}}$ genes consisted of two segments. The first segment harbored IS*26*-linked $bla_{\text{CTX-M-15}}$ like in Group F but lacked the orf gen between IS*26* and $bla_{\text{CTX-M-15}}$, and the second segment resembled the second segment of Group F2.

In Group H, the genetic environment of $bla_{\text{CTX-M-15}}$ was very similar in structure to Group G, but both segments were localized on plasmid.

In Group I, the IS*26*-linked orf- $bla_{\text{CTX-M-15}}$–*wbuC*-Tn2-IS*26* was present, and no other antibiotic resistance genes were detected in its environment.

For IS*Ecp1*-linked $bla_{\text{CTX-M-15}}$ (Group A–D), one copy of the Group A structure was found in one isolate (EC16–Cluster E), three copies of the Group B structure were found in one isolate (EC15–Cluster E), where one of the IS*Ecp1*-mediated insertion sequences was located in an opposite orientation. One copy of the Group C structure was observed in the K10 isolate (Cluster C). All four isolates in Cluster B (EC1, EC2, EC3, EC7) carried 2 copies of the IS*Ecp1*-linked $bla_{\text{CTX-M-15}}$ gene inserted into identical positions in their bacterial chromosome. One copy of $bla_{\text{CTX-M-15}}$ belonged to the Group B structure, while the second copy belonged to the Group D structure, which was located 84,842 bp downstream.

For IS26-linked $bla_{\text{CTX-M-15}}$ (Group E–I), one copy of the Group E structure was found in four isolates (K4, K6, K8, EC19–Cluster D), one copy of the Group F1 in one isolate EC25 (Cluster F), and one copy of the Group F2 in two (EC24 and EC22) isolates.

The K5 isolate formed Group G, the F31/F36:A1:B1-like plasmid integrated into the chromosome, where the IS26 mediated translocable unit was inserted upstream next to the 23S ribosomal RNA coding region.

The Group H structure was located on an F2:A1:B-like plasmid (91,036 bp), which was uniquely found in the K3 isolate (Cluster F). The Group I structure was observed in three isolates (EC10, EC18–Cluster D, and K7–Cluster F).

2.3.3. Localization and Genetic Environment of $bla_{\text{CTX-M-27}}$ in the C1-M27 Isolates (Region I–III)

All investigated $bla_{\text{CTX-M-27}}$ genes were located on IncF-like plasmids (Figure 3). The assembled plasmids were 133,550 bp for EC4, 98,865 bp for EC5, 134,637 bp for EC26, 96,681 bp for EC28, 80,412 bp for EC29, 137,256 bp for EC33, 105,606 bp for EC34, 100,608 bp for EC36, 127,181 bp for EC39, 139,893 bp for K9, and 110,045 bp for K17 in size. Nine out of eleven investigated isolates carried IncF(F1:A2:B20) plasmid, one carried IncF(F2:A2:B20), and another isolate had IncF(F1:A2:B-).

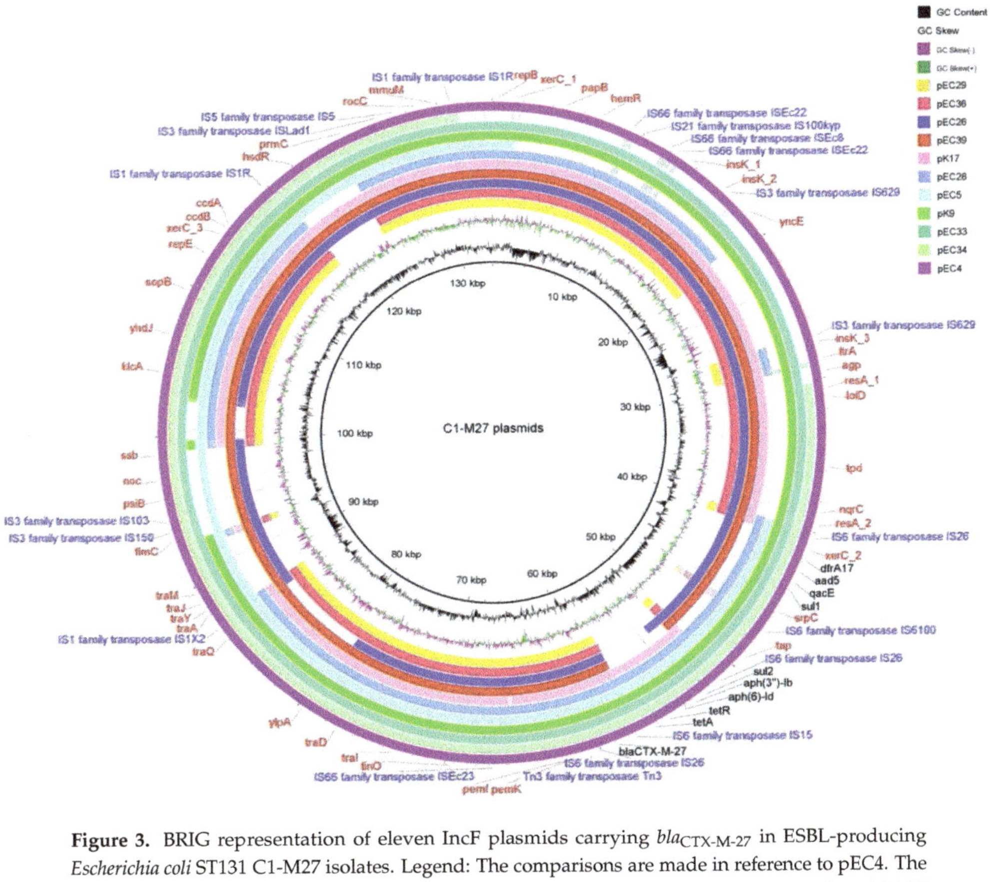

Figure 3. BRIG representation of eleven IncF plasmids carrying $bla_{\text{CTX-M-27}}$ in ESBL-producing *Escherichia coli* ST131 C1-M27 isolates. Legend: The comparisons are made in reference to pEC4. The inner rings show GC content (black) and GC skew (purple/green). The remaining rings show BLAST comparisons of eleven IncF plasmids carrying $bla_{\text{CTX-M-27}}$ of C1-M27 *E. coli*. The outer ring highlights the genes of pEC4, shown in different colors. The genomic features of annotated genes are indicated and color-coded in red. AMR genes are indicated in black and MGE in blue.

The antibiotic resistance genes were encoded in three regions of the IncF plasmids (Figure 4, Table 2): Region I. with $bla_{CTX\text{-}M\text{-}27}$ (10/11 was carried by *IS6* family-linked composite transposon); Region II. with *tetA*, *aph(6)-Id*, *aph(3'')-Ib*, *sul2* (six/seven was carried by IS6 family-linked composite transposon); Region III. with *dfrA17*, *aadA5*, *qacE*Δ*1*, *sul1*, *mph(A)* (six/eight was carried by IS6 family-linked composite transposon). EC29 and EC36 isolates carried Region I. alone, and K17 isolates possessed Regions I. and II. and EC26 and EC39 isolates possessed Regions I. and III. Six isolates (EC4, EC5, EC28, EC33, EC34, and K9) possessed Regions I., II., and III.

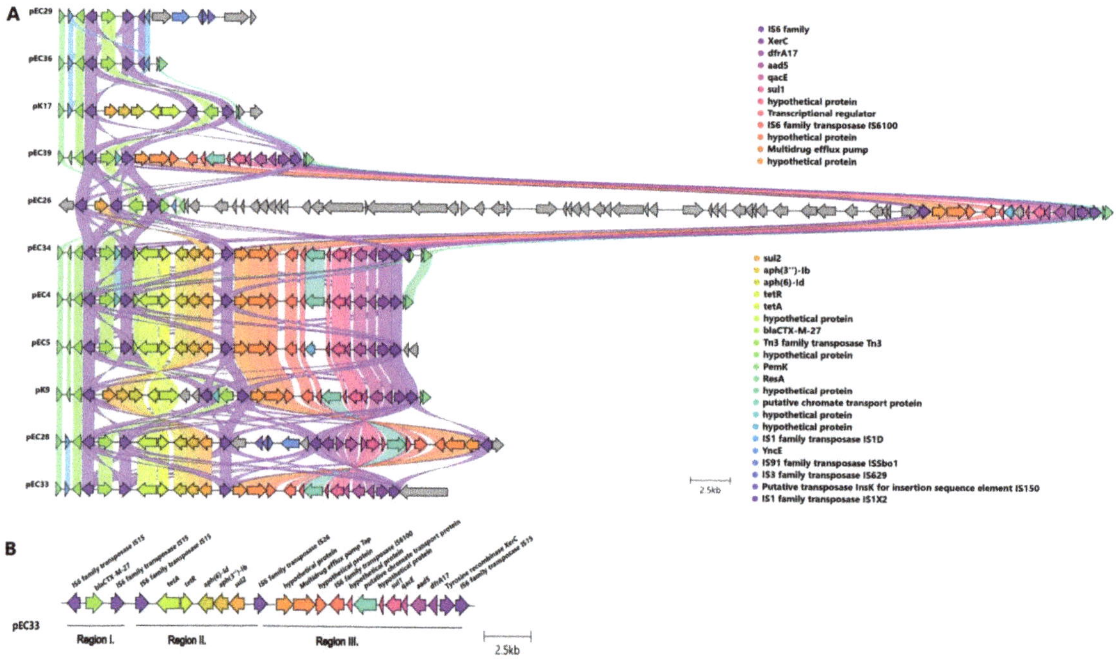

Figure 4. Linear sequence comparison of the genetic environment of $bla_{CTX\text{-}M\text{-}27}$. Legend: Genes are grouped and color-coded according to the function of the genes. The color-coded arrows indicated genes correspond to the circles in the top right corner of the figure. Grey arrows indicate genes with no similarity. Colorful wavy links represent the sequence identity of homologous gene groups. (**A**) Genetic composition of 11 plasmids containing $bla_{CTX\text{-}M\text{-}27}$. (**B**) Genetic environment of EC3 with Regions. Regions are indicated R on the figure. Three similar regions of resistance genes were present in the plasmid genetic context. Region I. consists of $bla_{CTX\text{-}M\text{-}27}$ and Region II. consists of *tetA*, *aph(6)-Id*, *aph(3'')-Ib*, sul2. Region III. consists of *dfrA*17, *aadA*5, *qacE*Δ*1*, *sul1*, and *mph(A)*.

Table 2. Main genetic characteristics of ST131 *E. coli* isolates with hybrid assembled genomes.

Subclade	Cluster Type	Cluster-Specific Virulence Gene	Localization of bla_{CTX-M}	IS Element Linked to bla_{CTX-M}	Copy of bla_{CTX-M}	Genetic Environment of bla_{CTX-M}: Categorization of the Genetic Environment of bla_{CTX-M}	1st Segment	2nd Segment	3rd Segment	4th Segment	Additional Antibiotic Resistance Genes	Corresponding Isolate (s)
C1-M27	A	*traT*, *senB*, *mcbA*, *pic*	plasmid	IS6 family	1	Regions I	$bla_{CTX-M-27}$	-	-	-	-	EC29, EC36
						Regions I–II		*tetA-aph(6)-Id-aph(3″)-Ib-sul2*	-	-		K17
						Regions I–III		*dfrA17-aadA5-qacE*Δ*1-sul1-mph(A)*	-	-		EC39, EC26
						Regions I–II–III		*tetA-aph(6)-Id-aph(3″)-Ib-sul2*	*dfrA17-aadA5-qacE*Δ*1-sul1-mph(A)*	-		EC4, EC5, EC28, EC33, EC34, K9
C2/H30Rx	B	*astA*, *east1*	chromosomal	ISEcp1	2	Group B, D	$bla_{CTX-M-15}$	*tmrB-aac(3)-Iia*	-	-	*qnrB19*, *sul2*, *aph(6)-Id*, *aph(3″)-Ib*	EC1, EC2, EC3, EC7
	C	-			1	Group C		Δ*catB3-*bla_{oxa-1}*-aac(6′)-Ib-cr5*	-	-	-	K10
	D	*hyl*, *cnf1*		IS26		Group I		-	-	-		EC10, EC18
						Group F2		Δ*catB3-*bla_{oxa-1}*-aac(6′)-Ib-cr5*	-	-		EC24, EC22
						Group F1		Δ*catB3-*bla_{oxa-1}*-aac(6′)-Ib-cr5*	Δ*catB3-*bla_{oxa-1}*-aac(6′)-Ib-cr5*	Δ*catB3-*bla_{oxa-1}*-aac(6′)-Ib-cr5*		EC25
						Group E		*tmrB*-aac(3)-Iia	Δ*catB3-*bla_{oxa-1}*-aac(6′)-Ib-cr5*	-		K4, K6, K8, EC19
	E	-		ISEcp1	3	Group B		-	-	-		EC15
		nfaE, *daaA-F*, *afa* operon, *draA-D*				Group A		-	-	-		EC16
	F			IS26	1	Group I		-	-	-		K7
			chromosomal			Group G		Δ*catB3-*bla_{oxa-1}*-aac(6′)-Ib-cr5*	-	-		K5
			plasmid			Group H		Δ*catB3-*bla_{oxa-1}*-aac(6′)-Ib-cr5*	-	-		K3

Legend: The symbol "-" indicates the absence of a genetic element.

3. Discussion

This genomic epidemiology study revealed for the first time the population structure of the C1-M27 and C2/H30Rx *E. coli* ST131 clones isolated from blood cultures in Hungary. This study was based on our previous observation study [11], which highlighted the dominance of the ST131 clone among ESBL-producing *E. coli* strains isolated from blood samples.

In Hungary, a CTX-M-15-producing ST131 clone was first identified in 2010, and later, the first ESBL-producing invasive *E. coli* isolates belonging to the C1-M27 subclade were detected in 2015. Since then, the number and rate of ESBL-producing ST131 clones and their subclades have gradually increased until 2018 [11]. In 2015–2018 and in 2021, a similar proportion of invasive ESBL-producing *E. coli* isolates investigated at the National Center for Public Health and Pharmacy belonged to the ST131.

In this study, all isolates proved to be resistant to ceftriaxone and ciprofloxacin, but the CTX-M-15-producing isolates showed higher resistance rates than the CTX-M-27-producing isolates to ceftazidime. Although CTX-M-27 also has the Asp240Gly amino acid substitution that is responsible for ceftazidime resistance in CTX-M-15, generally lower MIC values can be measured in vitro [12–14]. All isolates were susceptible to carbapenems. This finding is supported by data from EARS-Net, where only one carbapenem-resistant isolate was reported from Hungary between 2015 and 2021. The two colistin-resistant *E. coli* C1-M27 isolates (EC34 and EC35) (3.2% resistance rate in this collection) originated from the same healthcare institution in 2021. Similar low levels of colistin resistance were observed in a previous report from Hungary, where out of 146 investigated *E.coli* isolates, one isolate showed *mcr*-1 related colistin-resistance, and six isolates were colistin-tolerant in 2010–2011, and four isolates were colistin-tolerant in 2016 [15]. In this study, none of the isolates possessed any plasmid-mediated colistin resistance (*mcr*) genes. Only the EC34 had *pmrB* (L14R) mutation, which is strongly associated with colistin resistance [16]. Each of the 63 isolates possessed the same amino acid substitution in *pmrB* (E123D), which has been described in the context of chromosome-mediated colistin resistance [17,18]. However, they remained colistin-susceptible (except for two isolates). A Korean study also found *E. coli* isolates with *pmrB* (E123D) substitution as colistin-susceptible [19]. Thus, these results are still ambiguous and the origin of the resistance was not identified in one isolate and requires further investigation. The C2/H30Rx isolates showed lower susceptibility rates than the C1-M27 ones to amikacin, tobramycin, and gentamicin due to aminoglycoside-modifying enzymes [20,21]. Among the aminoglycoside-modifying enzymes, the *aac(6′)-Ib-cr* was present only in C2/H30Rx isolates, while *aac(3)-IIa* and *aph(6)-Id* were present both in C2/H30Rx and C1-M27 isolates. According to the scientific literature, *aac(3)-IIa* confers resistance to gentamicin and tobramycin, while *aac(6′)-Ib-cr* to amikacin and tobramycin [22]. In the study, the same associations were found between the presence of genes of these aminoglycoside-modifying enzymes and corresponding antibiotics. The identified *gyrA* and *parC* mutations are well known to confer resistance to ciprofloxacin. Additional resistance mechanisms such as *qnrA* or *aac(6′)-Ib-cr* may also contribute to fluoroquinolone resistance. Among the isolates of the C2 subclade, *qnrA* and *aac(6′)-Ib-cr* were present [23,24].

It has been shown that IncFII-type plasmids are mainly associated with the $bla_{CTX-M-15}$ gene [4]. In this study, among the 19 C2/H30Rx isolates, only one isolate from 2015 carried the C2-associated F2:A1:B- plasmid harboring $bla_{CTX-M-15}$. The remaining 18 isolates showed chromosomal integration of the $bla_{CTX-M-15}$ gene in one or several copies. The chromosomal integration of $bla_{CTX-M-15}$ has been described before in a few studies that reported a local distribution [25,26]. However, these 19 isolates were delivered from five clusters and sixteen healthcare institutions. These data indicate that C2/H30Rx ST131 clusters harboring chromosomal $bla_{CTX-M-15}$ might have emerged convergently and spread across Hungary.

There was a difference in the case of the type of IS, which was responsible for the translocation of the genetic environment of the $bla_{CTX-M-15}$. Two types of IS-mediated translocable elements were detected: IS*Ecp1* was located upstream of the *bla*$_{CTX-M-15}$ gene or IS*26* was incorporated upstream and downstream of the *bla*$_{CTX-M-15}$. There were also IS*26* or IS*Ecp1* transposon structures that consisted of overlapping IS*26*-based translocatable units. The IS*26* is one of the few IS types that have been shown to form fusions between two DNA sequences via replication and form cointegrates rather than move alone to a new location [27,28]. IS*26* could also be found in arrays, intercalated next to other transposable elements, and could form units able to undergo tandem amplification in drug-resistant plasmids or chromosomes [29]. Like IS*26*, the IS*Ecp1*-mediated transposition proceeds via one IS*Ecp1*, mediating genetic transposition events that involve homologous recombination. Both IS families could be responsible for the higher level of expression of the *bla*$_{CTX-M}$ genes, which play an important role in the dissemination of antibiotic-resistance genes among Gram-negative bacteria [30]. The phenomenon of IS*Ecp1* and/or IS*26*-mediated *bla*$_{CTX-M-15}$ transposition has been described before in a few studies [31–35]. Shawa et al. described the co-occurrence of IS*Ecp1*- *bla*$_{CTX-M-15}$ and other *catB3*, *bla*$_{oxa-1}$, *aac(6′)-Ib-cr5* genes on the chromosomal environment of CTX-M-15-producing ST131 *E. coli* [31,36].

In this study, the chromosomal integration of one or more copies of *bla*$_{CTX-M-15}$ and its genetic environment was mediated either by IS*Ecp1* or by IS*26*. Additionally, the IS26-mediated *bla*$_{CTX-M-15}$ gene was followed by three copies of Tn2-IS*26*-Δ*catB3*-*bla*$_{OXA-1}$-*aac(6′)-Ib-cr5*-IS26 segment in one isolate (Group F1).

Shropshire et al. described IS*26*-mediated amplification of *bla*$_{CTX-M-15}$ and *bla*$_{OXA-1}$ in the carbapenem-resistant ST131 *E. coli* genome and hypothesized that the IS*26*-*bla*$_{CTX-M-15}$-ΔTn*2* could drive the amplification of the genetic environment of *bla*$_{CTX-M-15}$ [34,37]. In this study, amplification mediated by IS*Ecp1* also occurred in four isolates.

The results suggest that the IncFII plasmid may have been progressively integrated into chromosomes in the 2010s and later was progressively lost in the genetic environment of *bla*$_{CTX-M-15}$. The Cluster B isolates with two copies of *bla*$_{CTX-M-15}$ probably emerged independently in Hungary.

Therefore, apart from the horizontal gene transfer of the plasmids encoding *bla*$_{CTX-M-15}$, the composite transposon-linked antimicrobial resistance genes had undergone several chromosomal insertion events mediated by IS*26* or IS*Ecp1*. The independent convergent appearance of various IS-mediated chromosomal integration of antimicrobial resistance genes suggests that this process may have an evolutionary potential. The benefit of the chromosomal integration of the *bla*$_{CTX-M-15}$ and other AMR genes might contribute to maintenance and dissemination under the selection pressure of the antimicrobial environment. The plasmid may impose a fitness burden on their hosts but also provides diversity, promoting the worldwide dissemination of successful clones [38–40].

Unlike the C2 subclade, in C1-M27 isolates all *bla*$_{CTX-M-27}$ and AMR genes were located on IncF plasmids. The majority of all C1-M27 ST131 isolates carried an F1:A2:B20 plasmid. The F1:A2:B20 plasmid type is strongly associated with the C1/H30R subclade [6]. Here, all the IS6 family-linked composite transposons carrying AMR genes remained on the plasmid, in line with many examples in the literature [41,42]. Similar Regions (I., II., III.) consisting of AMR genes had been found before in other F1:A2:B20 plasmids of C1-M27 isolates, and AMR gene regions without *bla*$_{CTX-M-27}$ were found in F1:A2:B20 plasmid with *bla*$_{CTX-M-14}$ [9,43]. In this collection, one isolate with F1:A2:B- plasmid harbored the same Region I., II., III. while there were F1:A2:B20 isolates harboring Region I. or I. and III.

In the study, the most common Col-like replicon type in the C1 and C2 subclades was Col156 (30 vs. 16), followed by Col(MG828) (29 vs. 9), respectively. The exact role of Col-like plasmids or replicons is not clear. Col-like plasmids or replicons are mobilizable vectors that have been described as promoting the spread of antibiotic resistance plasmids via horizontal gene transfer in the *Enterobacteriaceae* family [36,44,45].

4. Materials and Methods

4.1. Bacterial Collection

The putative ESBL-producing *E. coli* isolates obtained from blood samples have been submitted to the NCPHP for confirmation and molecular typing from the whole country. All isolates were identified using matrix-assisted laser desorption ionization time-of-flight mass spectrometry (MALDI Biotyper, Bruker, Bremen, Germany). The Double Disc Synergy Test (DDST) was used for confirmation of ESBL production: the DDST confirmation test was performed using the "ESβL Detection Disc Set" (MAST Diagnostica GmbH, Reinfeld, Germany) according to the manufacturer's instruction [46,47]. The clonal relationship was examined using *XbaI*-pulsed-field gel electrophoresis (PFGE) in all ESBL-producing *E. coli* isolates [48]. For the determination of ST131 subclades, the PFGE results and multiplex PCR were used as described by Matsumura [49].

Non-duplicate isolates were selected for further investigation based on the PFGE results and covering spatiotemporal distribution during 2015–2018 and 2021 (Supplementary Table S1).

4.2. Antimicrobial Susceptibility Testing

The antibiotic susceptibility testing of all isolates was performed using the disc diffusion method. Where the disc diffusion method was not recommended (e.g., colistin) or where MIC values gave additional significance to the disc diffusion results, MIC values were determined. The antimicrobial susceptibility testing to ceftriaxone, cefotaxime, fosfomycin, ceftazidime/avibactam, ertapenem, ciprofloxacin, imipenem, meropenem, gentamicin, amikacin, tobramycin, and tigecycline was performed by disk diffusion (Mast Diagnostica GmbH, Reinfeld, Germany), to ceftriaxone, cefotaxime, fosfomycin, ceftazidime/avibactam, ertapenem by MIC Test Strips (Liofilchem, Roseto degli Abruzzi, Italy), to colistin by MICRONAUT MIC-Strip (MERLIN Diagnostika GmbH, Bornheim, Germany) and interpreted using EUCAST guidelines [50]. ATCC 25,922 *E. coli* reference strain was used for quality control of antimicrobial susceptibility testing.

4.3. Molecular Characterization

4.3.1. DNA Extraction

The Bacterial DNA was extracted and purified with the DNeasy UltraClean Microbial Kit (Qiagen, Hilden, Germany) according to the manufacturer's instructions. The assessment of DNA quality was carried out using a TapeStation 2200 automated electrophoresis system (Agilent Technologies, Palo Alto, CA, USA) and DNA quantity was measured using a Qubit 3.0 Fluorometer (Invitrogen, Life Technologies, Carlsbad, CA, USA).

4.3.2. Short-Read Sequencing

Library preparation for short-read sequencing was performed with an Illumina DNA Prep kit (Illumina, San Diego, CA, USA). Whole genome sequencing of the isolates was performed on an Illumina MiSeq platform (150-bp paired-end sequencing). Raw data were processed by default using the EnteroBase pipeline [51]. EnteroBase provides automated pipelines that implement multiple functions that allow users to upload their own sequencing data for de novo assembly using a streamlined pipeline. Once the reads are received, EnteroBase provides the following workflow: First, automatic assembly with QAssembly. QAssembly consists of quality assemblies, including read pre-processing, trimming (QAssembly: Sickle), assembly (QAssembly: Spades), post-correction and filtering (QAssembly: BWA), and converting results to fasta format (QAtoFasta). This is followed by a quality assessment of the assemblies (QA evaluation). The fasta files were retrieved and analyzed using additional bioinformatics tools [51].

The quality of draft genome assemblies was considered appropriate for downstream analyses if the average sequencing depth of contigs exceeded 50-fold, their total size matched the expected genome size (5 ± 0.3 Mb), and N50 was longer than 100 kbp. Detailed genomic quality indicators are available in Supplementary Table S2.

4.3.3. Long-Read Sequencing and Genome Assembly

After short-read sequencing of the isolates, a phylogenetic tree was constructed. The antimicrobial resistance genes, virulence genes, mobile genetic elements, and plasmid replicon types were identified within the clusters. Based on these results, isolates were selected for long-read sequencing to represent the main characteristics and clusters proportionally, with their corresponding isolation years (2015–2018 and 2021) and healthcare institutions.

The long-read sequencing library of the selected isolates was prepared by using the Ligation Sequencing Kit (SQK-LSK109, Oxford Nanopore, Oxford, UK) according to the manufacturer's instruction for genomic DNA with barcoding (Native Barcoding Expansions 96 (EXP-NBD196, Oxford Nanopore, Oxford, UK) and the sequencing was performed on MinION Mk1C (Oxford Nanopore, Oxford, UK). Basecalling (with fast basecalling mode) and demultiplexing of the long-read raw sequences were performed with Guppy v5.1.12 as implemented in MinKNOW v 21.11.6. Raw long-read data processing was performed using the following GalaxyEU (https://nanopore.usegalaxy.eu) (accessed on 4 January 2023) toolkit with default parameters: Porechope v0.2.4 [52] was used for quality trimming and removing barcodes. We carried out multiple assemblies: first, we assembled the long-reads using Flye v2.9.1 [53] nd polished the assemblies, then used a short-read first hybrid assembly approach as implemented in Unicycler v0.5.0 [54] with default parameters. BWA MEM v0.7.17 [55] was used for short-read alignment, and Pilon v1.20 [56] was used for polishing. To decrease the mismatch rate, another round of genome polishing was applied. Polished Flye and polished Unicycler assemblies were analyzed simultaneously. Finally, those assemblies with less fragmentation were selected as representative genome assemblies of the isolates: 22 Flye assemblies and eight Unicycler assemblies were obtained for further analysis. Assembly statistics were retrieved with Quast v5.0.2 [57]. Prokka v1.14.6 was used for the genome annotation of the assemblies [58].

Clinker v0.0.27 [59] was used to compare and visualize gene cluster homology (genes were predicted using Prokka v1.14.6). The plasmid content and mobile genetic elements (MGE) were analyzed by PlasmidFinder v2.1 [60], pMLST v2.0 [60] and Nucleotide BLAST (https://blast.ncbi.nlm.nih.gov/Blast.cgi) (accessed on 9 January 2023) and MGE v1.0.3 [61] online tools. The sequence similarity of plasmid sequences was analyzed using BRIG v0.95 [62].

4.3.4. Phylogenomic Reconstruction and Clustering of Isolates

The genome annotation prepared with Prokka was used as input for Panaroo v1.1.2 [63] to reconstruct the core genome alignment of the isolates. The default values of Panaroo were used, except that the frequency threshold of genes for inclusion in the core alignments was increased to 99% (--core-threshold 0.99). A total of 3788 genes were included in the core genome. Phylogenetic relationships of the isolates were reconstructed using IQtree v2.2.3 [64], with automatic model selection with ModelFinderPlus turned on (-m MFP), and the robustness of the results was assessed using 1000 aLRT and 1000 ultrafast bootstrap replications. The fastBAPS v1.0.8 [65] R v4.2.2 [66] package was used to identify clusters in the population structure using optimise_baps priors. Instead of using a single threshold, the fastBAPS algorithm was applied that relies on statistical genetic models to effectively partition molecular variation [67] and is independent of the phylogenetic reconstruction. The "optimise.baps" BAPS priors were optimized with the optimise_prior function. Then, the fastp_baps and best_baps_partition functions implemented in the fastbaps R package were used to obtain the best clustering scheme describing the genetic clusters in the dataset. Both IQtree and fastBAPS used the multiple sequence alignment of the complete core genome as input. The results of fastBAPS clustering with the phylogenetic tree were visualized using the R package ggtree v3.6.2 [68], for which the phylogenetic tree was midpoint rooted using phangorn v2.11.1 [69]. The phylogenetic trees were visualized and annotated with sample metadata using the Interactive Tree of Life (iTOL v6.7.4) web tool [70].

4.3.5. Virulence and Antibiotic Resistance Genes Analysis

The antimicrobial resistance genes (AMR) in the complete genome sequences were identified using ResFinder v4.1 [71–73] CGE online tools (https://www.genomicepidemiology.org/) (accessed on 7 January 2023).

Virulence factor genes were retrieved using VirulenceFinder v2.0 [74] and Virulence Factors Database (VFDB v2020-Feb-28 through SeqSphere+ [75]) online tools. The virotypes A to D of the ST131 isolates were assigned according to the scheme developed by Blanco et al., 2013 [76].

4.4. Statistical Analysis

For statistical analysis, Fisher's exact test was performed in an online program (https://www.socscistatistics.com/tests/fisher/default2.aspx) (accessed on 15 March 2023). Three approaches were used for statistical tests: 1. testing for independence between two variables (presence or absence of virulence or resistance gene compared to two subclade isolates); 2. testing for independence between two variables (presence or absence of phenotypic antibiotic resistance compared to two subclade isolates); 3. testing for independence between two variables (phenotypic antibiotic resistance compared to presence or absence of antibiotic resistance genes). For each comparison, $p < 0.05$ was considered to be statistically significant.

5. Conclusions

These results indicate that C2 and C1 subclades may have emerged and spread in different ways in Hungary. The C1-M27 variant has formed one cluster since its appearance in Hungary around 2015–2016. This hypothesis is further supported by the fact that C1-M27 isolates showed high similarities in the virulome and resistome, and 93.9% of them harbored an F1:A2:B20 plasmid. In contrast, the C2/H30Rx clusters may have appeared independently in the country. In the meantime, the composite transposons or a part of the IncF-like plasmids may have been integrated into the chromosome and later progressively lost from the genetic environment of $bla_{CTX\text{-}M\text{-}15}$. By the time the C1-M27 clone appeared, C2/H30Rx clones with chromosomally encoded $bla_{CTX\text{-}M\text{-}15}$ were already present in Hungary. The C2 subclade could have undergone a convergent evolutionary process in Hungary, and these clonal lineages were still detectable in 2021. Although the composite transposons ensuring the presence of ESBL and other AMR genes in the C1 clonal lineage can be linked to *IS*26, as in C2, it has not been transferred into a chromosomal environment and underwent a clonal expansion different from C2. Also, neither C1 nor C2 subclades have been able to displace each other but have been able to stably coexist and spread. Within the C2 subclade, the Cluster B isolates with two copies of $bla_{CTX\text{-}M\text{-}15}$ showed local clonal expansion and probably emerged independently in Hungary.

Supplementary Materials: The following supporting information can be downloaded at https://www.mdpi.com/article/10.3390/antibiotics13040363/s1. The data set supporting the results of this article is included within the article and in Supplementary Data Sheet 1: Table S1: Thirty C2/H30Rx and thirty-three C1-M27 ESBL-producing *E. coli* ST131 were used in the study, including the source, ST131 sublineage, collection date, sex and age; Table S2:The genomic quality indicators for sequencing of the 63 ESBL-producing *E. coli* ST131;Figure S1: Maximum likelihood phylogeny of sixty-three, ESBL-producing *E. coli* ST131 isolates and their genetic characteristic, including virulome and virotype; Figure S2.: The results of phylogenetic reconstruction and fastBAPS clustering using the core genome alignment prepared by Panaroo.

Author Contributions: Study design, K.T., Á.T. and D.S.; writing—original draft preparation, K.T.; phenotypic analysis of strains, L.B., L.J., A.H. and K.T.; genotypic characterization of strains, K.T., Á.T., E.U., L.B., I.D. and V.L.; bioinformatic analysis, K.T., L.L. and Á.T.; drafting and revising the manuscript, K.T., Á.T. and D.S. All authors provided critical comments, contributed to the article, and approved the final manuscript. All authors have read and agreed to the published version of the manuscript.

Funding: This research was funded by the European Union's Horizon 2020 research and innovation program: Antimicrobial Resistance Surveillance (952491-AmReSu).

Institutional Review Board Statement: Ethics Committee approval was not required as the Hungarian legislation on handling of personal health information (Law no. 1997. XLVII.) empowers the National Center for Public Health and Pharmacy (NCPHP) to analyze data and to take necessary measures in the interest of public health. Healthcare institutions culture clinical specimens (e.g., blood samples) as part of routine diagnostic testing. They are by law required to send MDR isolates from different specimens to the NCPHP for further testing for public health purposes. Personal data have been handled in accordance with legal regulations and the Center's data protection rules.

Informed Consent Statement: Not applicable.

Data Availability Statement: The raw data of long-reads and short-reads are available on the Sequence Read Archive (SRA) database under the BioProject number PRJNA1003301.

Acknowledgments: We would like to thank colleagues at the Department of Bacteriology, Parasitology and Mycology and the National Center for Public Health and Pharmacy for their assistance with the study.

Conflicts of Interest: The authors declare no conflicts of interest.

References

1. Magiorakos, A.-P.; Srinivasan, A.; Carey, R.B.; Carmeli, Y.; Falagas, M.E.; Giske, C.G.; Harbarth, S.; Hindler, J.F.; Kahlmeter, G.; Olsson-Liljequist, B.; et al. Multidrug-resistant, extensively drug-resistant and pandrug-resistant bacteria: An international expert proposal for interim standard definitions for acquired resistance. *Clin. Microbiol. Infect.* **2012**, *18*, 268–281. [CrossRef] [PubMed]
2. Cassini, A.; Högberg, L.D.; Plachouras, D.; Quattrocchi, A.; Hoxha, A.; Simonsen, G.S.; Colomb-Cotinat, M.; Kretzschmar, M.E.; Devleesschauwer, B.; Cecchini, M.; et al. Attributable Deaths and Disability-Adjusted Life-Years Caused by Infections with Antibiotic-Resistant Bacteria in the EU and the European Economic Area in 2015: A Population-Level Modelling Analysis. *Lancet Infect. Dis.* **2019**, *19*, 56–66. [CrossRef]
3. Petty, N.K.; Ben Zakour, N.L.; Stanton-Cook, M.; Skippington, E.; Totsika, M.; Forde, B.M.; Phan, M.-D.; Gomes Moriel, D.; Peters, K.M.; Davies, M.; et al. Global Dissemination of a Multidrug Resistant *Escherichia coli* Clone. *Proc. Natl. Acad. Sci. USA* **2014**, *111*, 5694–5699. [CrossRef] [PubMed]
4. Johnson, T.J.; Danzeisen, J.L.; Youmans, B.; Case, K.; Llop, K.; Munoz-Aguayo, J.; Flores-Figueroa, C.; Aziz, M.; Stoesser, N.; Sokurenko, E.; et al. Separate F-Type Plasmids Have Shaped the Evolution of the H30 Subclone of *Escherichia coli* Sequence Type 131. *mSphere* **2016**, *1*, e00121-16. [CrossRef] [PubMed]
5. Pitout, J.D.D.; Finn, T.J. The Evolutionary Puzzle of *Escherichia coli* ST131. *Infect. Genet. Evol.* **2020**, *81*, 104265. [CrossRef] [PubMed]
6. Matsumura, Y.; Pitout, J.D.D.; Gomi, R.; Matsuda, T.; Noguchi, T.; Yamamoto, M.; Peirano, G.; DeVinney, R.; Bradford, P.A.; Motyl, M.R.; et al. Global *Escherichia coli* Sequence Type 131 Clade with Bla CTX-M-27 Gene. *Emerg. Infect. Dis.* **2016**, *22*, 1900–1907. [CrossRef]
7. Pitout, J.D.D.; DeVinney, R. *Escherichia coli* ST131: A Multidrug-Resistant Clone Primed for Global Domination. *F1000Research* **2017**, *6*, Rev-195. [CrossRef]
8. Mathers, A.J.; Peirano, G.; Pitout, J.D.D. The Role of Epidemic Resistance Plasmids and International High-Risk Clones in the Spread of Multidrug-Resistant Enterobacteriaceae. *Clin. Microbiol. Rev.* **2015**, *28*, 565–591. [CrossRef]
9. Hayashi, M.; Matsui, M.; Sekizuka, T.; Shima, A.; Segawa, T.; Kuroda, M.; Kawamura, K.; Suzuki, S. Dissemination of IncF Group F1:A2:B20 Plasmid-Harbouring Multidrug-Resistant *Escherichia coli* ST131 before the Acquisition of BlaCTX-M in Japan. *J. Glob. Antimicrob. Resist.* **2020**, *23*, 456–465. [CrossRef]
10. Pitout, J.D.D.; Chen, L. The Significance of Epidemic Plasmids in the Success of Multidrug-Resistant Drug Pandemic Extraintestinal Pathogenic *Escherichia coli*. *Infect. Dis. Ther.* **2023**, *12*, 1029–1041. [CrossRef]
11. Tóth, K.; Tóth, Á.; Kamotsay, K.; Németh, V.; Szabó, D. Population Snapshot of the Extended-Spectrum β-Lactamase-Producing *Escherichia coli* Invasive Strains Isolated from a Hungarian Hospital. *Ann. Clin. Microbiol. Antimicrob.* **2022**, *21*, 32. [CrossRef] [PubMed]
12. Poirel, L.; Gniadkowski, M.; Nordmann, P. Biochemical Analysis of the Ceftazidime-Hydrolysing Extended-Spectrum β-Lactamse CTX-M-15 and of Its Structurally Related β-Lactamase CTX-M-3. *J. Antimicrob. Chemother.* **2002**, *50*, 1031–1034. [CrossRef] [PubMed]
13. Castanheira, M.; Simner, P.J.; Bradford, P.A. Extended-Spectrum β-Lactamases: An Update on Their Characteristics, Epidemiology and Detection. *JAC-Antimicrob. Resist.* **2021**, *3*, dlab092. [CrossRef] [PubMed]
14. Bonnet, R.; Recule, C.; Baraduc, R.; Chanal, C.; Sirot, D.; De Champs, C.; Sirot, J. Effect of D240G Substitution in a Novel ESBL CTX-M-27. *J. Antimicrob. Chemother.* **2003**, *52*, 29–35. [CrossRef] [PubMed]

15. Juhász, E.; Iván, M.; Pintér, E.; Pongrácz, J.; Kristóf, K. Colistin Resistance among Blood Culture Isolates at a Tertiary Care Centre in Hungary. *J. Glob. Antimicrob. Resist.* **2017**, *11*, 167–170. [CrossRef] [PubMed]
16. Wan, F.; Xu, L.; Ruan, Z.; Luo, Q. Genomic and Transcriptomic Analysis of Colistin-Susceptible and Colistin-Resistant Isolates Identify Two-Component System Evgs/Evga Associated with Colistin Resistance in *Escherichia coli*. *Infect. Drug Resist.* **2021**, *14*, 2437–2447. [CrossRef] [PubMed]
17. Abavisani, M.; Bostanghadiri, N.; Ghahramanpour, H.; Kodori, M.; Akrami, F.; Fathizadeh, H.; Hashemi, A.; Rastegari-Pouyani, M. Colistin Resistance Mechanisms in Gram-Negative Bacteria: A Focus on *Escherichia coli*. *Lett. Appl. Microbiol.* **2023**, *76*, ovad023. [CrossRef] [PubMed]
18. Zakaria, A.S.; Edward, E.A.; Mohamed, N.M. Genomic Insights into a Colistin-Resistant Uropathogenic *Escherichia coli* Strain of O23:H4-ST641 Lineage Harboring Mcr-1.1 on a Conjugative IncHI2 Plasmid from Egypt. *Microorganisms* **2021**, *9*, 799. [CrossRef] [PubMed]
19. Choi, Y.; Lee, J.Y.; Lee, H.; Park, M.; Kang, K.J.; Lim, S.K.; Shin, D.; Ko, K.S. Comparison of Fitness Cost and Virulence in Chromosome- and Plasmid-Mediated Colistin-Resistant *Escherichia coli*. *Front. Microbiol.* **2020**, *11*, 798. [CrossRef]
20. Shaw, K.J.; Rather, P.N.; Hare, R.S.; Miller, G.H. Molecular Genetics of Aminoglycoside Resistance Genes and Familial Relationships of the Aminoglycoside-Modifying Enzymes. *Microbiol. Rev.* **1993**, *57*, 138–163. [CrossRef]
21. Ramirez, M.S.; Tolmasky, M.E. Amikacin: Uses, Resistance, and Prospects for Inhibition. *Molecules* **2017**, *22*, 2267. [CrossRef]
22. Ramirez, M.S.; Tolmasky, M.E. Aminoglycoside Modifying Enzymes. *Drug Resist. Updat.* **2010**, *13*, 151–171. [CrossRef]
23. Robicsek, A.; Strahilevitz, J.; Jacoby, G.A.; Macielag, M.; Abbanat, D.; Chi, H.P.; Bush, K.; Hooper, D.C. Fluoroquinolone-Modifying Enzyme: A New Adaptation of a Common Aminoglycoside Acetyltransferase. *Nat. Med.* **2006**, *12*, 83–88. [CrossRef] [PubMed]
24. Phan, M.D.; Peters, K.M.; Fraga, L.A.; Wallis, S.C.; Hancock, S.J.; Nhu, N.T.K.; Forde, B.M.; Bauer, M.J.; Paterson, D.L.; Beatson, S.A.; et al. Plasmid-Mediated Ciprofloxacin Resistance Imparts a Selective Advantage on *Escherichia coli* ST131. *Antimicrob. Agents Chemother.* **2022**, *66*, e02146-21. [CrossRef]
25. Hirai, I.; Fukui, N.; Taguchi, M.; Yamauchi, K.; Nakamura, T.; Okano, S.; Yamamoto, Y. Detection of Chromosomal BlaCTX-M-15 in *Escherichia coli* O25b-B2-ST131 Isolates from the Kinki Region of Japan. *Int. J. Antimicrob. Agents* **2013**, *42*, 500–506. [CrossRef]
26. Ludden, C.; Decano, A.G.; Jamrozy, D.; Pickard, D.; Morris, D.; Parkhill, J.; Peacock, S.J.; Cormican, M.; Downing, T. Genomic Surveillance of *Escherichia coli* ST131 Identifies Local Expansion and Serial Replacement of Subclones. *Microb. Genom.* **2020**, *6*, e000352. [CrossRef]
27. Harmer, C.J.; Hall, R.M. IS26 Cannot Move Alone. *J. Antimicrob. Chemother.* **2021**, *76*, 1428–1432. [CrossRef]
28. Harmer, C.J.; Hall, R.M. An Analysis of the IS6/IS26 Family of Insertion Sequences: Is It a Single Family? *Microb. Genom.* **2019**, *5*, e000291. [CrossRef] [PubMed]
29. Varani, A.; He, S.; Siguier, P.; Ross, K.; Chandler, M. The IS6 Family, a Clinically Important Group of Insertion Sequences Including IS26. *Mob. DNA* **2021**, *12*, 11. [CrossRef] [PubMed]
30. Lipszyc, A.; Szuplewska, M.; Bartosik, D. How Do Transposable Elements Activate Expression of Transcriptionally Silent Antibiotic Resistance Genes? *Int. J. Mol. Sci.* **2022**, *23*, 8063. [CrossRef]
31. Shawa, M.; Furuta, Y.; Mulenga, G.; Mubanga, M.; Mulenga, E.; Zorigt, T.; Kaile, C.; Simbotwe, M.; Paudel, A.; Hang'ombe, B.; et al. Novel Chromosomal Insertions of ISEcp1-Bla CTX-M-15 and Diverse Antimicrobial Resistance Genes in Zambian Clinical Isolates of Enterobacter Cloacae and *Escherichia coli*. *Antimicrob. Resist. Infect. Control* **2021**, *10*, 79. [CrossRef] [PubMed]
32. Biggel, M.; Moons, P.; Nguyen, M.N.; Goossens, H.; Van Puyvelde, S. Convergence of Virulence and Antimicrobial Resistance in Increasingly Prevalent *Escherichia coli* ST131 PapGII+ Sublineages. *Commun. Biol.* **2022**, *5*, 752. [CrossRef] [PubMed]
33. Awosile, B.B.; Agbaje, M. Genetic Environments of Plasmid-Mediated BlaCTXM-15 Beta-Lactamase Gene in Enterobacteriaceae from Africa. *Microbiol. Res.* **2021**, *12*, 383–394. [CrossRef]
34. Shropshire, W.C.; Aitken, S.L.; Pifer, R.; Kim, J.; Bhatti, M.M.; Li, X.; Kalia, A.; Galloway-Penã, J.; Sahasrabhojane, P.; Arias, C.A.; et al. IS26-Mediated Amplification of BlaOXA-1and BlaCTX-M-15 with Concurrent Outer Membrane Porin Disruption Associated with de Novo Carbapenem Resistance in a Recurrent Bacteraemia Cohort. *J. Antimicrob. Chemother.* **2021**, *76*, 385–395. [CrossRef]
35. Biggel, M.; Xavier, B.B.; Johnson, J.R.; Nielsen, K.L.; Frimodt-Møller, N.; Matheeussen, V.; Goossens, H.; Moons, P.; Van Puyvelde, S. Horizontally Acquired PapGII-Containing Pathogenicity Islands Underlie the Emergence of Invasive Uropathogenic *Escherichia coli* Lineages. *Nat. Commun.* **2020**, *11*, 5968. [CrossRef] [PubMed]
36. Stoesser, N.; Sheppard, A.E.; Pankhurst, L.; de Maio, N.; Moore, C.E.; Sebra, R.; Turner, P.; Anson, L.W.; Kasarskis, A.; Batty, E.M.; et al. Evolutionary History of the Global Emergence of the *Escherichia coli* Epidemic Clone ST131. *MBio* **2016**, *7*, e02162. [CrossRef] [PubMed]
37. Mahmud, B.; Wallace, M.A.; Reske, K.A.; Alvarado, K.; Muenks, C.E.; Rasmussen, D.A.; Burnham, C.-A.D.; Lanzas, C.; Dubberke, E.R.; Dantas, G. Epidemiology of Plasmid Lineages Mediating the Spread of Extended-Spectrum Beta-Lactamases among Clinical *Escherichia coli*. *mSystems* **2022**, *7*, e0051922. [CrossRef] [PubMed]
38. Yoon, E.-J.; Gwon, B.; Liu, C.; Kim, D.; Won, D.; Park, S.G.; Choi, J.R.; Jeong, S.H. Beneficial Chromosomal Integration of the Genes for CTX-M Extended-Spectrum β-Lactamase in Klebsiella Pneumoniae for Stable Propagation. *mSystems* **2020**, *5*, 00459-20. [CrossRef] [PubMed]
39. Cerquetti, M.; Giufrè, M.; García-Fernández, A.; Accogli, M.; Fortini, D.; Luzzi, I.; Carattoli, A. Ciprofloxacin-Resistant, CTX-M-15-Producing *Escherichia coli* ST131 Clone in Extraintestinal Infections in Italy. *Clin. Microbiol. Infect.* **2010**, *16*, 1555–1558. [CrossRef]

40. Palkovicova, J.; Sukkar, I.; Delafuente, J.; Valcek, A.; Medvecky, M.; Jamborova, I.; Bitar, I.; Phan, M.-D.; San Millan, A.; Dolejska, M. Fitness Effects of Bla CTX-M-15-Harbouring F2:A1:B− Plasmids on Their Native *Escherichia coli* ST131 H 30Rx Hosts. *J. Antimicrob. Chemother.* **2022**, *77*, 2960–2963. [CrossRef]
41. Matsuo, N.; Nonogaki, R.; Hayashi, M.; Wachino, J.-I.; Suzuki, M.; Arakawa, Y.; Kawamura, K. Characterization of Bla CTX-M-27/F1:A2:B20 Plasmids Harbored by *Escherichia coli* Sequence Type 131 Sublineage C1/H 30R Isolates Spreading among Elderly Japanese in Nonacute-Care Settings. *Antimicrob. Agents Chemother.* **2020**, *64*, e00202-20. [CrossRef] [PubMed]
42. Yasugi, M.; Hatoya, S.; Motooka, D.; Matsumoto, Y.; Shimamura, S.; Tani, H.; Furuya, M.; Mie, K.; Miyake, M.; Nakamura, S.; et al. Whole-Genome Analyses of Extended-Spectrum or AmpC β-Lactamase-Producing *Escherichia coli* Isolates from Companion Dogs in Japan. *PLoS ONE* **2021**, *16*, e0246482. [CrossRef] [PubMed]
43. Zhang, Y.; Sun, Y.H.; Wang, J.Y.; Chang, M.X.; Zhao, Q.Y.; Jiang, H.X. A Novel Structure Harboring BlaCTX-M-27 on IncF Plasmids in *Escherichia coli* Isolated from Swine in China. *Antibiotics* **2021**, *10*, 387. [CrossRef] [PubMed]
44. Cherubini, S.; Perilli, M.; Azzini, A.M.; Tacconelli, E.; Maccacaro, L.; Bazaj, A.; Naso, L.; Amicosante, G.; Cascio, G.L.; Piccirilli, A. Resistome and Virulome of Multi-Drug Resistant *E. coli* ST131 Isolated from Residents of Long-Term Care Facilities in the Northern Italian Region. *Diagnostics* **2022**, *12*, 213. [CrossRef] [PubMed]
45. Kondratyeva, K.; Salmon-Divon, M.; Navon-Venezia, S. Meta-Analysis of Pandemic *Escherichia coli* ST131 Plasmidome Proves Restricted Plasmid-Clade Associations. *Sci. Rep.* **2020**, *10*, 36. [CrossRef]
46. M'Zali, F.H.; Chanawong, A.; Kerr, K.G.; Birkenhead, D.; Hawkey, P.M. Detection of Extended-Spectrum I-Lactamases in Members of the Family Enterobacteriaceae: Comparison of the MAST DD Test, the Double Disc and the Etest ESBL. *J. Antimicrob. Chemother.* **2000**, *45*, 881–885. [CrossRef] [PubMed]
47. EUCAST The EUCAST Guideline on Detection of Resistance Mechanisms v 2.0. 2017. Available online: https://www.Eucast.Org/Fileadmin/Src/Media/PDFs/EUCAST_files/Resistance_mechanisms/EUCAST_detection_of_resistance_mechanisms_170711.pdf (accessed on 12 December 2022).
48. Tóth, Á.; Kocsis, B.; Damjanova, I.; Kristóf, K.; Jánvári, L.; Pászti, J.; Csercsik, R.; Topf, J.; Szabó, D.; Hamar, P.; et al. Fitness Cost Associated with Resistance to Fluoroquinolones Is Diverse across Clones of Klebsiella Pneumoniae and May Select for CTX-M-15 Type Extended-Spectrum β-Lactamase. *Eur. J. Clin. Microbiol. Infect. Dis.* **2014**, *33*, 837–843. [CrossRef] [PubMed]
49. Matsumura, Y.; Pitout, J.D.D.; Peirano, G.; DeVinney, R.; Noguchi, T.; Yamamoto, M.; Gomi, R.; Matsuda, T.; Nakano, S.; Nagao, M.; et al. Rapid Identification of Different *Escherichia coli* Sequence Type 131 Clades. *Antimicrob. Agents Chemother.* **2017**, *61*, e00179-17. [CrossRef]
50. EUCAST—The European Committee on Antimicrobial Susceptibility Testing. Breakpoint Tables for Interpretation of MICs and Zone Diameters. Version 11.0. 2021. Available online: http://www.Eucast.Org (accessed on 12 December 2022).
51. Zhou, Z.; Alikhan, N.F.; Mohamed, K.; Fan, Y.; Achtman, M. The EnteroBase User's Guide, with Case Studies on Salmonella Transmissions, Yersinia Pestis Phylogeny, and Escherichia Core Genomic Diversity. *Genome Res.* **2020**, *30*, 138–152. [CrossRef]
52. Wick, R.R.; Judd, L.M.; Gorrie, C.L.; Holt, K.E. Completing Bacterial Genome Assemblies with Multiplex MinION Sequencing. *Microb. Genom.* **2017**, *3*, e000132. [CrossRef]
53. Kolmogorov, M.; Yuan, J.; Lin, Y.; Pevzner, P.A. Assembly of Long, Error-Prone Reads Using Repeat Graphs. *Nat. Biotechnol.* **2019**, *37*, 540–546. [CrossRef] [PubMed]
54. Wick, R.R.; Judd, L.M.; Gorrie, C.L.; Holt, K.E. Unicycler: Resolving Bacterial Genome Assemblies from Short and Long Sequencing Reads. *PLoS Comput. Biol.* **2017**, *13*, e1005595. [CrossRef]
55. Li, H.; Durbin, R. Fast and Accurate Long-Read Alignment with Burrows-Wheeler Transform. *Bioinformatics* **2010**, *26*, 589–595. [CrossRef] [PubMed]
56. Walker, B.J.; Abeel, T.; Shea, T.; Priest, M.; Abouelliel, A.; Sakthikumar, S.; Cuomo, C.A.; Zeng, Q.; Wortman, J.; Young, S.K.; et al. Pilon: An Integrated Tool for Comprehensive Microbial Variant Detection and Genome Assembly Improvement. *PLoS ONE* **2014**, *9*, e112963. [CrossRef] [PubMed]
57. Mikheenko, A.; Prjibelski, A.; Saveliev, V.; Antipov, D.; Gurevich, A. Versatile Genome Assembly Evaluation with QUAST-LG. *Bioinformatics* **2018**, *34*, i142–i150. [CrossRef]
58. Seemann, T. Prokka: Rapid Prokaryotic Genome Annotation. *Bioinformatics* **2014**, *30*, 2068–2069. [CrossRef] [PubMed]
59. Gilchrist, C.L.M.; Chooi, Y.H. Clinker & Clustermap.Js: Automatic Generation of Gene Cluster Comparison Figures. *Bioinformatics* **2021**, *37*, 2473–2475. [CrossRef] [PubMed]
60. Carattoli, A.; Zankari, E.; Garciá-Fernández, A.; Larsen, M.V.; Lund, O.; Villa, L.; Aarestrup, F.M.; Hasman, H. In Silico Detection and Typing of Plasmids Using Plasmidfinder and Plasmid Multilocus Sequence Typing. *Antimicrob. Agents Chemother.* **2014**, *58*, 3895–3903. [CrossRef]
61. Johansson, M.H.K.; Bortolaia, V.; Tansirichaiya, S.; Aarestrup, F.M.; Roberts, A.P.; Petersen, T.N. Detection of Mobile Genetic Elements Associated with Antibiotic Resistance in Salmonella Enterica Using a Newly Developed Web Tool: MobileElementFinder. *J. Antimicrob. Chemother.* **2021**, *76*, 101–109. [CrossRef]
62. Alikhan, N.F.; Petty, N.K.; Ben Zakour, N.L.; Beatson, S.A. BLAST Ring Image Generator (BRIG): Simple Prokaryote Genome Comparisons. *BMC Genom.* **2011**, *12*, 402. [CrossRef]
63. Tonkin-Hill, G.; MacAlasdair, N.; Ruis, C.; Weimann, A.; Horesh, G.; Lees, J.A.; Gladstone, R.A.; Lo, S.; Beaudoin, C.; Floto, R.A.; et al. Producing Polished Prokaryotic Pangenomes with the Panaroo Pipeline. *Genome Biol.* **2020**, *21*, 180. [CrossRef]

64. Minh, B.Q.; Schmidt, H.A.; Chernomor, O.; Schrempf, D.; Woodhams, M.D.; Von Haeseler, A.; Lanfear, R.; Teeling, E. IQ-TREE 2: New Models and Efficient Methods for Phylogenetic Inference in the Genomic Era. *Mol. Biol. Evol.* **2020**, *37*, 1530–1534. [CrossRef] [PubMed]
65. Tonkin-Hill, G.; Lees, J.A.; Bentley, S.D.; Frost, S.D.W.; Corander, J. Fast Hierarchical Bayesian Analysis of Population Structure. *Nucleic Acids Res.* **2019**, *47*, 5539–5549. [CrossRef] [PubMed]
66. R Core Team. *A Language and Environment for Statistical Computing*; R Foundation for Statistical Computing: Vienna, Austria, 2018.
67. Willems, R.J.L.; Top, J.; van Schaik, W.; Leavis, H.; Bonten, M.; Sirén, J.; Hanage, W.P.; Corander, J. Restricted Gene Flow among Hospital Subpopulations of Enterococcus Faecium. *MBio* **2012**, *3*, e00151-12. [CrossRef]
68. Yu, G.; Smith, D.K.; Zhu, H.; Guan, Y.; Lam, T.T.Y. Ggtree: An R Package for Visualization and Annotation of Phylogenetic Trees with Their Covariates and Other Associated Data. *Methods Ecol. Evol.* **2017**, *8*, 28–36. [CrossRef]
69. Schliep, K.P. Phangorn: Phylogenetic Analysis in R. *Bioinformatics* **2011**, *27*, 592–593. [CrossRef] [PubMed]
70. Letunic, I.; Bork, P. Interactive Tree of Life (ITOL) v5: An Online Tool for Phylogenetic Tree Display and Annotation. *Nucleic Acids Res.* **2021**, *49*, W293–W296. [CrossRef]
71. Bortolaia, V.; Kaas, R.S.; Ruppe, E.; Roberts, M.C.; Schwarz, S.; Cattoir, V.; Philippon, A.; Allesoe, R.L.; Rebelo, A.R.; Florensa, A.F.; et al. ResFinder 4.0 for Predictions of Phenotypes from Genotypes. *J. Antimicrob. Chemother.* **2020**, *75*, 3491–3500. [CrossRef]
72. Zankari, E.; Allesøe, R.; Joensen, K.G.; Cavaco, L.M.; Lund, O.; Aarestrup, F.M. PointFinder: A Novel Web Tool for WGS-Based Detection of Antimicrobial Resistance Associated with Chromosomal Point Mutations in Bacterial Pathogens. *J. Antimicrob. Chemother.* **2017**, *72*, 2764–2768. [CrossRef]
73. Camacho, C.; Coulouris, G.; Avagyan, V.; Ma, N.; Papadopoulos, J.; Bealer, K.; Madden, T.L. BLAST+: Architecture and Applications. *BMC Bioinform.* **2009**, *10*, 421. [CrossRef]
74. Joensen, K.G.; Scheutz, F.; Lund, O.; Hasman, H.; Kaas, R.S.; Nielsen, E.M.; Aarestrup, F.M. Real-Time Whole-Genome Sequencing for Routine Typing, Surveillance, and Outbreak Detection of Verotoxigenic *Escherichia coli*. *J. Clin. Microbiol.* **2014**, *52*, 1501–1510. [CrossRef] [PubMed]
75. Chen, L.; Zheng, D.; Liu, B.; Yang, J.; Jin, Q. VFDB 2016: Hierarchical and Refined Dataset for Big Data Analysis—10 Years On. *Nucleic Acids Res.* **2016**, *44*, D694–D697. [CrossRef] [PubMed]
76. Blanco, J.; Mora, A.; Mamani, R.; López, C.; Blanco, M.; Dahbi, G.; Herrera, A.; Marzoa, J.; Fernández, V.; De La Cruz, F.; et al. Four Main Virotypes among Extended-Spectrum-β-Lactamase-Producing Isolates of *Escherichia coli* O25b:H4-B2-ST131: Bacterial, Epidemiological, and Clinical Characteristics. *J. Clin. Microbiol.* **2013**, *51*, 3358–3367. [CrossRef] [PubMed]

MDPI AG
Grosspeteranlage 5
4052 Basel
Switzerland
Tel.: +41 61 683 77 34

Antibiotics Editorial Office
E-mail: antibiotics@mdpi.com
www.mdpi.com/journal/antibiotics

www.ingramcontent.com/pod-product-compliance
Lightning Source LLC
Chambersburg PA
CBHW061514210926
51850CB00008B/1920
* 9 7 8 3 7 2 5 8 5 3 0 7 6 *